T0342464

Creating Wine

THE PRINCETON ECONOMIC HISTORY
OF THE WESTERN WORLD

Joel Mokyr, Editor

Creating Wine

THE EMERGENCE OF
A WORLD INDUSTRY, 1840–1914

James Simpson

PRINCETON UNIVERSITY PRESS

PRINCETON & OXFORD

Published by Princeton University Press, 41 William Street,
Princeton, New Jersey 08540

In the United Kingdom: Princeton University Press, 6 Oxford Street,
Woodstock, Oxfordshire OX20 1TW

press.princeton.edu

Library of Congress Cataloging-in-Publication Data
Simpson, James, 1953–
  Creating wine : the emergence of a world industry, 1840–1914 / James Simpson.
    p. cm. — (The Princeton economic history of the Western world)
  Includes bibliographical references and index.
  ISBN 978-0-691-13603-5 (hardcover : alk. paper) 1. Wine industry—Europe—
History. 2. Wine and wine making—Europe—History. I. Title. II. Series:
Princeton economic history of the Western world.
  HD9385.A2S56 2011
  338.8'8763209034—dc22   2011014516

British Library Cataloging-in-Publication Data is available

This book has been composed in Garamond Pro

Printed on acid-free paper. ∞

Printed in the United States of America

10 9 8 7 6 5 4 3 2 1

TO MARÍA JESÚS

Who gave me so many years of fun

IN THE OLD WORLD, WINE-MAKING IS AN ART; IN
AMERICA, IT IS AN INDUSTRY.

—André Simon, 1919:105

Contents

List of Illustrations

Maps

Figures

List of Tables

·

Acknowledgments

THIS BOOK would not have been written but for the help and encouragement I have enjoyed from a whole range of people and institutions. I owe a special debt to libraries and archives in many countries, including those of the Ateneo and Universidad Carlos III in Madrid; the British Library, Guildhall Library, and British Library of Political & Economic Science in London; the Archivo and Biblioteca Municipal in Jerez de la Frontera; the Bibliothèque and the Archives Départementales de la Gironde in Bordeaux; the Bibliothèque Municipal in Épernay; the Biblioteca Nazionale, Firenze; the FAO Library, Rome; Special Collections at the University of California, Davis; the North Baker Research Library, San Francisco; the State Libraries of South Australia and Victoria in Australia; the Biblioteca Nacional in Buenos Aires; and the Biblioteca Municipal and the Instituto Nacional de Vitivinicultura in Mendoza. However, the institution that has been crucial to this undertaking is without doubt the Viticulture and Enology Collection at UC Davis. The chief librarian, Axel Borg, deserves special mention: not only did he respond to all my demands for help, but he was active in putting me in contact with other scholars and people who have been of invaluable help to this work.

My colleagues at the Universidad Carlos III de Madrid have shown tremendous patience and support, especially over the past two or three years. Juan Carmona, Eva Fernández, and Joan Roses all took the time and trouble to read the manuscript at one moment or another and gave useful comments. Antonio Tena helped with the trade figures, and Leandro Prados de la Escosura provided sound advice and encouragement as always.

I am particularly grateful to Kym Anderson, Antonio Miguel Bernal, Luis Bértola, Hubert Bonin, Thomas Brennan, Jean-Michel Chevet, Carlos Coello, Luis Coria, Nicholas Faith, Valmai Hankel, Francisco Javier Fernández Roca, Lina Gálvez, Regina Grafe, Kolleen Guy, Colleen Haight, Pablo Lacoste, Pedro Lains, Ian McLean, José Miguel Martínez Carrión, Enrique Montañés, José María O'Kean, Vicente Pinilla, Ramón Ramón, Blanca Sánchez Alonso, Alessandro Stanziani, Steve Stein, Ron Weir, Nickolai Wenzel, and Bartolome Yun for their comments and help.

I enjoyed many visits to the London School of Economics, where I wish to thank especially Dudley Baines, Gerben Bakker, Janet Hunter, Tim Leunig, Colin Lewis, Patrick O'Brien, and Max Schultze. When not in London I seem to have been at UC Davis, where Julian Alston, Greg Clark, Peter Lindert, Alan Olmstead, and Alan Taylor all provided great encouragement. In Davis I enjoyed meals and conversations with Marta Altisent, Mary McComb, and JaRue

Manning. Darrell Corti showed great enthusiasm about my work and provided some outstanding wines for me to sample. In Mendoza, Ana Mateu and Lizzi Pasteris looked after me and showed me around the town, as did Jan Burbery in Melbourne.

I benefited greatly from long conversions with Jim Lapsley, who gave me important feedback on a number of chapters. Patricia Barrios kindly gave me access to some of her unpublished work in Mendoza and provided very useful criticism on an early draft of chapter 11. Paul Duguid was particularly helpful with his comments on the British trade. John Nye and Harvey Smith both read an earlier version with great attention and gave detailed notes on how to improve the manuscript. Begoña Prieto has never ceased to encourage me in this project and provided the means to organize a congress on wine economics in Silos, Burgos, in 2010.

I owe a special debt to Joel Mokyr, the series editor, for his enthusiasm and wise comments. At Princeton University Press I especially wish to thank Janie Chan, Dale Cotton, Lauren Lepow, and Seth Ditchik, and Dimitri Karetnikov for the maps.

The Spanish government generously funded a sabbatical at the London School of Economics (PR2008-0043) and provided research money for visits to numerous other academic institutions (ECO2009-10739, SEJ2006-08188/ECON, BEC2003-06481). I also thank Alan Olmstead for funding provided by the Institute of Governmental Affairs, and Alan Taylor at the Center for the Evolution of the Global Economy, both at UC Davis.

Various chapters in this book build on material previously published elsewhere. These include "Selling to Reluctant Drinkers: The British Wine Market, 1860–1914," *Economic History Review* 57 (1) (2004): 80–108; "Cooperation and Conflicts: Institutional Innovation in France's Wine Markets, 1870–1911," *Business History Review* 79 (Autumn 2005): 527–58; "Too Little Regulation? The British Market for Sherry, 1840–90," *Business History* 47 (3), (2005): 367–82; and "Factor Endowments, Markets and Vertical Integration: The Development of Commercial Wine Production in Argentina, Australia and California, c1870–1914," *Revista de Historia Económica. Journal of Iberian and Latin American Economic History* 29 (1) (2011).

A number of loyal and close friends have given invaluable support during this undertaking, including Pepe, Lindy, Kate, Madeleine, Miles, Nuala, Isabel, Elías, Gerardo, Lourdes, Maribel, Ramón and Sole. Azucena in particular was tremendous in my hour of need and greatly helped in getting this book finished.

My final debt is to those of my family who failed to see this project finished, namely, Ba, Mary, and, most tragically, both Gordon and María Jesús. María Jesús was a magnificent companion for almost thirty years, and without her I would never have come to Spain and become an academic. It is to her memory that this book is dedicated.

Map 1. The Wine and Brandy Districts of France. Source: Adapted from Robinson, 2006, p. 280

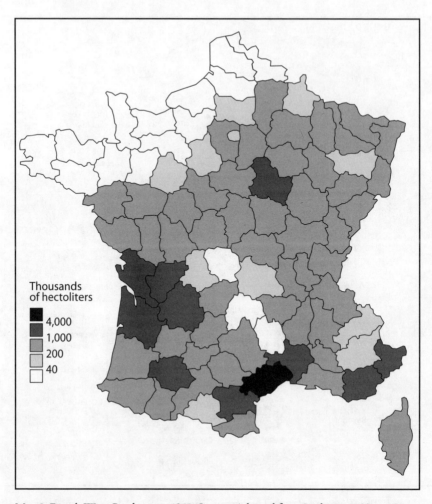

Map 2. French Wine Production, 1852. Source: Adapted from Lachiver, 1988, p. 389

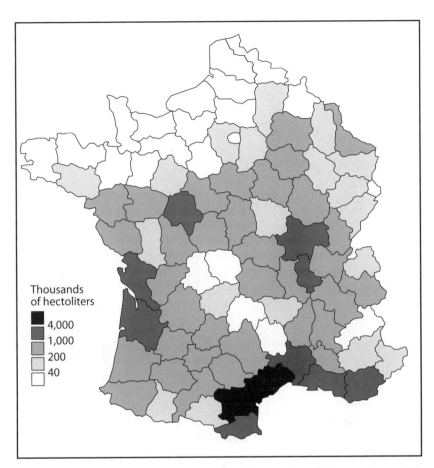

Map 3. French Wine Production, 1900–1909. Source: Adapted from Lachiver, 1988, p. 451

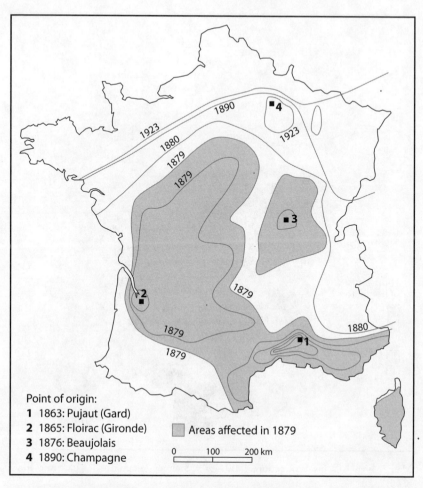

Point of origin:
1 1863: Pujaut (Gard)
2 1865: Floirac (Gironde)
3 1876: Beaujolais
4 1890: Champagne

Areas affected in 1879

0   100   200 km

Map 4. The Spread of Phylloxera in France. Source: Adapted from Braudel, 1986, p. 131

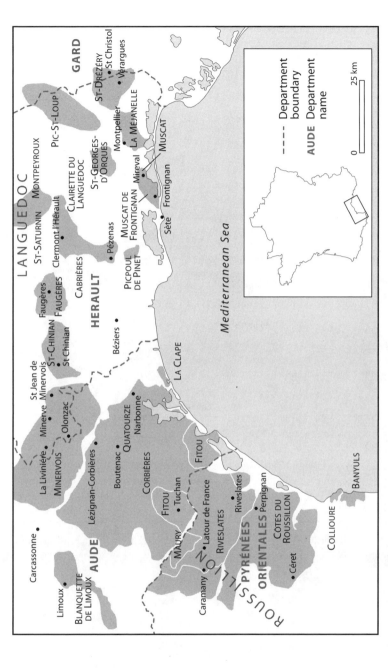

Map 5. The Midi (Languedoc-Rousillon). Source: Adapted from Robinson, 2006, p. 391.

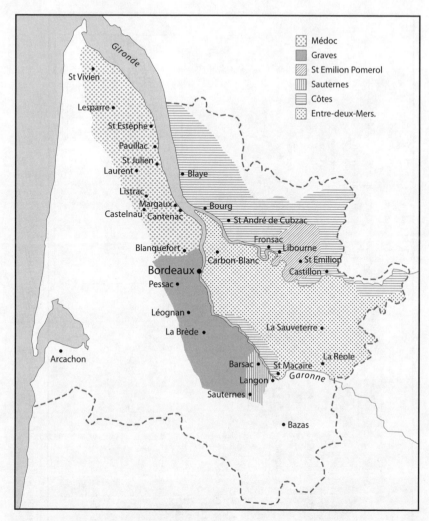

Map 6. The Bordeaux Wine Regions. Source: Adapted from France, Ministere de l'Agriculture, 1937

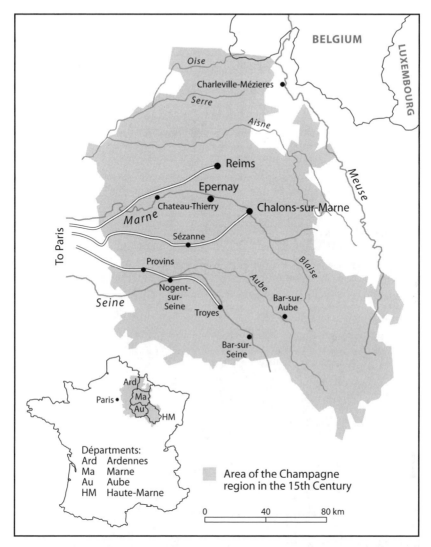

Map 7. The Champagne Region. Source: Adapted from Faith, 1988

Map 8. Portugal and the Porto Region. Source: Adapted from Viniportugal.pt

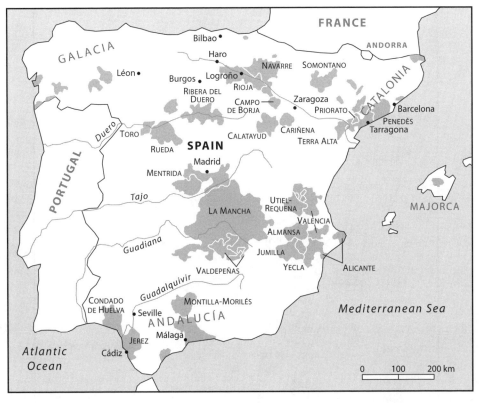

Map 9. The Wine Districts of Spain. Source: Adapted from Robinson, 2006, p. 653

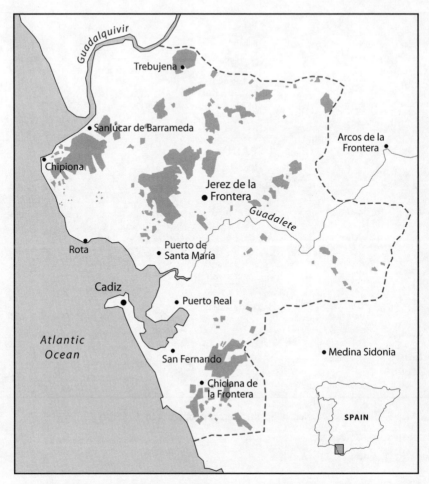

Map 10. The Jerez Region. Source: Adapted from Jeffs, 2004

Map 11. California Wine Districts. Source: Adapted from Wineweb.com

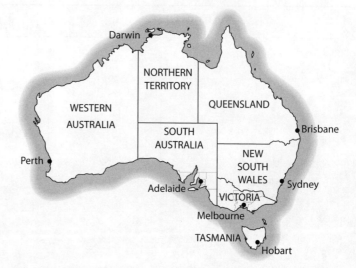

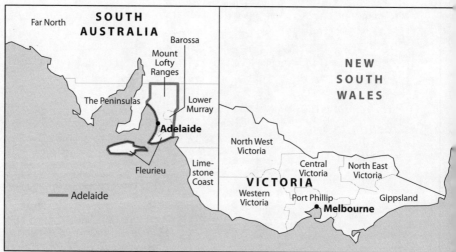

Map 12. The Australian States. Source: Adapted from Rightpundits.com

Map 13. Mendoza and San Juan Regions. Source: Adapted from Wineweb.com

Introduction

WINE IS NOT a homogenous product. A bottle from one of Bordeaux's first growths of the 2000 vintage today sells for several thousand dollars, much more than virtually any other wine, not just for that particular vintage, but also those produced on the same estates in previous and later years. Like most readers of this book, I have never drunk such a wine and probably never will, but I still get great pleasure from drinking more modest ones. A current favorite for everyday drinking is Tábula's Damana 2006 *crianza*, which my local shop sells for around fifteen euros, or twenty dollars. By the time this book is published it may be something else—perhaps because another winery will release a more competitively priced *crianza* next year; or the Tábula winery may have problems meeting demand for its wines and be forced to buy lesser ones to sell under its own brand (or instead may restrict the supply and raise the price above my budget); or my own tastes may change; or, but highly unlikely, my disposable income available for wines may increase to allow me to move upmarket. I also enjoy drinking Alfredo's *vino tinto*, especially as it is free and each glass comes with five or ten minutes of lively conversation (although whether supplies will be quite so forthcoming when Alfredo realizes that I have no influence to help his "export drive" remains to be seen). It is of course this huge diversity that makes wine interesting to so many different people. It also probably explains why few wine histories consider more than a single country, even though many regional studies exist. This book is an attempt to understand the dynamics of production and marketing of all wines during the period of major change that took place between the mid-nineteenth century and the First World War.

Wine production in Europe today is dominated by small family vineyards and cooperative wineries, while in the New World viticulture and viniculture are highly concentrated and vertically integrated. As a result, in the United States and Australia 70 percent of the wine is produced by the top five wine companies; Argentina and Chile, 50 percent; and in France, Italy, and Spain, only 10 percent.[1] In this book I argue that these fundamental organizational differences already existed by 1914 and were caused by six distinct but interrelated variables, namely, *terroir* (which can be explained roughly as production conditions), tradition (or path dependency), technology, the nature of market demand, political voice (especially of small producers), and political organization in each country. The result was that in Europe the functions of grape growing and wine making were integrated but marketing was a specialist activity, while in the hot climates of the New World the situation was reversed, as grape growing was a specialist

---

[1] Anderson, Norman, and Wittwer (2004), table 2.1. The figures exclude champagne.

activity, but the wine-making and marketing activities were often integrated. The ability of groups within the industry everywhere, from grape growers, winemakers, merchants, retailers to consumers, to influence government policy led to very different policy responses to the problems of periodic overproduction, adulteration, and mislabeling of wines, which further helped to determine the organizational structure. In particular, Europe's small family producers successfully lobbied governments and created a variety of new institutions, including regulatory bodies to control fraud, regional appellations, and cooperatives. By contrast, New World winemakers found a much more favorable political environment to create large, integrated businesses and trusts, investing heavily in advertising and brands to sell to distant consumers generally unaccustomed to wine drinking.

Yet despite these changes, about nine-tenths of wine in 1914 was produced in Europe, with France, Italy, and Spain alone accounting for almost three quarters of world output. Higher wages, urbanization, and the railways all helped contribute to the doubling of French consumption to more than 150 liters per person annually in the sixty or seventy years before the First World War. Technological change was rapid in both the vineyard and winery, and this led to the spread of commercial wine production to new regions, especially those with hot climates. However if a stable, good-quality dry wine could now be produced in places as diverse as Algeria, California, or South Australia, merchants everywhere had limited success in selling them to consumers who originated from outside Europe's traditional wine-producing regions. The period before 1914 saw the spread of commercial wine production to new geographical regions, but it is only in the past two or three decades that we can genuinely talk of a global wine market.

A number of different processes are required in the production and distribution of wine before it reaches the consumer. First the grower must prepare the land and choose and plant a suitable vine variety, followed by an annual cycle of cultivation (hoeing, pruning, harvesting) and wine making (crushing of the grapes and fermenting of the must). The wine is then racked, matured, and blended with others for sale. When the wine is marketed over long distances, merchants are needed to find suitable supplies in sufficient quantities and create distribution networks. Producers of fine wines such as claret, champagne, port, or sherry developed diversified commodity chains, but these wines accounted for only a small percentage of total production, even in their own countries.[2] Most wines were cheap, referred to in the literature by a variety of names, including *vin ordinaire*, *vino común*, jug, commodity, or beverage wines. Despite the huge variety of wines produced, the different steps involved in the commodity chain can be summarized relatively easily.

---

[2] This book follows the usual convention of using lowercase for the drink (champagne, claret, etc.) and uppercase for the region where it is produced.

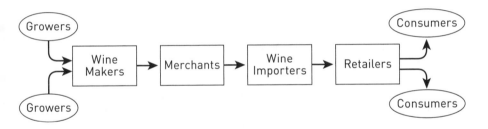

The commodity chain above suggests that different enterprises carried out each activity. On a few occasions this was indeed the case, but usually a single business integrated backward or forward into one or more activities. The particular business that controlled the chain and was ultimately responsible for quality and establishing the brand could vary significantly. Fine wines, for example, were sold by a variety of methods: under the name of the grower (Château Margaux), the manufacturer/exporter (Moët & Chandon), the shipper (Sandeman), or the importer (Victoria Wine Company). By 1914 there were also fundamental differences in the production and marketing of cheap table wines between the Old and New Worlds.

In much of Europe, growing grapes was risky. Hubert de Castella, a Swiss viticulturist who settled in Australia, noted that European growers faced "a long list" of potential damages: "First the spring frost, next the rain at blossoming time, the hail all through the summer. As to diseases and parasites, *le jaune, le collis, la pyrale, l'écrivain*—the sweeping oidium—and of very late years several new ones with ominous names, including peronospera, the antracnose."[3] Harvest size was not only up to four times more volatile than with cereals, but grape quality varied significantly from one plot to another, as well as from one year to the next. Grapes had to be processed rapidly after collection, and because growers could not depend on independent wineries purchasing them within such a short time frame, most had their own crushing and wine-making facilities. Therefore grape growing and wine making, with a few notable exceptions such as Champagne, were activities that were almost always integrated, eliminating any coordinating problems that would have occurred if two distinct businesses had existed instead. By contrast, European wine production and marketing were separate activities. Merchants played a crucial role in collecting the highly diverse wines from the thousands of small family producers and blending them to produce an acceptable drink for urban consumers. Even so, wine quality varied significantly, and there was virtually no attempt by merchants to sell their com-

[3] He concludes, "Considering the capital necessary to establish a vineyard in the old world, the time to wait for a first crop, it requires some courage to undertake it" (Castella 1886:17–18). Peronospera is downy mildew and antracnose, black spot.

modity using brand names. Therefore in Europe, where wine was the alcoholic beverage of choice for most consumers, it was the reputation of individual retailers in the thousands of taverns in cities such as Paris, Madrid, and Rome that determined which wines were drunk.

In Europe, viticulture and wine production were generally family businesses. Vines were widely cultivated despite the considerable output volatility because grapes could be successfully produced on land that was marginal to most other crops and was therefore cheap. Large quantities of labor were required to prepare the vineyards for planting, but much of the work could be carried out during periods of slack demand, when there was little alternative employment. Small plots of vines were therefore excellent vehicles for family producers with limited means to acquire a capital asset, which provided both a pleasant (and important) dietary supplement, as well as a possession that could be sold in times of difficulties. In France in the late nineteenth century, there were around one and half million producers, most of whom had only a very small area of vines and combined viticulture activities with others, both rural and urban.

The small-scale nature of production did not stop producers responding quickly to market upturns by adding to their vines, but they were much slower to uproot them at times of overproduction, creating a ratchet effect and more price volatility. During periods of low prices, growers preferred to dedicate more of their labor resources to other activities rather than destroy their vineyards. Traditionally the inadequate keeping quality of most wines limited the impact of a large harvest, and hence low prices, because stocks could not be carried over to the following year. By the late nineteenth century, however, production conditions had begun to change. In particular, increased scientific knowledge related to grape growing and fermentation allowed potentially higher yields and improved the keeping quality of wines. In addition, many growers, merchants, and retailers learned during the phylloxera-induced shortages of the 1870s and 1880s to adulterate their wines, using a variety of substances other than fresh grapes to increase volume and improve their keeping quality—techniques they were reluctant to give up when production recovered with phylloxera-resistant vines. By the turn of the century overproduction and low prices threatened to drive many producers to bankruptcy and seriously damaged some regional economies. France's Midi region was especially affected, and large and small owners there joined together in a series of demonstrations in the summer of 1907 that culminated in over half a million people protesting in Montpellier, which obliged the government to allow a growers' organization (the Confédération générale des vignerons du Midi) to monitor fraud and prosecute illegal wine-making activities and the sale of adulterated wines.

Although the protests in the Midi were largely organized by the new "industrial" producers, they underlined the growing negotiating strength of small growers everywhere with the state, and their ability to challenge the economic power of merchants. This new power was already clearly apparent by the First

World War in two distinct areas: the creation of cooperatives (which allowed family grape producers to benefit from the increasing economies of scale found in wine production) and regional appellations (which restricted the use of names such as "Bordeaux" and "Champagne" to local producers). Similar changes appeared in Europe's other wine-producing nations, although the timing differed, partly because political institutions initially were less favorable to family growers, and partly because wines had to achieve a sufficient popularity to encourage deception through mislabeling.

The European wine-producing tradition stretched over several millennia, and with the massive emigration of the late nineteenth century migrant laborers took their highly site-specific knowledge with them to the New World. There they often found climatic conditions much more favorable for viticulture, as fully mature and disease-free grapes could be produced virtually every year. Furthermore, by 1900 new wine-making technologies, which were especially suited to the hot climates of these countries, allowed producers to make consistently better wines that kept longer and were cheaper than those made using traditional methods. Consequently the new, large wineries depended on specialized growers for a significant quantity of their grapes, leading to different coordination problems from those found in Europe.

In the New World, unlike in Europe, the combination of good grape-growing conditions and the new fermentation technologies allowed large quantities of brandable homogenous wines to be produced. As a result, Australian wines were sold under the importer's brand in London, while many Argentine and California wines were sold under that of the winemaker. Major marketing problems existed, however, because production was located at considerable distances from large urban markets, and also because the consumers of California and Australian wines were traditionally beer or spirit drinkers. Even when wine was the alcoholic beverage of choice among consumers, such as in Argentina, price primarily determined demand, which encouraged producers to sacrifice quality for quantity and thereby reduced the value of the brand. Overproduction and adulteration were just as common in the New World as in Europe, but the political voice of growers was much smaller, and it was the winemakers and merchants who imposed solutions, in particular by exploiting the economies of scale in production and marketing and restricting competition.

This book is divided into four parts. Part 1 examines the transformation of Europe's cheap table wine industry. The period was dominated by the vine disease phylloxera, which was brought on vine stock from the United States and over the course of many decades would kill almost all vines in Europe and the rest of the world. The only permanent solution was to graft European varieties to phylloxera-immune American rootstock. Wine shortages caused by phylloxera in one market encouraged growers in others to increase output, as well as to produce artificial wines, while the new phylloxera-resistant vines were capable of producing significantly higher yields. Recovery in production and continued fraud resulted in markets becoming glutted by the turn of the twentieth century. Growers looked to avoid the consequences of their own success, namely, over-production, low prices, and the need to exit the industry. They demanded help from the government to resolve collective-action problems associated with con-trolling the production of artificial wines and to create wine-making coopera-tives to allow them to improve wine quality and cut production costs.

Part 2 looks at export failure and the difficulties of selling in a market where wine, for most consumers, was not the alcoholic beverage of choice. For centu-ries wine had been an important export commodity, and in 1850 it still ac-counted in terms of value for half of all Portugal's exports, a quarter of Spain's, and one-fifteenth of France's. Much of this was fine wine destined for the British luxury market. The potential of this market changed dramatically with Glad-stone's legislation of the early 1860s that specifically sought to create a mass mar-ket for table wines. Britain had a large and wealthy population and was open to world trade, importing over half of its food and beverage needs by 1914. Al-though wine imports tripled between the late 1850s and the mid-1870s, there was no permanent change in drinking habits. Consumption drifted lower for the rest of the century, even though the population continued to increase, living standards rose, and the duties on wines fell in real terms. On the eve of the First World War, per capita consumption was no greater than it had been a hundred years earlier. The failure to create a mass market for cheap table wines among consumers who traditionally preferred beer and spirits was caused, at least in part, by the difficulties associated with developing cheap, impersonal exchange mechanisms such as brands because of the major annual fluctuations in wine quality and the ease with which the drink could be adulterated.

Part 3 looks at the response by local producers and exporters in Bordeaux, Champagne, Porto, and Jerez to the growth in the generic use of words such as "port" or "champagne" for wines of a certain type, regardless of where they had been produced. Local growers had long claimed that "claret" or "sherry" consti-tuted collective trademarks, and they wanted to restrict the supply of grapes to a designated area by creating regional appellations. This not only led to confronta-tions between local growers and those located outside the appellation but was also opposed by merchants who claimed that it was they, not the growers, who created a wine's reputation, and that they needed the freedom to purchase out-side wines for blending to compensate for the vagaries in the local harvest and to sell at competitive prices. Local producers argued that regional appellations were essential to maintain wine quality and provide guarantees for consumers, but their opponents claimed they served only to create geographical monopolies and shift rents to the local producer. Collective action on the part of growers therefore needed the backing of the state. Although the sixty thousand growers in Bordeaux might oppose the production of wines made from raisins or the sell-ing of cheap Midi wines as claret, they were happy to bend the rules when it suited them. Only when a grower believed that a system could adequately identify and punish cheats were they likely to respect the rules themselves. The creation of an appellation led to bitter and sometimes violent conflicts as the economic livelihood of the excluded growers was threatened.

Outside France, political institutions were often less favorable to small growers. In Spain, for example, authorities routinely dismissed demands by growers in Jerez for a regional appellation and the creation of a local bank because they went against the interests of a small number of powerful merchants. By contrast, in Porto, vine growers found a considerably more sympathetic state because the foreign shippers, unlike those in Jerez, still retained their British nationality, which placed them at a distinct disadvantage when negotiating with the Portuguese state.

The final section, part 4, looks at the development of the industry in the New World, comparing the experiences of California, Australia, and Argentina. In these areas factor endowments were reversed, as labor was scarce and land abundant. The potential of these countries to become major producers was quickly appreciated, but both the technology and virtually all the world's wine drinkers were found in the Old World. During the half century prior to 1914, technology was transferred and adapted to the needs of the new producers, and millions of potential consumers migrated from Europe. Change was rapid and the New World's reputation for poor-quality wines on account of faulty fermentations in the 1890s had disappeared in some areas by 1914. Good wine production under these conditions required heavy investment and significant technical skills, leading to a separation between grape production and wine making. From the final decade of the nineteenth century, the problems associated with overproduction and adulteration and the huge distances between the areas of specialized viticulture and their major markets led to the creation of very different organizational structures from those found in Europe. Winemakers successfully lobbied for tariff protection, but the small, family growers had very little political influence compared with their European counterparts. The leading producers were huge compared to those in Europe, and they learned to standardize production and integrate forward into distribution and marketing of their brands in distant urban centers, and to different degrees to manipulate prices. In both Australia and the United States, the difficulties of selling over long distances to consumers unacquainted with wine at a time of increasing interest in prohibition limited per capita consumption to less than a tenth of that found in Argentina or Chile. Technological change in wine-making technologies was especially impressive in Australia and California, but the difficulties of creating new markets for these table wines led to a switch to fortified dessert wines after 1900, and as late as the 1960s wines such as "sherry" or "port" made up at least half of California's and Australia's wine production.

Weights, Measures, and Currencies

Gay-Lussac—alcohol strength measured by the number of liters of pure ethanol in 100 liters of wine
14.8 degrees Gay Lussac = 26 degrees proof Sykes
24.5 degrees = 40 degrees proof
1 pound sterling = 20 shillings (s.); 1 shilling = 12 pence (p.)

MEASURES OF CAPACITY

1 pint = 0.125 imperial gallon = 0.567 liter
1 imperial gallon = 4.546 liters = 6 bottles
1 U.S. gallon = 3.788 liters = 5 bottles
1 hectoliter = 100 liters = 22 imperial gallons = 26.4 U.S. gallons
Port: pipe = 522.5 liters
Sherry: butt = 480 liters
Bordeaux: hogshead = 224 liters; tun = 900 liters
Champagne: demiqueue = 183 liters; butt = 200 liters
California: barrel = 190 liters (50 U.S. gallons)
Australia: hogshead = 270–295 liters (60–65 imperial gallons)

MEASURES OF AREA

1 hectare = 2.47 acres
1 aranzada (Jerez) = 0.45 hectare = 1.05 acres

Acronyms and Abbreviations

Arch. Gironde	Archives Départementales de la Gironde
AMA	Archivo del Ministerio de Agricultura, Spain
AMJF	Archivo Municipal de Jerez de la Frontera
Association syndicale	Association syndicale autorisée pour la défense des vignes contre le phylloxera
AVC	Association viticole champenoise
BOIV	Bulletin de l'office international du vin
CGV	Confédération générale des vignerons du Midi
Companhia	Companhia Geral da Agricultura das Vinhas do Alto Douro
CVN	Centro vitivinícola nacional
CWA	California Wine Association
CWMC	California Wine-Makers Corporation
Fédération	Fédération des syndicats viticoles de la Champagne
ISTAT	Istituto centrale di statistica, Italy
IIA	International Institute of Agriculture
Ligue	Ligue des Viticulteurs
MAIC	Ministero di agricoltura, industria e comercio, Italy
PWSR	*Pacific Wine & Spirits Review*
Ridley's	*Ridley & Co.'s Wine and Spirit Trade Circular*
Syndicat du commerce	Syndicat du commerce des vins de Champagne

PART I

Technological and Organizational Change in Europe, 1840–1914

FOR THOUSANDS OF YEARS wine has been widely produced and drunk in western Europe. Traditional preindustrial economies often had high levels of underemployment, which made the vine attractive as it provided considerable employment opportunities during "all seasons, to all ages and both sexes."[1] According to the French historian Le Roy Ladurie, the "classic response of Mediterranean agriculture to a rise in population" was to "plant trees or vines on old or new assarts, thereby increasing the returns from agriculture by more intensive forms of land utilization."[2] This essentially Malthusian vision assumes that output at best grew with population growth and implies little or no change in labor productivity that might have helped improve living standards.[3] In fact, long before 1840 growers were also quick to extend their vines when grape prices increased and new opportunities to trade opened.[4] Some of Europe's most dynamic agricultural regions included Bordeaux and Champagne, which specialized in viticulture, and the key to their success, as the highly observant James Busby noted in the early 1830s, was the substitution of capital for land and labor. Yet in most regions prior to the railways during the second half of the nineteenth century, production remained labor-intensive, and the wine produced was very ordinary.

The production and marketing of wines in Europe were transformed beyond recognition in the period between 1840 and 1914. Improved production methods and declining transport costs led to more integrated markets, regional spe-

[1] Gasparin (1848, 4:595). In southern Spain in the 1920s, for example, cereals provided only 25 days employment a year, olives between 31 and 62 days, but viticulture between 44 and 237 days. The figures for cereals do not include land left fallow in alternative years. Extensive and intensive irrigation created 175 and 375 days employment, respectively (Carrión 1932/1975:324, 341–42).

[2] Le Roy Ladurie (1976:56–57).

[3] Lewis (1978, chap. 8) argued that unlimited areas of suitable land and underemployed labor in the late nineteenth century resulted in the supply increasing without productivity changes, producing stagnant or declining living standards among producers of tropical farm commodities.

[4] For example, the historian Pierre Vilar (1962), in his classic account of eighteenth-century Catalonia, argued that the expansion of local viticulture increased the demand for manufactured goods, helping to stimulate economic growth.

cialization, and a major increase in consumption in producer countries. The appearance of new vine diseases led to the need for cash payments for chemicals and pesticides. Wine shortages also increased farm gate prices, which encouraged the spread of commercial viticulture to new areas and stimulated a huge amount of scientific research on vine growing, leading in turn to higher yields. Unfortunately for wine producers, high prices also encouraged widespread adulteration, so that when domestic wine consumption recovered by the turn of the century, prices collapsed.

This part looks at the growth in output of cheap table wines in Europe, the impact of the vine disease phylloxera, and how the problems caused by overproduction and adulteration in domestic markets were resolved. The French market was central to events, as not only was it responsible for about a third of the world's production in 1909–13, but it also accounted for 60 percent of world imports. The rapid spread of new vine diseases, such as oidium (powdery mildew), phylloxera, and downy mildew, was resolved mainly by French scientists, while the contributions of Chaptal and Pasteur marked the beginning of a proper understanding of fermentation and wine making. By the First World War a major distinction had developed between traditional and modern viticulture, with the former defined by growers using labor-intensive production methods and surplus labor to increase the size of their holdings, and the latter dependent on capital markets.

Chapter 1 looks at the nature of grape production and wine making on the eve of the railways. Chapter 2 examines the impact of phylloxera and the widespread changes in viticulture and wine-making technologies, while chapter 3 looks at the response of growers to overproduction, fraud, and declining wine prices at a time of rising labor costs. By 1914 French winegrowers had established a degree of political influence, which required the state to intervene at periodic intervals when adverse movements of supply and demand threatened growers' livelihoods.

CHAPTER 1

European Wine on the Eve of the Railways

> Wine-making is an art which is subject to important modifications
> each year.
> —Nicolás de Bustamente, 1890:103

WINE WAS AN INTEGRAL PART of the population's diet in much of southern Europe. In France on the eve of the railways, there were reportedly over one and a half million growers in a population of thirty-five million. High transport costs, taxation, and poor quality all reduced market size, and most wines were consumed close to the place of production. Volatile markets also forced most growers to combine viticulture with other economic activities. Alongside the production of cheap table wines, a small but highly dynamic sector existed that specialized in fine wines to be sold as luxury items in foreign markets. In particular, from the late seventeenth century foreign merchants and local growers combined to create a wide range of new drinks, primarily for the British market, and in the 1850s wine still accounted for about half of all Portugal's exports, a quarter of Spain's, and one-fifteenth of France's.[1]

The process of creating wine followed a well-determined sequence: grapes were produced in the vineyard; crushed, fermented, and sometimes matured in the winery; and blended (and perhaps matured further) in the merchant's cellar; finally, the wine was drunk in a public place or at home. This chapter looks at the major decisions that economic agents faced when carrying out these activities. It examines the nature of wine and the economics of grape and wine production, market organization, and the development of fine wines for export before 1840.

WHAT IS WINE?

The *Oxford Companion to Wine* defines wine as an alcoholic beverage obtained from the fermentation of the juice of freshly gathered grapes.[2] Grapes are cut from the vine and crushed to release their sugars, which are then fermented into alcohol by the natural yeasts found on the grape's skin. In expert hands and in

---

[1] Port wine and madeira represented 38 percent and 7 percent, respectively, of all of Portugal's exports in the 1850s, while sherry accounted for 20 percent of Spain's exports between 1850 and 1854 (Prados de la Escosura 1982:41; Lains 1992:126 ; France. Direction Générale des douanes).
[2] Robinson (2006:768).

favorable years, this simple process might produce an excellent wine, but until recently, quality in most years was often poor. Much of the wine produced was drunk with water, especially as this reduced the risks posed by water-borne diseases, which helps to explain the seemingly high levels of drinking in producer countries.[3] In some regions resin, honey, or herbs were added to hide the deficiencies in wine making. There was also a long history of the use of more dangerous substances, such as lead and lead compounds, to balance the wines and give them a slightly sweet taste, but by the second half of the eighteenth century these additives had been correctly identified as the cause of severe abdominal pains (the colic of Poitou) and banned.[4] Adulteration was carried out by any of several economic agents along the commodity chain wishing to gain financially from the illegal activity. For example, growers in the Douro on occasion used the skins of dried elderberries to give more color to port wine; the Midi wine producers added sugar and water to make "second" and "third" wines; Spanish wine merchants used industrial alcohol produced from potatoes to fortify their wines before exporting; and Paris retailers added water to increase volume.

A combination of urbanization, which implied that consumers no longer knew the origins of the wine that they drank; scientific progress, which had the negative externality of making it easier for would-be fraudsters; and greater consumer awareness brought about by the rapidly expanding local press led to the adulteration of food and beverages being a major issue in all countries by 1900. Therefore, although romantic poets appear never to have drunk a poor wine, one popular French saying of this time suggested that it took three to drink a bottle: one to hold the person, a second to pour it down his throat, and finally the victim himself. An Argentine joke had a wine merchant on his deathbed confessing to his son that wine could be produced "even" from grapes. National laws, such as France's Griffe Law of 1889, might classify wine as being made from fresh grape juice, but additional legislation was used to control additives considered legitimate for the wine-making process, such as those required to increase sugar or acidic content, antiseptics such as sulfur dioxide, fining agents, or yeast cultures. Exactly what was permissible and what not varied both over time and between countries, and inevitably what was considered as adulteration depended on who did it and whether this information was made available to the consumer prior to purchase.

Wines are highly diverse and can be classified by their color (red, white, rosé); by alcohol content (table wines usually range from 9 to 15 percent alcohol); by sweetness; by age (young or old); as still or sparkling; or by whether alcohol has been added during the wine making (dessert wine such as port or sherry). For perhaps 95 percent of wine this type of classification is sufficient, but for fine

---

[3] People in northern Europe drank tea much earlier and in much greater quantities than did those in the wine-producing areas farther south.

[4] Gough (1998:82).

wines the list is far too simple. The characteristics of a wine produced from a single grape variety such as cabernet sauvignon differ significantly according to the soil's physical and chemical characteristics, hydrology, topography, micro climate, and so on. Wine produced on one part of a vineyard might sell for many times that of wine from another part, even though grape varieties, cultivation techniques, and wine-making technologies are broadly similar. The price of a fine Bordeaux claret, such as Château Latour or Château Margaux, depends on a number of characteristics, some of which are exceedingly difficult to classify, such as color, body, tannin content, bouquet, and freshness. Equally important, and especially in northern Europe, quality varies significantly with each harvest because of weather conditions and vine diseases: the price of Château Latour fluctuated from a minimum of 550 francs a tun to a maximum of 6,250 francs, with a standard deviation of 2,088 in the 1860s.[5] Consequently connoisseurs of fine wine look at two variables in particular as indicators of quality: the growth (a vineyard's geographical location) and vintage. Today's technology and producers' skills have greatly reduced the influence of weather, although perfect conditions remain a rarity, as suggested by Bordeaux's 2005 vintage, which was described as "the deckchair vintage" because it was sufficiently easy that owners could afford to "spend the summer sunning themselves."[6]

Fine wines sold for high prices, and were frequently exported (and thereby provided a country with an important source of foreign exchange), but they were insignificant in terms of volume compared with total production. Chaptal estimated in 1811 that approximately three quarters of all French wines sold for less than 20 francs a hectoliter, or less than a tenth of what the leading 2 percent of wines fetched, although the latter accounted for over 20 percent of the total value of wine production (table 1.1). In Spain ordinary table wines still accounted for almost 90 percent of volume in the 1930s.[7]

A major distinction in this book is therefore between fine and commodity wine production, as they were two discrete industries in the period under discussion, each following quite different objectives. Quality was crucial for fine wines, and growers enjoyed significantly higher incomes when this variable was achieved than when harvests were large. By contrast for commodity producers quality was limited to their ability to produce a sound wine that could be sold to merchants for blending.[8] Both groups were market oriented, but for commodity producers there was no attempt to improve vineyard management practice to produce a better wine, as this would reduce yields and consequently profits. Be-

[5] Pijassou (1980).

[6] Robinson (2009). In 2008, by contrast, "they had to spend most of the time in the vineyard fighting rampant mildew and fending off rot."

[7] Ministerio de Agricultura (1933:128–29).

[8] Figures for the Côte d'Or in Burgundy between 1820 and 1879 show that fine-wine producers maximized their incomes when quality was high and harvests were small, but the reverse was true for commodity wine producers (Loubère 1978:125–26).

TABLE 1.1
Output and Value of French wine, 1804–8

Francs per hectoliter	Quantity produced (millions of hectoliters)	Value (millions of francs)	Output (as % of total)	Value (as % of total)
7.5	5.4*	40.2	15.3	5.6
7.5	10.5	78.8	29.8	11.0
10.0	4.6	46.0	13.1	6.4
15.0	3.4	51.0	9.6	7.1
20.0	2.3	46.0	6.5	6.4
25.0	2.0	50.0	5.7	7.0
30.0	1.7	51.0	4.8	7.1
35.0	1.6	56.0	4.5	7.8
40.0	1.5	60.0	4.3	8.3
50.0	1.6	80.0	4.5	11.1
200.0	0.6	160.0	1.7	22.2
	35.2	718.9		

Source: Chaptal (1819, 1:175–77).
*Wine destined to be distilled

tween these two major extremes there was a small group of favorably located growers in places such as Bordeaux or Champagne who chose vineyard management techniques to limit production and improve quality, and who therefore needed to sell their wines at better prices than commodity producers to cover their higher production costs. These premium producers were the ones that demanded appellations at the turn of the twentieth century.

An important characteristic of Europe's commodity wine production was that it was produced by hundreds of thousands of growers in very small quantities and then blended by merchants to meet consumer demand. The advantages of blending were considerable, as André Simon, a leading authority, noted in 1920:

> The blending of wine is both legitimate and necessary. It is a perfectly legitimate manner of improving different wines, of improving them without tampering with any fundamental law of Nature. It is sometimes necessary, in order to render more readily saleable wines which, in spite of being sound, might be otherwise difficult to sell. Blending is resorted to by honest people, who deal in none but honest wines, as and when it is their best chance of supplying better value. Blending is the only sound method of improving the quality and of lowering the cost price of most wines.
>
> The general object of blending may be said to consist in giving better value to the consumer and greater profits to the trader; the means to that end being the standardization of quality and prices.[9]

[9] Simon (1920:77).

Simon uses words such as "legitimate" and "honest" because blending clearly provided plenty of opportunities for fraud and adulteration. Fraud involved selling wine under the label of a private brand such as Moët & Chandon, or collective regional brands (Bordeaux or Champagne) when it had been produced elsewhere. Adulteration, by contrast, consisted of adding ingredients that were considered illegal or "unnatural" to wine and the wine-making process.

As late as the First World War, Europe still produced and consumed approximately 90 percent of the world's wine, with France and Italy alone accounting for three-fifths (table 1.2). Algeria, Argentina, Chile, and the United States, by contrast, produced little more than 10 percent, although production in these countries was growing rapidly. Production in Algeria, which was an administrative district of France at the time and consequently enjoyed free access to this market, reached a million hectoliters for the first time in 1885, a figure that the United States achieved in 1887, Argentina in 1900, and Chile in 1901.

FAMILY PRODUCERS

The virtues of the vine for traditional small-scale family farms are best summarized by Arthur Young, a British observer who was otherwise generally highly critical of French farming. Young noted in his travels through France in the late 1780s that the cultivation depended "almost entirely on manual labor ... demanding no other capital than the possession of the land and a pair of arms; no carts, no ploughs, no cattle."[10] The vine required no fallow and little if any manure, and it adapted to all kinds of soils, including those "which produce nothing but useless thorns and briers."[11] Because the vine was much more productive than most other crops, a couple of hectares in ideal conditions could maintain a family.[12] Yet few relied exclusively on it because vineyards in one year "yield nothing: in another, perhaps, casks are wanted to contain the exuberant produce of the vintage: now the price is extravagantly high; and again so low, as to menace with poverty all who are concerned in it."[13] In France harvest size had a coefficient of variation more than four times greater than other crops.[14]

According to Young's calculations, vines in France yielded an annual average of £9 per acre, compared to the £6 or £7 for the best land in England, which also

[10] Young (1794, 2:25). The version used here is the second chapter, on vines, of the Bury St. Edmonds edition.

[11] Gasparin (1848:595).

[12] Lachiver (1988:245). Augé-Laribé (see below) suggests 5 hectares as a minimum for full-time growers in southern France.

[13] Young (1794, 2:203). Labrousse (1933, 1:269–76) noted that in eighteenth-century France, "the cyclical fluctuations [of wine prices] are ... superior to those of all other products." Cited in Brennan (1988:97).

[14] Toutain (1961:154), cited in Loubère (1978:121).

TABLE 1.2
Leading Wine-Producing Countries before 1914

	Wine production (millions of hectoliters)			% of total in 1909–13
	1865–74	1885–94	1909–13	1909–13
France	55.4	31.9	46.4	31.4
Italy	23.6	31.9	46.0	31.2
Spain	17.1	21.9	14.9	10.1
Austria-Hungary	3.2	7.7	7.9	5.4
Portugal	2.1	4.3	4.8	3.3
Greece	0.2	1.8	3.2	2.2
Germany	2.5	2.5	1.8	1.2
Russia	3.3	3.5	1.4*	0.9*
Rumania	0.1	2.8	1.4	0.9
Bulgaria			0.8	0.5
Other European			0.9	0.6
European total			129.5	87.8
Algeria	0.2	3.1	7.9	5.4
Argentina	n.a.	n.a.	4.4	3.0
Chile	n.a.	n.a	2.0	1.4
United States	0.3	1.0	1.9	1.3
Russia (Asia)			0.9	0.6
South Africa	0.7	0.3	0.3	0.2
Tunisia			0.3	0.2
Australia	n.a	0.1	0.2	0.1
Other countries	5.1	15.7	0.2	0.1
Non-European producers			18.1	12.3
World	113.7	125.7	147.6	100.0

Sources: 1865–74 and 1885–94: Morilla Critz (1995:303); 1909–13: International Institute of Agriculture (1915b:110–11, and Mitchell (1995:240). For Portugal, Lains and Sousa (1998:965). For Argentina, table 11.2. For Australia, fig. 10.1. For the United States, Shear and Pearce (1934, table 10).

Refers to European Russia.

required an expensive fallow.[15] As a result Young claimed the vine produced "more than sugar pays in the West Indies, which is usually supposed the most profitable cultivation in the World."[16] His estimates of an annual net return of 7–10 percent would be much higher if the wine was stored six months after the vintage. High returns, however, were achieved with high production costs. On France's vineyards on the eve of the revolution, average annual cultivation costs

[15] Young (1794, 2:21).
[16] Ibid., 22.

were between £4 10s. and £5 17s. an acre, of which £2 12s. 6d. was for labor, a figure "about thrice as high as that of common arable crops."[17] Indirectly, however, labor costs were even higher, as most of the capital costs were associated with planting the vineyard, another labor-intensive activity.

Land inequality was common in many parts of Europe, resulting in a few individuals owning significantly more land than they could farm themselves with their own labor, and much larger numbers having to find additional employment. For the large landowners, some types of agriculture lent themselves much better than others to either direct farming using wage labor or some form of rental agreement. In general, and apart from exceptional circumstances such as when farm prices were very high, the use of cash rental contracts or wage labor was unsuitable with most tree or bush crops such as the vine. The explanation is not difficult to find because vines can be easily and permanently damaged if the pruning, plowing, and hoeing operations are badly carried out. According to the Spanish agronomist Esteban Boutelou in 1807, "no crop suffers more from the omission or poor quality of work, requiring many years to recover from the abuses of a single year."[18]

The problem was therefore one of moral hazard, namely, wage laborers or tenants treating the vines with less care and attention than their owners required. On some occasions landowners did succeed in developing an incentive structure that aligned the laborers' interests with their own. Fine wine producers in Bordeaux developed a contract that provided good working conditions and pay, precisely because they wanted workers to return each year and acquire specialized, site-specific knowledge of a particular plot of vines (chapter 8).

Yet Bordeaux is the exception that fits the rule. Most landowners could not use wage labor because the poor-quality wine they produced made it unprofitable to pay a sufficiently high wage to ensure good-quality work. With the exception of the harvest, most vineyards were therefore worked by their owners, although by the late nineteenth century some landowners found ways to create large wine estates in parts of the Midi and Algeria, for reasons that will be explained in chapter 2.

Tenancy arrangements were also rare in viticulture as landowners found it extremely difficult to determine whether a poor harvest was the result of climatic factors or the agent's idleness, a problem made more difficult by the inverse relationship between harvest size and quality.[19] A short-term rental agreement gave incentives for tenants to prune the vines to maximize harvests, but this short-

---

[17] Ibid., 20, assumes a production £9 per acre and a net profit of £3 3s.–£4 10s.

[18] Boutelou (1807:66). Monitoring wage labor was especially costly on old vineyards that had developed in haphazard fashion over the years. For a general discussion of viticulture and labor monitoring, see Galassi (1992:78–83), Hoffman (1984), and Carmona and Simpson (forthcoming).

[19] Bardhan (1984:161); Galassi (1992:82); and Hayami and Otsuka (1993:3).

ened their productive life. Even with long-term contracts, the landlord ran the risk of receiving back a dying vineyard once the tenant's lease expired.[20] One alternative was a sharecropping contract, where the landowner provided the land and the tenant the labor, and the harvest was divided in a predetermined way, often fifty-fifty between the two parties. In this case incentives often coincided, especially if the contract was considered indefinite, as both landowner and sharecropper benefited from exceptional harvests or jointly suffered the losses from a poor one.[21] However, sharecropping was rare outside a few specific areas, such as Burgundy, Beaujolais, Tuscany, or Catalonia. There were several problems of using sharecropping, especially the practical difficulties of dividing the harvest. In the regions noted above, wine quality was often better than average, the grapes were usually brought to the central farm, and the landowners themselves were heavily involved in wine making and marketing. Most landowners did not want to get involved in production, especially if only average quality wines were produced.[22] Another problem was that because contracts divided output according to the relative value of land and labor inputs at the time that they were originally agreed, major conflicts occurred from the very late nineteenth century in places such as Tuscany and Catalonia when wine and land prices declined and wages rose.[23]

Most vineyards were therefore small and worked by the owner's family. France in 1907 had 1,662,000 hectares of vines and 1,612,000 growers, or an average of about one grower per hectare, although these figures are distorted by the very large number of miniscule plots owned by part-time producers who only made wine for home consumption. The 1892 census for land use is inaccurate, but the more reliable one of 1924 shows almost 70 percent of French vines being on holdings of less than 5 hectares, with only 20 percent found on those farms with more than 20 hectares. France's southern Midi region had a number of large estates; if these four *départements* are excluded, 91.5 percent of all vines were on holdings of less than 10 hectares, and 83 percent on less than 5.[24] By contrast, 40 percent of the Midi's vines were on vineyards of 10 hectares or more. According to Augé-Laribé, writing in 1907, vineyard owners

[20] Evidence suggests that in Catalonia it took about five years to exhaust the vines. Carmona and Simpson (1999:292).

[21] Some contracts, such as the *Rabassa Morta* in Catalonia, were explicitly long term or indefinite, while others, such as the *Mezzadria* in Tuscany, were renewed annually, although in reality the farm frequently passed from one generation to the next. To ensure that sharecroppers supplied sufficient labor, they were expected to be married, and all the family's labor was to be dedicated exclusively to the landowner's land.

[22] Sharecropping was also used with multicropping. These issues are discussed in Carmona and Simpson (forthcoming).

[23] The appearance of new off-farm inputs such as fungicides and taxes had to be allocated between the two parties and could, at least in theory, compensate for these adverse shifts in the relative price of land and labor (Carmona and Simpson 1999:301).

[24] See appendix 1.

in the Midi with less than 5 hectares usually worked their vines with hand hoes rather than plows. On larger properties, the vines were likely to be plowed, and those over 25 or 30 hectares needed hired labor and perhaps a manager. Vineyards of 80 hectares or more took on the characteristics of an industrial enterprise.[25]

The higher labor inputs compared with most other crops, especially cereals, implied that population densities were greater in areas of viticulture, and the villages larger. The more equitable property distribution led the historian Ernest Labrousse to claim that there were fewer conflicts than with other farming and livestock regions, while Marcel Lachiver notes that "one finds more homogeneity in the wine-growing regions, less submission, more democratic spirit, more fraternity."[26] However, if this was indeed true for earlier historical periods, the half century prior to the First World War saw plenty of conflicts between local growers and outside interests.

THE PRODUCTION OF GRAPES PRIOR TO PHYLLOXERA

Contemporary descriptions of grape growing were similar for much of the world. What distinguished production, for example, on a major Bordeaux château from that on a small patch of vines on the edge of some poor isolated village was not the sequence of activities, but rather the level of care and their timing. Production costs in Bordeaux were very high because yields were low and labor inputs considerably greater than elsewhere. Yet everywhere viticulture was considered a business in the sense that growers were fully aware that the care and time spent on their vines would affect the size and quality of the future harvest, and therefore the level of attention depended on both the outlook for future wine prices and the opportunity cost of the grower's labor. High wine prices (or the lack of alternative and more remunerative employment) increased the care that growers gave to their vines.

Traditional viticulture consisted essentially of land and labor and required little capital expenditure. Before the appearance of powdery mildew in the 1850s, a family needed only a few farm tools and the means to transport grapes to be crushed and turned into wine. Contemporary estimates for cultivation costs, such as those shown in table 1.3, therefore consist essentially of labor.

Considerable amounts of labor also were required to establish a vineyard. Du Breuil suggested in his *Vineyard Culture* that vines be planted at a depth of 0.30 meter in northern Europe and twice that in the South because "a dry and warm soil must be dug deeper than a rich, substantial, and somewhat cold soil, because

---

[25] Augé-Laribé (1907:134–37).

[26] See Labrousse (1944:596) for the late eighteenth century and Lachiver (1988:246), cited in Brennan (1997:10).

TABLE 1.3
Production Costs in Aude (the Midi), 1829 (francs per hectare)

	By spade	By plough
Depreciation of cost of vineyard (5%)	23	23
Replacement of vines	6	12
Pruning	12	12
First digging—24 days' work @ 1 franc + wine	36	—
Second digging—15 days' work @ 1 franc + wine	18	—
First plowing—7 days' work @ 4 francs	—	28
Second plowing—6 days' work @ 4 francs	—	24
First hoeing of vines—20 days' work (female) @ 0.50 franc	—	10
Second hoeing of vines—16 days' work (female)	—	8
Harvesting, transport, and treading of grapes	22.5	17
Transferring wine from the vat and pressing the marc	2.5	2
Upkeep of equipment	2	2
Land tax	12	12
Total cost in francs	134	150

Source: Archives Départementales de l'Aude, 13 M 61, cited in Valentin (1977, 2:108). My estimate for depreciation.

the root must penetrate deeper in the first than in the last."[27] To achieve a depth of 0.45 meter, Du Breuil recommended working the soil first with an ordinary plow to 0.15 meter, and then a further 0.30 meter using a strong plow pulled by six horses or oxen along the same furrow.[28] When the land was inaccessible, or the farmer lacked a plow team, the preparatory work had to be done by hand and labor inputs were much greater, especially if the land had to be cleared or terracing constructed. Augé-Laribé estimated 210 days labor needed to uproot a hectare of dead vines and then dig it to the depth of half a meter, while a study for Valencia gives 340 days to achieve a depth of 0.4 meter and a further 18 days to dig holes and plant the vines.[29] As much of this labor could be supplied during periods of high seasonal unemployment, its opportunity cost was low for family growers. In reality, however, the amount of time used to prepare the land often had less to do with the advice found in technical books than with the price of wine and the immediate profitability of viticulture.[30]

[27] Du Breuil (1867:29). The first edition was published in French in 1863.
[28] Ibid., 30.
[29] Augé-Laribé (1907:139) and Dirección General de Agricultura Industria y Comercio (1891:xv–xvi).
[30] Thus in Valencia (Spain) during the prosperous years between 1880 and 1885, the

cultivation of the vine advanced in intensity year by year, the plantation being made with great care after deep plowings, the digging of spacious holes for the plants, and abundant manuring. Today [1889], circumstances have unfortunately changed . . . , and those cares and plowings,

There are many different grape varieties, so growers had to search for those most suited to their vineyard. Many small growers planted a selection to reduce the risk that their whole harvest would be lost, as varieties differed in their susceptibility to extreme weather conditions or the presence of disease and pests.[31] In addition, as the grapes often ripened at different moments, diversity allowed the harvest to be spread over a longer period and thus reduced the demand for outside labor. A well-balanced wine was far more likely to be produced in Europe from a "harmonious blend" of different grapes than from those of one kind of vine, "however fine some of its qualities may be." This, according to Maurice Tait, was because "one wine may give 'musts' with an excess of acidity, another with superlative colour, a third a specially delicate 'nose,' a fourth may be good in dry years, and a fifth may be better in rather cool shady situations."[32] This was just as true for the classic Bordeaux clarets, where growers looked for a desired balance between cabernet sauvignon, merlot, and cabernet franc, as it was for the cheapest beverage wine.

Vines were sometimes propagated by cuttings, which in France was considered the "simplest," gave the best results, and was "the one most in vogue."[33] The number of specialist nurseries increased with the appearance of new vine diseases but the large number of plants required in some regions and the long time before a reasonable harvest was produced were considered disadvantages. Instead, many growers resorted to layering (*provignage*), which involved taking a long cane from a mature plant and burying it under the soil to where a new vine was required. The following year the vine was detached from the mother plant. If grapes were not produced in the first year, they appeared in the second.[34] In many regions layering was the chosen method to replace dead or diseased vines, allowing vineyards to be kept in production indefinitely. When vine density was to be high in a new vineyard in a region such as Champagne or Burgundy, perhaps a third or half of the total number of vines were planted with cuttings or roots, and then each of these was used to produce one or two additional vines by layering.[35] Grafting was also carried out in some regions, such as in Hérault (the Midi), Champagne, or the Loire, especially to reduce the time required for certain varieties to produce their first crop, but in Spain it was rare outside of Barcelona prior to phylloxera.[36]

---

while still taking place with some regularity, do not reach such a degree of perfection on account of the lack of resources (Spain, Dirección General de Agricultura Industria y Comercio 1891:xv)

[31] Bustamente (1890:28) suggests a maximum of five or six varieties, as weather conditions that were unfavorable for one were favorable for others.

[32] Tait (1936:17–18).

[33] Du Breuil (1867:67–68).

[34] Ibid., 248.

[35] Ibid., 113.

[36] Ibid., 85. Grafting was rare in the Gironde prior to phylloxera (Cocks and Féret 1883:38). In 1877 for Barcelona it was noted that "the grafting of vines is widespread in this province; whether this is because of climatic reasons, the intelligence of the farmer, or for other unknown reasons, the

The number of vines planted per hectare varied significantly, from as many as 60,000 per hectare in the Champagne region to 4,000 in Hérault, falling to 2,000 in Spain's arid La Mancha.[37] The vine density therefore had a tendency to be greater in areas with higher rainfall, and where land was at a premium. It made layering easier and reduced the need for weed control, but it was highly labor intensive.

Pruning was the single most important activity in the vineyard and, if badly carried out, could both ruin the following year's harvest and permanently damage the vine. As one French proverb notes, "anyone can dig me, but only my master can prune me."[38] If too many fruitful buds were left on the vine, the fruit matured slowly and quality was impaired. If too few were left, the quality was generally good but the yield small. As Amerine and Singleton write, "Much skill and considerable luck are required to prune every year to obtain the maximum yield consistent with optimum quality, since grape variety, vine health, the weather, berries set per cluster, and the wine to be made all contribute in determining the optimum crop in any one season for a certain vineyard." Not unnaturally, these authors continue that "economics and human nature conspire to produce more errors on the 'too many' side."[39] Indeed, most of Europe's growers were advised in the mid-nineteenth century to "sacrifice quality for quantity," as the "price will never be sufficiently high to compensate for the diminished yield."[40]

In the low-density, dry vineyards of southern Europe the goblet method of pruning was preferred. Here the vines were left freestanding at about 0.40 to 0.50 meter above the ground, and the grapes grew without supports. This saved the expense of staking, and plowing was less necessary because the ground was covered with the branches, choking the weeds.[41] Elsewhere other methods of pruning were used, and stakes to support the grapes were required in sixty of the seventy-six districts where vines were cultivated in France.[42] At the extreme, in the Champagne region, some sixty thousand wooden stakes were driven into the ground each spring and the vines attached to them, only to be removed again after each harvest so that the wood would not rot during the winter.

In the Médoc the vines were trellised on long laths, which accounted for a

---

graft, which is not only difficult but almost completely unknown in Castile, is done here with amazing ease and the vine grower changes the type of vine when it pleases him." In particular this was the case with the malvasia because, unless grafted with a fast-growing rootstock, it took ten to twelve years to produce fruit (Exposición Vinícola Nacional 1878–79:293).

[37] Du Breuil (1867:96); Spain, Dirección General de Agricultura Industria y Comercio (1891).
[38] Cited in Brennan (1997:17).
[39] Amerine and Singleton (1977:42).
[40] Du Breuil (1867:56, 58).
[41] Ibid., 129.
[42] Ibid., 156.

fifth of annual cultivation costs in the early 1880s.[43] The advantage of trellises is that vines could be grown in straight lines, making it much easier to cultivate between them. A further development, namely, the training of the vines along wires attached to large posts every four or five meters, significantly reduced the number of stakes and labor requirements. Du Breuil describes one system used by Michaux as early as 1845, but it was only with the development of an efficient method for tightening the wire to allow the trellises to remain in position permanently with the vines attached that this method gained in popularity. The savings associated with training vines along wires rather than the annual staking was 88 percent in Champagne and 50 percent in Bordeaux.[44] However, the diffusion of wire trellises on old vineyards was slow, as vines were rarely found in neat, long rows, and many owners adopted the system only when they were forced to replant after phylloxera.

Horse plows could be used when sufficient space existed between the vines, allowing growers to cultivate a larger area. A minimum of two plowings were considered necessary: in February or March after the pruning and fastening of shoots to their supports, and again at the end of the spring.[45] A plow worked by a single animal needed at least a meter of space to work between the rows, so they were common on almost all the vineyards in southern France but were much rarer farther north, where the vines were more densely planted. Hoes were used where the rows were too narrow or the vines no longer grew in straight lines because of layering. When wine prices weakened everywhere in Europe from the late 1880s, wages continued to increase, and reducing tillage costs became critical for some regions to remain competitive. In the Priorat region in Catalonia, for example, the steepness of the terrain made it difficult to use the plow, and the attraction of high-wage employment in the growing urban market of nearby Barcelona led to a rapid decline in viticulture after 1900.[46]

The harvest was the one time of year when even family farms might employ wage labor, and the risk of disease or inclement weather made growers begin as soon as possible. Traditionally municipal ordinances fixed the start of the harvest, which helped reduce theft, but even after the ordinances were abolished, the fact that many proprietors made the wine from a dozen or so different grape varieties mingled together implied that some were inevitably unripe and "too often" imparted tartness.[47]

The significant annual fluctuations in yields encouraged growers to diversify, and in some areas this took the form of cultivating other crops among their

---

[43] Ibid., 166; Cocks and Féret (1883:52). Refers to ordinary wines (Palus region).
[44] Du Breuil (1867:170, 174–76).
[45] Producers of fine wines in Bordeaux carried out four in the 1860s (Cocks and Féret 1868:45).
[46] Perpinyá i Grau (1932:10–11).
[47] Redding (1851:41). In some areas, such as the Côte-d'Or, the ordinances continued during much of the nineteenth century. Lachiver (1988:209–14). For Spain, Pan-Montojo (1994:33).

vines. This was especially common in Italy, where in 1913 some 76 percent of the vines in the north of the country were intercropped (*coltura promiscua*), 85 percent of those in the center, and only in the drier south and the islands did it fall to 12 percent.[48] In France, intercropping (*culture à la Provençal*) was common south of the Loire, at least prior to phylloxera.[49] Where summer droughts restricted intercropping, growers often used only part of their land for vines. An alternative to crop diversification was for growers to work part-time as wage laborers. The large estates in both Bordeaux and the Midi, for example, required significant quantities of skilled labor for specific tasks such as pruning, while elsewhere some family labor was employed as domestic servants or in industry. In addition, small growers who owned a work animal or two could hire them out, and in the Midi as late as the 1950s this helped growers keep vineyards of less than 7 hectares viable.[50] Vineyard ownership helped smooth incomes, and Gérald Béaur has shown how laborers in the cereal land around Chartres in the eighteenth century bought small plots of vines when economic conditions were favorable and sold them again when they were adverse.[51] It could also help social mobility, and in central France it was noted in the late 1870s that land "is still rather cheap, and there is a very large profit derived from putting it under vines; it is a speculation which is undertaken upon a large scale, to buy land, to plant the vine, and to sell the vineyard when the vine begins to produce, which means in about three or four years."[52]

Growers therefore adapted to short-term price movements by varying the quantity and quality of labor inputs, which affected wine output. In the medium term supply was less flexible. While growers were quick to plant more vines at times of shortage, market conditions had often changed by the time these became productive four years later. Growers were then very reluctant to uproot healthy vines because of the large amounts of labor invested in establishing a vineyard, equivalent to about four times the annual demand in cultivation. In addition, the land was usually poor, and the area too small, to cultivate most other crops efficiently. This created a ratchet effect, with growers being quicker to increase supply in response to higher prices than to reduce it at times of falling prices. For much of the nineteenth century this produced only limited problems, as the poor wine-making technologies ensured that periods of overproduc-

---

[48] Table 2.7.

[49] Loubère (1978:9–10). For Bordeaux, see chapter 8. The system was often criticized because the vines had to compete for the nutrients from the soil, and tillage on occasions was delayed because of the other crops (Du Breuil 1867:93–94).

[50] Études et Conjoncture (1953:530).

[51] (Béaur 1998). In Zaragoza, Spain, during the boom of the 1880s, a large number of wage earners planted vines on scrubland, using the profits to purchase small irrigated plots (Rivera y Casanova 1897:93–94).

[52] United Kingdom. Parliamentary Papers (1878–79), F. R. Duval, no. 5517, p. 272.

tion were short because little wine could be carried over to the next harvest. By 1900 conditions had changed.

TRADITIONAL WINE-MAKING TECHNOLOGIES

The highly variable weather conditions facing most European growers implied that grapes needed to be collected quickly, crushed immediately, and the juice fermented. With the exception of fine wines such as port or champagne, or where climatic conditions were sufficiently hot and dry to allow good-quality grapes to be produced most years, virtually all growers preferred to make their own wines. This gave them much more time to find a buyer (or drink the wine themselves) than if they had tried to sell the grapes.[53] There were few economies of scale in traditional wine making, and although the cellar and wine-making equipment, such as presses or storage vessels, were expensive, they were often used over several generations, so production costs per hectoliter of wine differed little between small and large producers.

According to Joslyn and Cruess, "if the fermentation is skillfully conducted, fair wine may be made from relatively inferior grapes, while mistakes at this stage will spoil wine made from the best grape."[54] As the chemical composition of the grapes and the weather during fermentation varied even on a daily basis, producers had to know how to make the necessary adjustments in the fermenting vat. Before Pasteur showed that the spoilage of wine was due to aerobic microorganisms producing acetic acid, which could be avoided by careful wine-making techniques, there was limited knowledge of why wines were good in some years but undrinkable in others. Wine making was often poorly carried out, and Maynard Amerine, for many years chairman of the department of Viticulture and Enology at the University of California, Davis, thought that in the mid-nineteenth century at least 25 percent of the wine spoiled before fermentation was complete, and much of the wine was of a very poor quality.[55] The roles of yeasts, bacteria, enzymes, sugar, and oxygen were largely unknown, and white wines were "usually oxidized in flavor and brown in color; most red wines were high in volatile acidity and often low in alcohol."[56] L. Roos, director of the enological station at Hérault at the turn of the twentieth century, noted that wine produc-

---

[53] Growers would not know the quality of their grapes until they were actually at the winery door, giving excessive power to the winery owner during the price negotiations. In addition, after a large harvest, a winery owner might be unwilling to buy grapes. The relatively high prices of port or champagne allowed winery owners to offer growers a guaranteed market for their grapes. The appearance of specialized grape growers in hot climates is discussed in part 4.

[54] Joslyn and Cruess (1934:12).

[55] Amerine and Singleton (1977:21).

[56] Arata (1903:200–1); and Amerine (1973:63), cited in Pinney (1989:354).

ers "acted without method" and asked: "Are there many vignerons who are able to recall the behaviour of particular vatfuls of the preceding year, the diverse phases of their fermentation, or who possess such a stock of observations as to enable them to deduce the best conditions for the vinous fermentations? They are *raræ aves*."[57]

The first step in wine making is to collect the grapes at the correct stage of ripeness, as this has a major influence on the character of fermentation and wine quality. Growing grapes was difficult in northern Europe's damp and cool climate, but the lower temperatures during fermentation made wine making easier, and the grape's greater natural acidity helped keep the wine free of bacteria. The wines in general were weaker in alcohol content but were cleaner, had more bouquet, and lasted longer. From the eighteenth century some writers recommended the addition of sugar to increase the alcohol content and thus help preserve the wine, and Chaptal gave this widespread publicity in his *Traité théorique et pratique sur la culture de la vigne* (Paris, 1801). Chaptal suggested grape concentrates, partly no doubt because cane sugar at this time was so expensive. Technical improvements in the refining process from the 1820s encouraged the spread of sugar beet production, but prices remained too high for it to be used in commercial wine production on any great scale until the late nineteenth century.

By contrast in hot climates, grapes developed relatively free from disease and with high sugar content, allowing a natural strength in excess of 12 percent. Fermentation, however, was liable to stop prematurely as the wine yeasts weakened when the temperature of the must increased above 35°C, which created favorable conditions for dangerous bacteria to develop, producing volatile acids and mannite that made the wines unstable, imparted disagreeable flavors, and resulted in persistent cloudiness.[58] These wines also suffered because of their lack of acidity, and gypsum (calcium sulfate or plaster of Paris) was often added in southern Europe during fermentation. By the late nineteenth century a number of scientists and wine writers argued that this practice caused a health risk, and its use was banned or restricted in some wine regions.[59] Tartaric acid was recommended instead, either being added to the grapes in the crusher, or spread over the vat while it was being filled.[60]

As on the vineyards, labor was the producer's major cost, and this was mostly supplied by the family. Grapes were crushed by treading (*foulage*), a tedious and expensive operation, which in Jerez, for example, resulted in a worker

<hr />

[57] Roos (1900:11).

[58] Bioletti (1905:20).

[59] In France it was effectively banned by the 1891 law that limited the amount that could be added to two grams per liter (Roos 1900:148). For the use of calcium sulphate in Spain, see Archivo Ministerio de Agricultura, leg. 81–82; Exposición Vinícola Nacional (1878–79:392, 645); and Hidalgo Tablada (1880:173–74).

[60] Roos (1900:148) argued that "*it is better to completely reject this method.*" Emphasis in the original.

crushing only between 4 and 6 tons a day.[61] However, except on the smallest holdings or in areas of fine wines, mechanical crushers had replaced treading in most French wineries by 1914.[62] The simplest and best known consisted of two cylinders operated by hand, which crushed three tons an hour, but it was heavy work and required four men who took turns to operate it.[63] Mechanical crushers were rare among fine wine producers because they crushed the grape seeds and released excessive amounts of tannin into the wine.[64] Thus in Bordeaux it was noted in 1883 that "many machines have been invented for crushing the grape, but none have replaced, *in our vineyards*, the foot of man, the weight of whose body is heavy enough to crush the grains, and the sole of the foot sufficiently flexible to avoid crushing the sour grapes or breaking the stones, which later impart a very disagreeable taste to the wines. This is the only means employed in Medoc."[65]

Red wines were made with the juice, skins, and pips, and in some regions, such as Bordeaux, with stem fragments. Grape stems had also been present during wine-making in southern France, and, according to Coste-Floret in the 1890s, this led to the production of poor red wine that met the "depraved taste of a certain class of consumers who formerly drank our common wines . . . but who appear to have now abandoned us." By contrast, Roos believed it was poor wine-making methods, rather than the presence of stems, that caused the region's inferior wines at this time.[66]

The crushed grapes were fermented in vats. A good fermentation requires healthy, abundant, and vigorous yeasts, which are found naturally on the grape's surface, as well as the presence of oxygen and a favorable temperature of between about 25 and 35 degrees. Fermentation converts sugar to alcohol and gives off carbon dioxide, which pushes the solid material to the top of the vat to form a cap (*chapeau*), where the alcohol and heat produced during fermentation extract color and tannin. The cap has to be kept submerged to avoid acetic acid forming, and various devices were introduced within the vat to automatically keep it below the surface, although Paul Pacottet argued that it was only the labor shortages from 1890 in southern France that caused them to

[61] Ibid., 56. Marcilla Arrazola (1949–50, 2:69–70).

[62] Mandeville (1914:72). In Spain, the two principal surveys in the late nineteenth century (Exposición Vinícola Nacional 1878–79; Spain, Ministerio de Fomento 1886) rarely mention mechanical crushers. See also Elías de Molins (1904:102).

[63] Roos (1900:58). Grape yields were also low. See also Coste-Floret (1894:27).

[64] Mechanical crushers were still rare in the 1960s in the production of port and sherry (Croft-Cooke 1957:72); Jeffs 2004:170). Another difficulty was the imperfect aeration in cylinder crushers, although this could be resolved by allowing the grapes to travel along an open shuttle (Roos 1900:73).

[65] Cocks and Féret (1883:59). My emphasis.

[66] Roos (1900:68–70). Grape stems contain tannin and, if broken during the fermentation, leave wine bitter and astringent, as well as slightly reducing alcoholic content and color (Robinson 2006:227).

become widespread.[67] White wine can be made from either white or red grapes (without the presence of the solid parts), with the former being easier to produce, but the latter producing a better-quality wine.

The size of the vats needed to approximate the quantity of grapes collected in a single day, but very large vats were already being rejected in some places by the 1880s because of the excessive temperatures that large quantities of fermenting wine generated.[68] The choice of the material used was less important than the need for it to be easily cleaned so as not to impair the taste or the chemical composition of the wine, as well as to maximize heat loss caused by the fermentation. In Spain, for example, where wood was often scarce, wine was fermented in stone, cement, or brick tanks, sometimes lined with tiles or plaster, and large earthenware containers (*tinajas*).[69]

The time needed for fermentation varied greatly, depending on the temperature in the vat, whether white or red wines were being made, and according to "custom and experience."[70] As a long fermentation increased the threat of oxidation and the wine turning sour, it was usually brought to an end after six or eight days.[71] The length of time the must remained in contact with the marc (the residue of the grapes) in the fermenting vat also varied significantly according to "the nature of the wine it is proposed to make, to the *cépage* used, to the method of fermentation adopted, the temperature of the vat, and the manipulation the must undergoes during fermentation." In southern France it was usually three or four days but lasted "eight days when the temperature does not exceed 30°C."[72] The wine was then racked (separated from the lees or solid matter) into casks and stored in the cellar to mature, and occasionally fined, before being sold.[73]

---

[67] Pacottet (1924:169). Chaptal had argued for fermentation to be carried out in airtight vats, and although the larger growers in Bordeaux and the Midi began to do this, many specialists questioned the need when fermentation was carried out rapidly. See especially Loubère (1978: 99–101).

[68] In Bordeaux we read in 1883 that "The vats of 20 tuns are becoming very uncommon, as expert workmen have recognized that the capacity of 10 tuns, in large vineyards, is that which harmonizes best with the conditions of a good vinification, which cannot be guaranteed unless the vats have been filled in 24 hours, so that fermentation may not be disturbed" (Cocks and Féret 1883:55–56). This passage is absent in the second French edition (1868).

[69] The advantage of the *tinajas* were that they were cheap and easily cleaned, and the temperature of the must reduced as heat loss was high. They were fragile, however, and there were difficulties in removing the wine from the lees (Marcilla Arrazola 1949–50, 2:163).

[70] Manuel de l'agriculteur du Midi (1831:141). Today red wine fermentations are usually complete within four to seven days, but white wines, which are fermented at a much lower temperature, may take several weeks (Robinson 2006:770).

[71] Loubère (1978:98). The *Manuel de l'agriculteur du Midi* (1831:141) gives a minimum of three and a maximum of nine days. However, the same source notes that a few growers left the wine in the vat for as little as 24 hours, while others left it a couple of weeks.

[72] Roos (1900:145, 147).

[73] A variety of products were used for fining wines, including the whites of eggs, fish glue, and animal blood, and in regions such as Valencia or Barcelona it was carried out by middlemen or exporters rather than individual producers (Spain, Ministerio de Fomento 1886:83–87).

Most wines had to be drunk within about nine months, and because the wooden barrels in which they were traditionally stored and shipped were porous, "wines evaporated from within, air penetrated from without, and unless the barrels were 'topped up' (kept full) the wines, especially the more fragile varieties quickly deteriorated."[74] Many wine producers, especially the smaller ones, were anxious to sell the wine as quickly as possible. By selling it directly from the vat, the grower avoided the costly duplication of storage capacity, which in many instances they did not possess.[75]

The addition of water and sugar to the marc allowed a second, thin wine (*piquette*) to be produced. In the mid-nineteenth century most of the presses were bulky and made of wood and required several men to operate, but by the early 1880s an increasing number were of iron and worked by animals.[76] The second wines were traditionally reserved for farm laborers and family, but during the phylloxera shortages they were often marketed as wine.

Finally, distilling of surplus or poor wines was traditionally widespread in wine-producing countries. Stills were simple and often portable, but by the mid-eighteenth century technical change, improved communications, and growing demand encouraged specialization for cheap spirits in a number of regions, including the Midi and Catalonia.[77] However, the prices paid for wines used for distilling were low, and the process virtually disappeared during the very high wine prices in the 1870s and 1880s.

MARKETS, INSTITUTIONS, AND WINE CONSUMPTION

High transport costs in the prerailway age meant that wines were produced over large geographical areas and consumed locally. Those remaining in the cellar on eve of the harvest were thrown out to make way for the more valuable new wine, and stocks played only a very small role in smoothing out supply from one year to the next. Per capita consumption therefore fluctuated with the size of the local harvest. The harvests of 1865 and 1866 were so large in the Midi, for example, that consumers in *les cabarets* bought their wine by units of time—one *sou* an hour—rather than by the quantity they drank.[78] By contrast, Richard Ford noted in the 1840s that the Spaniard "drinks the wine that grows in the nearest vine-

---

[74] Gough (1998:80).

[75] Richard Ford noted for Valdepeñas (La Mancha) in the 1840s that "the red blood of this 'valley of stones' issues with such abundance, that quantities of old wine are often thrown away, for the want of skins, jars, and casks into which to place the new" (1846/1970:161). In Barcelona in the 1880s, for example, "only in some coastal regions and in Villafranca de Penadés were there containers where the wine could be conserved, and generally the growers left it in the fermenting vats (*trujales* or *cubas*) until they sold it" (Spain, Archivo Ministerio de Agricultura, leg. 81–83).

[76] Cocks and Féret (1883:62).

[77] For the Midi, see Augé-Laribé (1907:36–42) and Pech (1975); for Catalonia, Vilar (1980: 234–35), Segarra Blasco (1994), and Valls Junyent (1996).

[78] Augé-Laribé (1907:84). For price elasticities of wine, see chapter 2.

yards, and if there are none, then regales himself with the water from the least distant spring."[79] Yet local consumption did not necessarily imply peasant self-sufficiency. In France in 1852, when the country still possessed only 3,654 kilometers of railways, 56 percent of wine was sold outside the arrondissement of production, compared with 31 percent of meat and 13 percent of wheat.[80] Even for wheat it has been argued that if most was consumed locally, "we can be sure that perhaps 70 percent of net output was marketed."[81] In Italy, household budget surveys for a slightly later period also suggest that farmers bought and sold extensively in markets.[82]

A major demand restriction was the low level of urbanization, with only 8.8 percent of France's population living in urban centers in 1800, 14.6 percent of Italy's, and 11.1 percent of Spain's.[83] Even these modest figures encouraged some market specialization, although the production and sale of wines were often carefully regulated in the ancien régime. In 1577 the *parlement* of Paris prohibited local retailers (*débitants*) from buying wines within a zone of 20 leagues (88 kilometers) of the capital, a measure that was abolished only in 1776.[84] The aim in part was to encourage trade and hence increase royal revenues, and for two centuries the Île-de-France ceased to supply the capital with wines and the land was used for cereals and sheep farming. The population of Paris increased from 130,000 in 1550 to 430,000 a century later, reaching 576,000 in 1750. By 1800 it totaled 581,000, six times the size of the next largest French city, and was the focus of the nation's long distance trade.[85] The small, family growers of northern France—Champagne, the upper Loire, and lower Burgundy—responded to the Parisian demand:

> The wine produced in the north of France had long enjoyed a reputation for its quality, but the populace of Paris demanded vast quantities of cheap wine. The good wine had been the result of the pinot noir or, in the upper Loire, the related auvernat grapes, grown on the best slopes, which produced a low yield of rich, complex wine. In response to Parisian demand, the regions within reach of Paris turned increasingly to high-yielding grapes, such as the gamay or gros noir, that could grow even on flat land and make an abundance of cheap, unsophisticated wine.[86]

[79] Ford (1846/1970:159).
[80] Demonet (1990:215). See also Postel-Vinay and Robin (1992:496). An arrondissement was typically twenty miles across."
[81] Postel-Vinay and Robin (1992:497).
[82] Federico (1994).
[83] De Vries (1984:45). Urban centers were considered to have ten thousand inhabitants or more.
[84] For administrative purposes the exclusion area was not circular but connected different towns at approximately this distance. A number of exemptions implied that only about three quarters of the capital's wine was supplied outside this radius (Garrier 1998:169).
[85] De Vries (1984:273–75). For the Paris wine market, see Dion (1977, chap. 16); Lachiver (1982); and Brennan (1988).
[86] Brennan (1997:6–7).

By contrast, the incentives for change in those areas outside the direct influence of the Parisian market were much less. The historian Roger Dion explains the survival of Chablis' wines into the twentieth century, almost alone among all the ancient wines of lower Burgundy, by the disadvantage the region suffered from being so far from the Yonne River:

> Beyond a certain distance from water routes, the cost of carting to the nearest river port passes a certain price, the exportation ceases to be advantageous for wines of little value and can only be contemplated for wines of quality. Hence the vineyards of Chablis, inaccessible to commercial shipping, remained faithful to the pinot blanc grape, a vine of quality that many of the vignerons closer to the river routes had abandoned in order to respond to the demands of Parisian commerce.[87]

The same was also true concerning the sale of wines from France's major port, Bordeaux. Growers over a large area of southwestern France were linked by navigable rivers to Bordeaux, and from there to the export markets of northern Europe. From an early date Bordeaux's local wines enjoyed important privileges— the *sénéchaussée privilégiée*—in both the city and export markets.[88] Details changed over time, but the preamble to the Royal Edict of 1776, which abolished these privileges, provides a good indicator of their general nature: wines from outside the city could not be sold before Christmas, and "thus the wine growers of the Haut Pays cannot profit from selling their wine at the most profitable time."[89] When wines were admitted, they could be sold only with numerous restrictions, to the distinct advantage of the Bordeaux growers. The debates concerning privileges of this nature were resurrected in the early twentieth century with the demand for regional appellations (chapter 5).

Wines everywhere were taxed by local, and sometimes national, authorities. The duty was usually levied by volume and consequently, as a percentage of the final price, was heaviest on cheaper wines.[90] Most wines were sold directly from the barrel, and bottles, except for those taken by consumers to taverns to be refilled, were irrelevant for much of the trade. Unlike fine wines, commodity wines were not sold under brand names, but in a world where many economic decisions were carried out face to face, reputation played a major role in the trade. Consumers might be unable to tell exactly what had been added to the wine they were drinking or where it had been produced, but they could make a choice of which retailer to frequent, being influenced no doubt by that person's reputation for fairness, quality, price, or good company.

---

[87] Dion (1977:553), cited and translated by Brennan (1997:133–34).

[88] Kehrig (1884:4).

[89] Dion (1977:393). The potential trade from the Midi with the opening of the Canal des Deux-Mers in 1681 was limited therefore. Legal restrictions also existed limiting the spread of vines at the expense of cereal in eighteenth-century France.

[90] Franck, Johnson, and Nye (2010).

THE DEVELOPMENT OF FINE EXPORT WINES

Most wines were considered items for everyday consumption, and their markets were limited by their poor keeping quality, high transport costs, and taxation. Yet long before the mid-nineteenth century, a handful of wines had been created in select regions such as Bordeaux, Porto, or Jerez that were very different. In particular, between the mid-seventeenth and late eighteenth centuries, the production of specialized fine wines underwent changes that were as spectacular as any of those that took place in Britain during the so called Agricultural Revolution over the same period. André Jullien, in his classic *Topographie de tous les vignobles connus*, shows that the major areas of fine wine production were already well established by 1816, and it was only with the appearance of fine wines in the New World over the final couple of decades of the twentieth century that this map was radically changed. However, the nature of these fine wines was completely transformed over the seventy or eighty years prior to the First World War, as consumers' preferences switched away from sweet and heavy wines toward lighter and drier ones, together with the development of vintage ports and champagnes.

Fine wines can be divided into two major categories: dessert wines, such as sherry, port, malaga, madeira, or tokay, which are fortified during production and whose additional alcoholic strength allows them to be kept for several years and withstand rough treatment and transportation without being ruined; and nonfortified ones, including claret, burgundy, and champagne. The production of quality distilled wines, such as cognac, also dates from the same period. In fine wine production, grapes had to be selected from shy-bearers grown in a few highly favorable sites, and producers needed considerable grape-growing and wine-making experience and large quantities of capital. Not only was Britain the most important market for many of these wines before 1914, but merchants contributed directly to the development of their styles in response to demands in that market.[91]

The addition of brandy stopped fermentation, and the remaining glucose left the wine sweet, while the high alcoholic content kept it stable, allowing it to be transported long distances without deteriorating. Port was a drink developed by British merchants, and the trade benefited greatly from the Methuen Treaty of 1703, which restricted duties on Portuguese wines to a maximum of two-thirds of the custom duties paid for French wines in the British market.[92] Merchants experimented to find the best moment and quantity of brandy to be added during fermentation. According to Ralph Davis, until the mid-eighteenth century nearly all wines were drunk during the year of making but, "between 1780 and

[91] This is dealt with in more detail in part 3.
[92] Nye (2007); Simpson (2010).

1830 the occasional practice of allowing long maturity for port and sherry became general. . . . [In 1780] very little wine was kept for more than two years; by the latter, it was taken for granted that port should be kept in a bottle for ten to fifteen years to develop the genuine aroma of the wine, and that sherry needed to be mellowed by at least five years in cast, and did not attain its full perfection for fifteen to twenty years."[93]

Madeira's producers benefited from favorable trade winds that brought American-bound ships to its port, and the tax privileges granted in the British Sugar Act (1764). If the wine they produced in 1640 was cheap and simple, "made from a base of white must to which growers and exporters added varying amounts of red must in order to give it color and flavor," by 1800 it was "a complex, highly processed, expensive, and status-laden beverage."[94] One drawback of adding lower-density alcohol such as brandy was that it produced "a harsh, bracing, and often uneven taste," which required wines to be matured for much longer. Shippers observed, however, that this process was speeded by long sea voyages, and the improvement was even more rapid when the journey was to a hot climate. Wine was deliberately sent via the West Indies as early as 1749, and via the East Indies in 1772, and a circuit of "floating ovens" had been established by 1775. The American Revolutionary War led to producers and exporters shifting the process of heating to Madeira itself, with the creation of stoves, ovens, and hothouses. Steam engines were installed to artificially create the conditions of a ship's movement. According to David Hancock, "a wine that would be palatable to Americans only after four or five years in England, three years in Madeira, or one year in the East or West Indies could be readied in a stove in three or six months."[95]

Sherry also became popular, especially from the early nineteenth century. In 1772 Jean Haurie successfully challenged the growers' guild and removed the institutional restrictions to shippers making and storing their own wines, and the development of the solera system of production gave them considerable flexibility to allow merchants to blend wines of a consistent nature and quality, and for British retailers to establish their own brands.[96] As with port and madeira, it was the ingenuity of the producers and shippers, mainly British, who changed wine-making procedures in response to market demand in their home and colonial markets. The capital requirements for production and trade soared with the increasingly complexity of these wines from the late eighteenth century.

By contrast, the best and most expensive nonfortified wines were produced in northern Europe.[97] Before the late seventeenth century, when bottles and corks began to be used to protect the wine from air, wines had to be drunk within a

---

[93] Davis (1972:96).

[94] Hancock (2009:73).

[95] Ibid., 89, 92–93.

[96] For the solera, see chapter 8.

[97] Berget (1908) links the presence of fine wines and difficult growing conditions.

year of their production, and most Bordeaux wines were shipped, if not before Christmas, then in the early spring.[98] The development of fine wines in the Bordeaux region can be dated to between about 1650 and 1740. In 1647 the highest prices paid for wines were for those produced in the Palus region, but by 1730 in real terms these had changed very little, while the best Médoc wines now sold for five or six times more.[99] Perhaps the most famous wine ever drunk was the Ho Bryan that Samuel Pepys enjoyed on April 10, 1663, which he noted "hath a good and most perticular taste that I never met with."[100] By the early eighteenth century, all Bordeaux's first growths appeared to have been drunk with a certain frequency in Britain, which was a more important market than Paris. In 1797 Christie's sold its first vintage claret, six hogsheads of "first-growth claret" of 1791.[101] Among Bordeaux's thousands of small vineyards producing common table wines, there were now approximately 250 vineyards, often owned by the nobility, that produced 80,000 hectoliters of fine wines and covered some 3,200 hectares.[102]

The creation of burgundy and champagne has been explained by the response of growers and négociants to lost markets for their traditional wines.[103] The improvements in water transportation in the seventeenth and eighteenth centuries opened up the rapidly growing Parisian market to the wines of lower Burgundy, which in the case of Beaujolais "was practically created" by this trade.[104] However, this reduced the competitiveness of the wines from northern Burgundy along the Côte d'Or, and it encouraged growers to produce better-quality wines for foreign markets. The same was true in the Champagne region. The late seventeenth and early eighteenth centuries saw major advances in knowledge concerning the production of sparkling wines. This trade was very risky, not least because of the difficulties in controlling the second fermentation and the large losses through broken bottles. Thomas Brennan suggests that it was only the difficulties surrounding unsold young wines for the Paris market that encouraged some growers to turn to bottling. Therefore, "in both Burgundy and Champagne, commercial difficulties pushed growers and brokers away from mass markets, which were all too stagnant from growing com-

---

[98] The inability to store wines for any length of time followed the replacement of sealed amphorae by the use of wooden barrels at the end of the second century AD (Unwin 1991:13).

[99] Lachiver (1988:294–302).

[100] Latham and Matthews (1970, 4:100).

[101] The *London Gazette* of 1707 advertised wines from the châteaux "Margaux," "La Tour," and "Lafitte" as "New French Prize Clarets" (Pijassou 1980:372–79). Walpole consumed large quantities of Margaux, Lafite, and Pontac (Ray 1968; and especially Penning-Rowsell 1973:108–10).

[102] Lachiver (1988:303).

[103] Brennan (1997:240). Theoretically négociants worked for themselves whereas brokers worked for others, although Brennan (p. 209) argues there was often some overlapping of their functions.

[104] Ibid., 210.

petition, and toward elite markets."[105] To sell the expensive wines, local brokers were forced to look for new markets instead of waiting for buyers to appear.[106] Once again, by controlling sales in distant markets, producers were able to create brand names that have lasted until today.

Finally, this process of specialist wine production was accompanied by the rapid growth from the eighteenth century in the distilling of wines. The ageing of spirits had become well understood in Cognac by the 1720s, and growers were switching from producing sweet white wines to acidic ones. Unlike most wines, these needed to be distilled only twice, resulting in a superior brandy to those produced elsewhere.[107] Important local families, such as Otard, Dupuy, Hennessy, and Martell, which would dominate the future trade, were already present by 1760. If Dutch buyers had originally turned what had previously been a small, widely diffused activity into a commercial enterprise, by the late eighteenth century England had established itself as the major market for quality brandy. According to the merchant James Delamain, the best judges of good cognac were to be found in London in 1788.[108] As with all successful beverages, by the turn of the nineteenth century cognac was attracting an increasing number of imitations.[109]

European producers over the centuries learned to specialize and develop wines suitable for their particular terroir and markets, and major geographical shifts to new areas occurred well before the railways, as Alexander Henderson noted in *History of Ancient and Modern Wines*, published in 1824:

> In tracing the history of French wines, we are struck with the fact, that many vineyards, which have now little or no repute, were renowned in former times for the excellence of their growths; while those which, of late years, have maintained the greatest celebrity, were then unknown, or almost unnoticed. Thus, the wines of Orleans and of the Isle of France were at one time in greater estimation than those of Burgundy and Champagne; and even Mantes, which is on the borders of Normandy, was famed for the produce of its vines.[110]

The railways would give growers the opportunity to plant new vineyards on land that had previously been considered unsuitable for commercial wine production and to compete with the old traditional areas. As Europe's population grew during the first two-thirds of the nineteenth century, so too did the area under vines. In France it is estimated that the area increased from about 1.6 million hectares in the late eighteenth century to a maximum of 2.5 million in

---

[105] Ibid., 240.
[106] Ibid., 209.
[107] Faith (2004:63); Cullen (1998:45).
[108] Cullen (1998:4, 17, 95–96).
[109] Lachiver (1988:582–83).
[110] Henderson (1824:146).

1874.[111] In Spain no reliable figures exist, but the area of vines perhaps increased from just under a million hectares in 1800 to just under two million in the mid-1880s.[112] In Italy, the widespread use of intercropping in the center and north of the country makes it difficult to establish accurately the area of vines even at the end of the period, but contemporaries speak of a similar growth.

The appearance of new commercial opportunities coincided with scientific research that began to unravel the secrets of the wine-making process. Yet even with limited scientific knowledge, some fine wines were produced and sold at what contemporaries considered astronomical prices in the early nineteenth century. As growers looked to plant vines in new areas, they naturally were attracted to the fame and high prices achieved by the top producers in places such as Bordeaux or Reims. Indeed, the poor quality of much of the imported wines found in the United States or Argentina that purportedly came from the leading wine estates led local growers in places such as California and Mendoza to believe that they could successfully compete. However, the challenge of producing a fine wine that could be sold at a significant price premium was considerable. James Busby noted in the 1830s that three major factors explained why a few wines were consistently better and sold at high prices. In the parlance of the twenty-first century, these were location, human capital, and physical capital. Busby was essentially interested in the vine's prospects for the New World (and especially New South Wales), so his greater emphasis on human and physical capital than on terroir is of special interest. His comments are worth quoting in full:

> The limited extent of the first rate vineyards is proverbial, and writers upon the subject have almost universally concluded that it is in vain to attempt accounting for the amazing differences which are frequently observed in the produce of vineyards similar in soil and in every other respect, and separated from each other only by a fence, or a footpath. My own observations have led me to believe, that there is more of quackery than of truth in this. In all those districts which produce wines of reputation, some few individuals have seen the advantage of selecting a variety of grape, and of managing its culture so as to bring it to the highest state of perfection of which it is capable. The same care has been extended to the making, and subsequent management of their wine, by seizing the most favourable moment for the vintage,—by the rapidity with which the grapes are gathered and pressed, so that the whole contents of each vat may be exactly in the same state, and a simultaneous and equal fermentation be secured throughout,—by exercising equal discrimination and care in the time and manner of drawing off the wine, and in its subsequent treatment in the vats or casks where it is kept,—and lastly, by not selling the wine till it should have acquired all the perfection which it could acquire from age, and by selling, as the produce of their own vineyards, only such vintages as were calculated to acquire or maintain its

[111] Lachiver (1988:582).
[112] See Pan-Montojo (1994:384–93); Simpson (1986).

celebrity. By these means have the vineyards of a few individuals acquired a reputation which has enabled the proprietors to command almost their own prices for their wines; and it was evidently the interest of such persons, that the excellence of their wines should be imputed to a peculiarity in the soil, rather than a system of management which others might imitate. It is evident, however, that for all this a command of capital is required which is not often found among proprietors of vineyards; and to this cause, more than to any other, it is undoubtedly to be traced, that a few celebrated properties have acquired, and maintained, almost a monopoly in the production of fine wines.[113]

Yet for the great majority of wine producers, to improve quality and make a profit were two quite different and often incompatible objectives. While some producers might call their wines claret or sherry, the economic incentives for the vast majority were to attempt to maximize production. Vineyard location, grape choice, pruning method, and a whole host of other variables were chosen to produce a wine that the consumer wanted, namely, one that was cheap. Only in the last couple of decades of the twentieth century did wine-making technology change sufficiently to allow fine wines to be made outside a few chosen areas, such as the Médoc, and were there sufficient consumers willing to pay high enough prices to make it economically sensible to invest heavily in regions such as Carneros or the Ribera del Duero. In this respect perhaps the surprising feature of the so-called Judgment of Paris blind wine tasting in 1976 was not that some of France's leading connoisseurs preferred California wines to those from their own country, but rather that a winery such as Stag's Leap Wine Cellars had been operational for only four years. All this is a far cry from the mid-nineteenth century, when 99 percent of producers were happy just to be able to produce wine that was sufficiently drinkable that it could be sold.

Finally, taxes on alcoholic beverages seriously distorted markets, with tariffs in most producer countries limiting imports to just fine wines. In Britain, taxes on all types of alcohol contributed 36 percent of national revenue in 1898–99, but they were also 19 percent in France (1898), 18 percent in Germany (1897–98), and 28 percent in the United States (1897–98).[114] In addition, wine provided significant revenues for local governments in producer countries, further restricting demand and market specialization. Merchants talked of consumer taste, but then, as today, this was often shaped by the prevailing tax regime.

---

[113] Busby (1833:106–7). Charles Higounet (1993: 110) notes that "from the eighteenth century onwards what differentiates the leading Médocain wine château from the small or average sized peasant of *bourgeois* property is the ability and the willingness to invest often considerable amounts of capital to improve the *terroir*, the natural source of the product."

[114] Tax figures given in *Ridley's Wine and Spirit Trade Circular*, April 12, 1900, p. 258, hereafter cited as *Ridley's*.

CHAPTER 2

Phylloxera and the Development of Scientific Viti-Viniculture

> The Midi's high yielding viticulture is perhaps the most industrial of France's farming systems, caused as much by the high levels of capital used as the production systems employed: a monoculture; the degree of mechanization in the vineyard and winery; the use of piece work and the creation of a salaried proletariat; and labor disputes between property owners and unions.
>
> —Michel Augé-Laribé, 1907:11

EUROPE'S GROWERS, winemakers, and merchants had to adapt to a series of important exogenous shocks in the half century prior to the First World War. On the demand side, the decline in transport costs produced by the railways, rapid urbanization, and rising incomes led to per capita wine consumption in France reaching more than 160 liters in the 1900s, and there were significant increases in other countries, such as Italy, Portugal, and Spain. The growth in consumption was all the more impressive given that the vine disease *phylloxera vastatrix* destroyed large areas of Europe's vineyards. The destruction initially was greatest in France, and rising prices encouraged wine merchants to look for new sources of supply in previously neglected regions in countries such as Spain or Algeria, while in the New World they spurred growers to increase production and become less dependent on imports from the Old World. The boom ended with the recovery in French domestic production and by the early 1900s prices collapsed, leading to protests by growers, especially in the Midi. Other countries, such as Spain, suffered less from the low prices because phylloxera was now destroying their own vines in large quantities and thereby reducing the supply of wine.

The only long-term solution to phylloxera was to uproot the dead vines and replant using American, disease-resistant roots stock, and grafting European scions. Considerable scientific research was undertaken to find suitable vines that produced both a drinkable wine and adapted successfully to the site-specific characteristics of each vineyard. The new vines produced higher yields, but they were more susceptible to other diseases, which required heavy expenditure on chemicals, so small growers had either to spend valuable cash or risk heavy crop losses. Scientific advances were just as spectacular in the wineries, with Pasteur's work on fermentation providing an understanding of the causes of fermentation and how to keep wines in good condition. One major break-through of the period was the ability to produce good, cheap wines in hot climates, which allowed a rapid reallocation of production to these new regions, so that by 1914 the Midi and Algeria produced the equivalent of about half of French wine consumption. Montpellier became the world's center for new wine-making technologies, but advances were quickly reported in other regions with hot climates thanks to individuals such as Frederic Bioletti (University of California), Arthur Perkins (Roseworthy, South Australia), and Raymond Dubois (Rutherglen, Victoria).

The new, modern wineries were capital intensive, and because of the economies of scale, production costs of a liter of wine were lower and wine quality better than in the small, family-operated wineries. They also needed large quantities of grapes if they were to be worked at full capacity, and in the Midi and Algeria, unlike in the New World, these were usually produced by winemakers integrating backward into grape production. Large, high-yielding vineyards were planted on the fertile plains and valley floors, and owners used labor-saving plows and introduced new work practices to coordinate wage labor in vineyards. Increasing wine output in these regions no longer depended on small growers using their underemployed labor to slowly extend their vineyards to create new capital assets; instead, producers looked to commercial banks for credit to plant vineyards on a major scale and create modern wineries. While small, family vineyards producing grapes remained competitive, they found it increasingly difficult to compete with the scale-dependent methods used in wine making.

This chapter looks at the growth in wine consumption in the second half of the nineteenth century and shows the impact of phylloxera on the French market, and how the stimulus of higher international prices led to a wine boom in Spain. Finally it discusses the development of scientific viticulture and wine making, and the appearance of large-scale wineries in the Midi and Algeria.

THE GROWTH IN WINE CONSUMPTION IN PRODUCER COUNTRIES

The railways transformed Europe's economy. As early as 1858 Paris was connected to all the country's major wine regions, and prior to phylloxera in the late 1860s French railways annually transported three million tons of wine and spirits, a figure that had tripled by 1913.[1] Transport costs fell by four-fifths in the Midi, helping the region to become *"une veritable monoculture de la vigne"* by 1900.[2] The railways pushed Europe's wine frontier into regions long known to contemporaries as being especially suitable for the vine—the Midi (France), La

---

[1] Price (1983:245, 296).

[2] Lachiver (1988:410). For example, after 1858 the cost of transporting a *muid* of wine from Montpellier to Lyon fell from 50 to 7 francs (Degrully 1910:324). The cost from Montilla (Córdoba) to Jerez fell by 75 percent. See chapter 8.

TABLE 2.1
The Growth of Railways in France, Italy, and Spain, 1850–1910 (thousands of kilometers)

	1850	1880	1910	1910 as % of 1930
France	2.9	23.1	40.5	96
Italy	0.6	9.3	18.1	82
Spain	—	7.5	14.7	86

Source: Mitchell (1992:655–59).

Mancha (Spain), and Puglia (Italy)—allowing them to specialize for the growing urban markets.[3]

The railways, by helping to integrate markets, encouraged urbanization. The percentage of the population living in French towns and cities of more than ten thousand increased from less than 10 percent in 1800 to 15 percent by 1850 and 25 percent by 1890, with Paris having more than 2.5 million inhabitants in 1896.[4] By 1910, 27 percent of France's population lived in centers of more than twenty thousand, and 59 percent worked outside the agricultural sector. In Spain and Italy, the urban population was 23 and 28 percent, respectively, and the nonagricultural labor force was 34 and 41 percent in 1910.[5]

Living standards also improved significantly. Between 1850 and 1913 gross domestic product (GDP) per capita doubled in France, Italy, and Spain.[6] Real wages of unskilled urban workers increased in France by about two-thirds between 1850 and 1910, and a similar improvement seems likely for Italy and Spain.[7] Engel's law suggests that consumers devote a smaller proportion of their income to food when living standards improve, but demand elasticities behave very differently according to the food item. Rising incomes in late nineteenth-century London, for example, resulted in a rapid growth in the consumption of fresh fruit, vegetables, milk, and meat but a falling demand for bread. In Europe, rising incomes led to an increase in wine consumption. In France, expenditure elasticities in 1852 for alcoholic beverages (wine, beer, and cider) among farm

---

[3] Water transport remained the most economical, and as late as 1903–5 some 40 percent of the wine that entered Paris's bonded warehouses came by boat, compared to 53 percent brought by rail and 7 percent by road (Richard 1934:20). My calculations.

[4] In France the figure was 8.8 percent in 1800; in Italy, 14.6 percent; and in Spain, 11.1 percent. By 1890 in France the figure was 25.9 percent; in Italy, 21.2 percent; and in Spain, 26.8 percent (De Vries 1984:45–46). For Paris, Pinchemel (1987:146–47).

[5] Simpson (1995, table 8.5).

[6] Maddison (1995:104–8). The increase was 109 percent in Italy, 107 percent in France, and 97 percent in Spain.

[7] Williamson (1995:164–66). For Italy, wages increased by two-thirds between 1880 and 1910. Rosés and Sánchez-Alonso (2004:407) give an increase of 53 percent for nonskilled urban wages between 1854 and 1914 in Spain.

TABLE 2.2
Per Capita Wine Consumption in Producer Countries, 1840s–1900s (liters)

	France	Italy	Spain	Portugal
1840s	106			
1850s	79			
1860s	127		61	
1870s	133			
1880s	97	99	72	
1890s	111	92	86	92
1900s	153	120	83	92

Sources: France, *Annuaire statistique année 1938* (1939:463); my calculations; and Simpson (1985a: 122).

laborers were strongly positive, and an increase in household income of one franc produced a growth in consumption of one and a half francs. Price elasticities were equally strong at −1.33, so that "expenditure on beer and wine increased rapidly when income improved, but quickly retreated when these drinks became more expensive."[8] In France wine consumption rose from 76 liters per capita in 1850–54 to 108 liters in 1890–94, peaking at 168 liters in 1900–1904.[9] The quantity consumed by producers and their families (and therefore exempt from taxes) grew from 5 to 9 million hectoliters between 1850–54 and 1900–1904, while the increase in off-farm consumption jumped from 18 to 42 million hectoliters.[10] Information for the other major producer counties, such as Spain, is much less reliable, but lower levels of urbanization and living standards help explain why consumption remained lower than in France on the eve of the First World War (table 2.2).[11] One major restriction in all producer countries was indirect taxes, which, because they were levied by volume rather than ad valorem, increased in relative terms during periods of low wine prices.

The growing market integration had an impact not just on the quantity of cheap wine produced but also on the quality. In central France it was noted:

> there have been great improvements in the making of wines going on for some years. The thing is very easy to understand; formerly we had no railways, and frequently very poor roads, so the wine was only drunk on the spot, now we are beginning to send the wine upon a large scale to Paris, and some rather distant districts of France; therefore the wine must be able to bear the journey, and to travel in any temperature, and so on.[12]

[8] Postel-Vinay and Robin (1992:503–4).
[9] Nourrisson (1990:321).
[10] Calculated from Degrully (1910:320–21).
[11] For Spain, see *Archivo Ministerio de Agricultura*, leg. 68, exp. 1, cited in Pan-Montojo (1994: 41).
[12] United Kingdom. Parliamentary Papers (1878/79), F. R. Duval, no. 5545, p. 273.

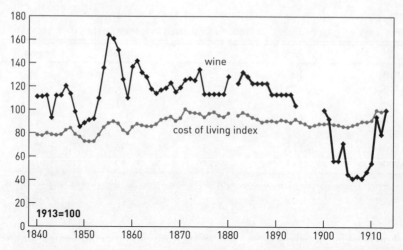

Figure 2.1. Price movements in Paris, 1840–1913. Source: Singer-Kérel (1961:452–53 and 472–73)

Wine quality and harvest size varied considerably, and, as noted in chapter 1, a major function of merchants was to reduce price volatility for urban consumers by blending wines from a variety of sources to compensate for these significant local annual fluctuations. As a result, prices peaked in Paris in 1855 with the appearance of powdery mildew but then remained remarkably stable until the end of the century when they fell sharply, despite the collapse in domestic wine production during the 1870s and 1880s. The movements in wine prices in Paris reflect those of the general price index, with the exception of the significant jump in the mid-1850s and fall in the 1900s (fig. 2.1). Farm-gate prices, by contrast, rose sharply, first during the powdery mildew epidemic and then during the 1870s and 1880s because of phylloxera (fig. 2.2).

PHYLLOXERA AND THE DESTRUCTION OF EUROPE'S VINES

The activities of botanists in studying and classifying local varieties, and growers in improving them, encouraged the movement of plants from one country to another.[13] With the faster shipping times in the Atlantic trade and increased trade in plants, European farmers suffered from a number of new and devastating diseases, such as potato blight, pébrine, powdery mildew, phylloxera, and foot-and-mouth. Governments often responded by banning imports of plants

[13] One of the first volumes of any importance was Simon de Roxas Clemente y Rubio, *Ensayo sobre las variedades de la vid común que vegetan en Andalucía* (1807). For a general background of the connection between the "Industrial Enlightenment," technological change, and economic growth, see Mokyr (2004).

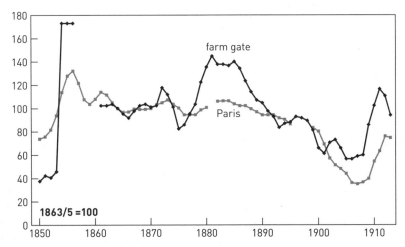

Figure 2.2. French wine prices, 1850–1913. Sources: France. *Annuaire statistique année 1938* (1939:62–63) and Singer-Kérel (1961:472–73)

and live animals, but once the diseases had breached national boundaries policies were needed to eradicate or control their spread.

The first major new vine disease was powdery mildew, which appeared in the 1840s, although its economic impact was delayed by a decade or so. Powdery mildew caused French production to slump to just 17.6 million hectoliters between 1853 and 1856, and output between 1851 and 1861 surpassed 41.7 million hectoliters only once, a figure that the country had averaged during the decade between 1832 and 1841 (fig. 2.3).[14] Wine prices doubled in the 1850s, causing French farm laborers, whose household income increased by 53 percent and food expenditure by 44 percent, to actually cut drink consumption by 25 percent,[15] while nationally per capita wine consumption fell by 21 percent between 1849–51 and 1859–61. The shortages were short-lived, however, as it was found that dusting the vines with sulfur checked the spread of this fungal disease, and the recovery in French output by the late 1850s is an indicator of the widespread use of chemicals.[16] Powdery mildew had now become endemic, reappearing especially in damp, warm years, increasing growers' annual production costs.

Some European growers imported vines that were immune to powdery mildew from eastern and southern regions of the United States, and by doing so

---

[14] Calculated from Lachiver (1988:582). No production figures exist for 1842–49.

[15] Meat increased by 27 percent and bread by 7 percent (Postel-Vinay and Robin 1992:506–7).

[16] If this took place too close to the harvest it would affect the taste of the vine, and in warm weather excessive spraying made workers ill. The Médoc châteaux were reluctant to use sulfur on a large scale before 1857–58 (Loubère 1978:79).

they inadvertently introduced a new disease.[17] Phylloxera, which was first noticed in 1863, spread much more slowly than powdery mildew, but its long-term economic consequences were considerably greater. This tiny aphid attacked and destroyed the root system of Europe's *Vitis vinifera* species, usually killing the plant within a couple of years. In time phylloxera destroyed virtually all Europe's vines, and permanent barriers to its devastation existed only in a few areas, such as on the sandy soils of the Camargue.[18] In France between 1868 and 1900, some 2.5 million hectares of vines were uprooted at an estimated cost of fifteen billion francs, while chemicals, imports of vines, replanting, and grafting added another twenty billion.[19] Yet the speed of this devastation varied significantly across the continent, as suggested in map 3. In Spain, some 277,000 hectares were infected fifteen years after the disease had been first identified in 1878, but the following fifteen years saw a million hectares of vines destroyed and another 125,000 infected.[20] Even so, the major wine-producing region of La Mancha only became infected in the 1930s. The spread of phylloxera was even slower in Italy, where as late as 1912 less than 10 percent of the nation's vines had been infected.[21]

In France, wine output, which had averaged 57.4 million hectoliters in 1863–75, fell to 31.7 million in 1879–92 before recovering to 52.5 million in 1899–1913 (fig. 2.3).[22] The French government offered a prize of 300,000 francs in June 1873 for an effective remedy to save the nation's vines, and although some 696 suggestions had already been studied by October 1876, it was never awarded.[23] A number of temporary solutions halted phylloxera's march, but all were expensive as they had to be applied annually. In 1873 the flooding of vineyards was shown to be successful, but this required relatively large and compact holdings on level ground close to good supplies of cheap water. Two chemical solutions were also used: injecting the vines' roots with liquid carbon bisulfide or spraying with sulfocarbonate. Although costly, they allowed growers to remain in production while wine prices were strong in the 1870s and 1880s and offered

[17] As early as 1872 Jules-Émile Planchon of Montpellier University suggested that the large imports of plants, rather than cuttings, had provided the medium for introducing the phylloxera aphid into Europe (Campbell 2004:108). By contrast, Ordish (1972:5) argues that the faster shipping times allowed the aphid to survive on plants imported by scientists and growers. Total imports of all plants to France jumped from 460 tons in 1865 to 2,000 tons by the 1890s (Robinson 2006:522).

[18] For the very few isolated areas in Europe that survived phylloxera, see Campbell (2004:275–77). Prices for the better land in the region of Aiguesmortes rose after the outbreak of phylloxera from 100 francs a hectare to 3,000 (Ordish 1972:94–96).

[19] Galet (1988), cited in Paul (1996:16). Trebilcock (1981:157) suggests a "final bill" in excess of £400 million (10 billion francs), equivalent to 37 percent of the average annual GDP for 1885–94.

[20] García de los Salmones (1893, 1:14); Spain, Ministerio de Fomento (1909:192–93). In 1893 some 14,871 hectares had been replanted, compared with 323,858 hectares in 1909.

[21] Loubère (1978:175).

[22] Calculated from Lachiver (1988:582–83).

[23] Ordish (1972:68, 70).

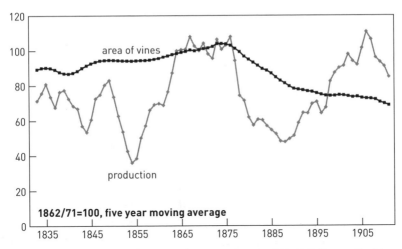

Figure 2.3. Area of vines and wine production in France, 1833–1911. Source: Lachiver (1988:582–83)

fine wine producers more time to adapt. Indeed, as chapter 5 shows, the phylloxera epidemic actually witnessed a major increase in fine wine production in Bordeaux. From 1884 the effects of phylloxera in France slowed, in part because of the widespread destruction that had already taken place, and in part because of the introduction of quarantine and restrictive measures in unaffected areas (table 2.3)[24]

As Jules-Émile Planchon, professor of botany at Montpellier, remarked, the phylloxera paradox was that the source of the disease, namely, certain indigenous American vines, was also its long-term cure.[25] Planchon had argued as early as 1877 that by grafting European scions (the chosen grape variety) to American phylloxera-resistant rootstock, both vines would retain their own characteristics: the rootstock its immunity to phylloxera and the scion its traditional wine quality.[26] However, French scientific opinion split into two distinct camps: the "chemists," who wanted to save the country's ungrafted *vinifera* vines, and the *américainistes*, who demanded the import of large quantities of American phylloxera-resistant vines.[27] Prior to the mid-1880s the French government sided with the chemists, passing legislation for local authorities to control the movement of vines, destroying infected vineyards, and providing grants for chemicals. However, from the very late 1870s the use of grafted vines became increasingly popular, and the declaration of a four-year tax morato-

[24] *Pacific Rural Press* (1901:372).
[25] Cited in Guy (2003:89–90).
[26] Campbell (2004:160).
[27] For details, see Pouget (1990); Paul (1996); and Campbell (2004).

TABLE 2.3
Phylloxera and the Response of French Growers, 1878–97 (thousands of hectares)

	Diseased	Destroyed	Submersion	Carbon bisulfide	Sulfo-carbonate	Grafted American vines	Total area of vines	Wine price (francs per hectoliter)
1878	243.0	373.4	2.8	2.5	0.8	1.4	2,296	29
1880	454.3	558.6	8.1	5.5	1.5	6.4	2,209	43
1884	664.5	1000.6	23.3	33.4	6.3	52.8	2,041	40
1888			33.5	66.7	8.1	214.7	1,844	30
1890			32.4	62.2	9.4	436.0	1,817	29
1894			35.3	50.5	8.7	663.2	1,767	20
1896			37.4	40.2	10.2	797.1	1,728	26
1897						962.0	1,689	24
1905							1,669	14
1910							1,618	39

Source: Galet (2004, 1:24–25). Wine prices: France, Annuaire statistique année 1938 (1939:63).

rium on vineyards replanted with American vines in 1886 marks the turning point of the French government in its switch in support to the américainistes. The slow response of the central government to replanting can be explained by the demand of local growers to keep foreign vines out of disease-free villages and regions, especially while there was still hope that chemists might suddenly discover a miracle solution that would make replanting unnecessary, just as they had done earlier with powdery mildew and potato blight. Finally, the quality of the wine produced using American vines was initially atrocious.

The new vine diseases not only caused interruptions in wine supplies but also radically changed the organization and skills required in viticulture. In prephylloxera viticulture there were few barriers to entry for growers, as production, as noted in chapter 1, consisted essentially of two inputs: land, which was often marginal for other crops, and labor. This now changed, as the uprooting of dead vines and replanting implied heavy capital costs. Deeper plowings were required to prepare the land if the new, postphylloxera vines were to maximize their output, and this required expensive steam plows, which produced considerable savings over hand labor but were impractical on small or fragmented holdings.[28] A distinction appeared in some areas between intensive, high-yielding "capitalist" viticulture, and low-yielding, less-intensive "peasant" farming.[29] Vineyards were

[28] Deep plowing facilitated the rooting and growth of young vines and economized hand labor. In the words of one contemporary, "spend more money and labor in getting your land into good order and you will have to spend less in trying to keep it in good order" (Bioletti 1908:54).

[29] Carmona and Simpson (1999:307) describe this in Catalonia.

no longer self-sufficient, as growers could not replace vines by layering but had to purchase from nurseries the American rootstock that was suitable for the conditions on their own vineyard and compatible with the chosen European scions. Some combinations performed much better than others, and this information was not easily available to growers. The new vines were more delicate and susceptible to fungus diseases, requiring expenditure for sulfur for powdery mildew, or "Bordeaux mixture" (copper salts) for downy mildew. Finally, the economic life of the vines was between twenty and thirty years, considerably shorter than for traditional vines. There were some benefits, however, as the destruction caused by phylloxera forced governments to make significant contributions to scientific research, which led to a much greater understanding of viticulture, and its divulgence in a number of classics on the subject, such as Gustave Föex's *Cours complet de viticulture* (1886) and Viala and Vermorel's seven-volume *Traité général de viticulture ampélographie* (1910). The new vines also came into production earlier, and the development of hybrids allowed significantly larger yields, forming the basis for a new "industrial viticulture."

Hybrids can occur naturally, but those of the late nineteenth and early twentieth centuries were the result of the deliberate crossing of two different species in an attempt to "combine the desirable wine quality of European *vinifera* varieties with American vine species' resistance" to American pests and diseases, especially the phylloxera louse.[30] As direct production hybrids required less care and chemicals than *vinifera* vines, they became the "easiest way to obtain cheap wine in difficult times" and were quickly adopted by commodity wine producers.[31] According to the historian Harry Paul, much of the bad reputation for direct producers came from the indiscriminate spread of a number of infamous varieties, which the French government would eventually ban.[32]

The increase in capital requirements to replant and combat vine diseases was accompanied by rising labor costs, as nominal farm wages increased by about 20 percent over the final quarter of the nineteenth century in France, and similar increases were found in parts of Italy and Spain. However, until about 1885 or 1890 growers were more than protected from rising production costs by high wine prices. This situation changed dramatically from the final decade of the century when the two indices diverged, with wages continuing to increase in many areas but wine prices falling everywhere. Growers now faced a sharp drop in profits unless they could reduce labor inputs or increase yields, as suggested by the drop in competitiveness of almost a half in France between the 1870s and 1900s, a third in Italy, and a quarter in Spain (table 2.4, fig. 2.4).

[30] Robinson (2006:351).

[31] Paul (1996:75). Because hybrids were not officially encouraged, much of the research was undertaken on private initiative, and growers received technical information from nurserymen, whose "chief interest was to sell" (p. 101).

[32] Ibid., 103, 105. Restrictions began in 1919, and all new plantations were banned in 1975.

TABLE 2.4
Changes in Relative Wine Prices and Wages in Europe, 1870–1910

	France	Italy	Spain
1870–79	100	100	100
1880–89	132		122
1890–99	85		81
1900–1909	54	64	75

Sources:

France: wages—Lévy-Leboyer (1971:490, table 11); wine prices—France, *Annuaire statistique année 1938* (1939:62–63) and Pech (1975:511–13).

Italy: wages—1870–79 taken as 1.5 lira; Arcari (1936:270–71); wine prices—Italy, Istituto Centrale di Statistica (1958:178).

Spain: wages—Bringas Gutiérrez (2000); wine prices for Sant Pere de Ribes (Barcelona)—Balcells (1980:337–39).

Note: Wine prices have been divided by wages.

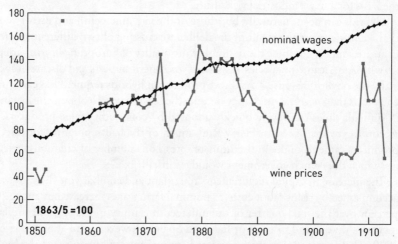

Figure 2.4. Movement in nominal wages and wine prices in France, 1850–1913. Sources: wages—Bayet (1997); wine prices—France. *Annuaire statistique année 1938* (1939:62–63)

As the phylloxera epidemic was spread over many decades, the moment when a particular vineyard was infected was often crucial in determining how growers responded. The Midi was one of the first major regions hit, allowing replanting to take place during a period of high wine prices. Others were not so lucky, and failed to replant, so that the area of vines in France declined by almost a million hectares, equivalent to 40 percent, from its peak in 1874 to 1,535,000 hectares in 1914, although higher yields implied that output did not change significantly (fig. 2.3).[33]

[33] The area reached 1.66 million hectares in the mid-1930s before slowly declining once more.

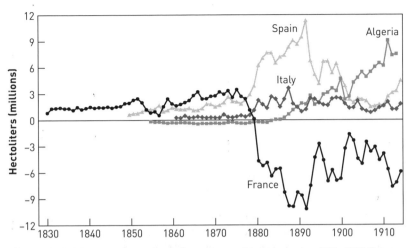

Figure 2.5. Net exports of wine from the major producer countries, 1831–1913. Source: national trade statistics

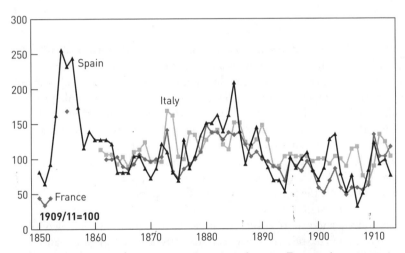

Figure 2.6. International wine prices, 1850–1913. Sources: France. *Annuaire statistique année 1938* (1939:62–63); Italy. Istituto Centrale di Statistica (1958:178); and Balcells (1980:337–39)

PHYLLOXERA AND THE INTERNATIONAL RESPONSE IN SPAIN AND ITALY

The rapid and early destruction of vines by phylloxera in France led to major changes in the international wine market, as the country switched from being a net exporter to an importer in 1879, requiring imports equivalent to 22 percent of total domestic wine supply between 1886 and 1895 (fig. 2.5). The presence of common exogenous shocks—shortages caused by new vine diseases and greater market integration—resulted in wine prices in France, Italy, and Spain moving in similar directions (fig. 2.6). Spanish producers benefited most from France's domestic shortages because their vineyards were closer to the major wine centers of Bordeaux and Cette than were those of Italy, and their wines better matched the French requirements for blending.

Powdery mildew ruined Europe's harvests in the 1850s, and overnight Spain became a major force in the international market for cheap table wines, as the incidence of the disease was less there. Wine prices in the Mediterranean provinces such as Alicante tripled between 1851 and 1855, as low transport costs allowed producers to export to France. By contrast, while there were plenty of potential wine-producing regions in Spain's interior, high transport costs made production unprofitable.[34] As one contemporary wrote:

> Large as is the extent of country in Aragon and Navarre cultivated with vineyards, it is small in comparison with what it might be if the demand for the wines of those provinces should continue, and what it certainly will be when the railroads now in course of construction are completed to the French frontier, as well as to Bilbao and Barcelona, which lines will be of equal benefit to the vineyards of Old and New Castile, many of which, like those of Aragon, have been as little known to the rest of Spain as they are to the rest of Europe.[35]

When phylloxera devastated French harvests and international prices rose again from the 1870s, the railways allowed many more Spanish growers to respond than had been possible a couple of decades earlier. The export boom drove up wine prices (table 2.5), leading to higher wages and land prices and producing widespread regional prosperity.[36] The French market required wines of good color and an alcoholic content of up to 15 percent to mix with their low-strength wines of the South and Southwest. In fact, exporting strong wines made economic sense as French import duties and transport costs were the same for all wines up to this strength, and fortifying wines with alcohol before shipping helped overcome the persistent problem of their poor keeping quality. The high prices, however, significantly reduced the wine available for distilling, and Spanish producers and exporters turned instead to foreign "industrial" alcohol produced from potato spirits or sugar beets. Imports jumped from less than 150,000

---

[34] Spain suffered less from powdery mildew than either France or Portugal. Wine prices in Alicante jumped from 10–15 dollars per pipe (100 gallons) in 1851 to 35–50 dollars in 1855, while exports increased from just 200 or 300 pipes to 23,767 from the port of Alicante (United Kingdom. Parliamentary Papers 1859:12–13). For a discussion on supply elasticities during powdery mildew and phylloxera shortages, see especially Nye (1994).

[35] United Kingdom. Parliamentary Papers (1859:31). Small quantities of Navarra wine did reach France at this time (Lana 2002:165–96).

[36] In landlocked Navarra, wine prices averaged only 60 percent of those in Barcelona between 1841 and 1865, but the railways helped to narrow the gap to 84 percent during the period between 1875 and 1894. It then opens to 77 percent between 1894 and 1900 as Navarra had greater difficulties in adapting to the loss of the French market (Balcells 1980; Lana 1999:211–14).

TABLE 2.5
Exports of Spanish Table Wines, 1861–1915 (millions of liters)

	Total exports	Exports to France	Exports to other countries	Exports to France (% of total)	Spanish wine price[1] (pesetas per 120 liters)	Index of exports by value[2]
1861–65	88.7	9.9	78.8	11	23.2	100
1866–70	111.5	13.0	98.5	12	19.4	105
1871–75	160.6	30.8	129.8	19	20.2	158
1876–80	302.4	180.1	122.2	60	26.0	382
1881–85	682.7	547.9	134.8	80	35.5	1,178
1886–90	825.6	698.0	127.6	85	26.0	1,043
1891–95	636.4	444.5	191.9	70	16.5	510
1896–1900	534.6	372.0	162.6	70	19.3	501
1901–5	208.4	143.3	65.1	69	20.7	210
1906–10	140.1	42.7	97.4	30	15.8	108
1911–15	269.9	141.8	128.1	53	27.4	359

Sources: Spain, Dirección General de Aduanas; Balcells (1980:375–79).

[1] Wine price in San Pere de Ribes (Barcelona)

[2] Index of exports obtained by multiplying total exports by wine price (1861–65 = 100).

hectoliters a year during much of the 1870s to over a million hectoliters in 1886, when it was noted in the important wine province of Tarragona (Catalonia) that "the commerce of true wines has been greatly diminished for some time in this area, as a considerable quantity of those that are exported have only a small base of wine, the rest is composed of water, foreign alcohol, colouring materials and tartaric, citric and sulphuric acids, the last of which is harmful to the health."[37]

Industrial spirits were also used for the home market, and one contemporary estimate suggested that if Spain exported 8 million hectoliters of wine, the 12.5 million hectoliters left for domestic consumption was augmented by a further 4 million of wines fabricated using foreign alcohol.[38] Complaints concerning the absence of "good" wines and the presence of adulterated ones became as common in Spain as elsewhere, a subject we shall return to in the next chapter.

Spanish growers responded to increased demand in a variety of ways, reflecting in particular local resource endowments. Perhaps the most dramatic change was La Mancha, a huge central plain that was ideally suited to the vine once the region was connected by rail to the country's major urban markets and ports. Yields were low, just 10 hectoliters in Ciudad Real in the 1880s compared with 24 hectoliters in Barcelona, but production costs were 25 percent less per hectoliter.[39] Costs were low because plows could be used between the widely spaced vines, and the hot, dry conditions permitted the goblet pruning system, removing the necessity of using stakes or trellises. The region was also phylloxera free until the 1930s and suffered less from other diseases because of climatic con-ditions, allowing growers to continue using traditional vines and planting systems.

Increasing output by using more land and labor in response to high prices and export demand was the most common, but not the only, response of Spanish producers. From the late eighteenth century the small but highly dynamic sherry export sector witnessed significant investment in cellars and wine-making equipment and a division of labor between grape production, maturing wines, and exporting (chapter 8). The rapid growth in export demand for cheap table wines now led to the creation of industrial bodegas, or large wineries, which were established, often by foreigners, around the Spain's Mediterranean ports.[40] Because there were few economies of scale to be achieved in wine making until the 1890s, these bodegas acted primarily as depositaries for collecting wines for blending, creating special wine types such as port, or adding industrial alcohol or artificial

[37] *Consejo Provincial de Agricultura, Industria y Comercio*, in Crisis Agrícola y Pecuaria (1887–89, 3:132:26).

[38] Antúnez (1887:16); Simpson (1985a:346–52). There are no reliable figures for the area of vines or wine output until the late 1880s, just when the export boom was ending. One estimate suggests that the area under vines grew by a third between 1858 and 1888, with the greatest growth occurring during the last decade (Pan-Montojo 1994:384–93).

[39] Simpson (1995:211).

[40] Pan-Montojo (2003). *Ridley's*, March 1882, pp. 73–74, regretted that there were not more, as it claimed that winegrowers attempted to maximize yields by gathering the harvest too early to avoid grapes falling to the ground.

coloring. The poor quality of many local wines encouraged some of these houses to integrate backward into wine making to ensure a better commodity or encourage winegrowers to make improvements. Producers along the Mediterranean littoral in Catalonia and Valencia copied production systems found in southern France, using sulfur to clean utensils and alcohol to strengthen wines, racking the wine from the lees, and maturing it in wooden casks. Iron crushers for grapes and presses for the pomace also became more common.[41] However, Spain's industrial wineries in the 1870s and 1880s enjoyed few economies of scale, capital was tied up in stock rather than equipment, and wine in the large fermenting vats ran the risk of overheating.[42]

By contrast, the attempts to produce premium table wines were less successful. One pioneer was the Marqués de Riscal in the Rioja region, who in the 1860s hired an enologist (Jean Cadiche Pineau) and imported fine vine varieties (cabernet sauvignon, malbec, and sémillon rouge), as well as copying the bordelaise wine-making methods. Riscal was never going to compete with Bordeaux's grand crus, but a major market existed in the late nineteenth century for blending wines to create good, ordinary claret that could be shipped from Bordeaux to Britain at £4 a hogshead (see chapter 4). When this market disappeared at the turn of the century, Riscal reinvented itself by selling much smaller quantities of bottled wine under its own brand name. Yet prices were low and remained stable from one harvest to the next, a clear indication that consumers were insensitive to annual changes in quality and suggesting that the problem of producing premium wines was as much one of demand as one of inadequate technology or lack of wine-making skills.[43]

The rapid response of thousands of small growers to high wine prices illustrates the market-oriented nature of Spanish viticulture rather than a simple response to growing population pressure, as suggested for an earlier period by Le Roy Ladurie and others. It did not to last, and the boom was brought to an end by the increase in French tariffs in 1892 in response to rising domestic production in that country, resulting in a 75 percent decline of Spanish exports between 1886–90 and 1901–5 (table 2.5).[44] A second shock was phylloxera, which destroyed a third of Spain's vines between the late 1880s and First World War. Be-

[41] Navarro Soler (1875:184–201), quoted in Pan-Montojo (1994:90–91). The quality of these cheap wines improved, so that, according to one report, "in former times these Spanish red wines were abominable, because they were put in skins instead of being put in wooden casks, and the consequence was that they were almost undrinkable in France, whereas now they are much better." United Kingdom. Parliamentary Papers (1878/79), Lalande and Guestier, no. 5191, p. 250.

[42] This was especially true for sherry in Jerez (Maldonado Rosso 1999:228–57; Montañés 2000; and Pan-Montojo 2003:316).

[43] For the Marqués del Riscal, see especially González Inchaurraga (2006). In the Rioja region, the Riscal bodega was founded in 1862 and Murrieta in 1872, while Vega Sicilia in Valladolid dates from 1864 (Pan-Montojo 1994:82–97).

[44] Prosperity for domestic growers remained linked to the French market. In 1923 it was noted that "if France takes 3 million hectolitres, wine is sold at 40 pesetas a hectolitre in Spain; if she takes no more than a million, it is sold at 15" (*El Progreso Agrícola y Pecuario* 1289, April 7, 1923, p. 220).

cause of the depressed international market, even the phylloxera-induced shortages failed to increase domestic prices in the early 1900s. The growers' response to the collapse in demand varied. In Barcelona, the divergent movement of wine prices and wages created considerable social tensions. Sharecropping, which had been considered an integral part of the success of local viticulture, was now regarded as an instrument of exploitation by landowners, and there were widespread demands for land reform.[45] By contrast, the decline in French demand for wines strengthened the industrial bodegas in Spain. The laws of 1887 and 1892 severely restricted imports of foreign alcohol, which encouraged local producers to develop a range of domestic brands of alcohol-based drinks, such as liquors or brandy, that enjoyed economies of scale in production and marketing, thereby benefiting the industrial wineries.[46]

The experience of Italy in the half century prior to 1914 was similar in many ways to that of Spain, although a growing population and rising wages increased domestic demand to a greater extent, and exports peaked in 1891–95 at just 8 percent of output, significantly less than Spain in absolute and relative terms (tables 2.5, 2.6). France was the major export market until it was closed to Italian wines in 1888 and then replaced by Austria-Hungary until 1904. Figure 2.6 shows wine prices rising until the late 1880s, but the area of vines increased by 42 percent and yields by 45 percent between 1880–84 and 1909–13.[47] The late unification of the country resulted in viticulture being widely practiced in each of the old states: 38 percent of wine was produced in the North in 1909–13, 22 percent in the center, 29 percent in the South, and 12 percent in the islands, although two-thirds came from six regions–Sicily, Puglia, Tuscany, Campina, Emilia, and Piedmont.[48] Phylloxera first appeared in Italy in 1879, but as late as 1914 perhaps only 7–10 percent of the vines were dead or dying, the great majority of these being in Sicily and the South, areas where the vine was predominantly cultivated on its own, and not intercropped.[49]

Sicily had long been famous for its strong, sweet wines, most notably marsala, which traced its history back to the arrival of the British merchant John Woodhouse in 1770. The rapid growth in wine output in the late nineteenth century, from 4.2 million hectoliters in 1870–74 to 7.7 million in 1879–83, however, was linked to cheap table wines. Phylloxera then devastated the island, and as wine

[45] Carmona and Simpson (1999).

[46] Pan-Montojo (2003:323–27).

[47] Calculated in Simpson (2000, table 3).

[48] MAIC (1914:24–27). By contrast, 45 percent of the population lived in the North, 17 percent in the center, 25 percent in the South, and 13 percent in the islands.

[49] Loubère (1978:178). According to this author, when grape production was accompanied by other farming activities on the same plot of land, the "vines hung on trees and in widely spaced rows, did not encourage the aphid" (p. 175). Measuring the area of vines and wine production in Italy in this period was very difficult as there were officially 3.5 million hectares of vines found in polyculture, and only 0.89 million as the sole crop in 1913.

Table 2.6
Production, Trade, and Consumption of Italian Wine, 1884–1911 (thousands of hectoliters)

	Production	Exports	Imports	Domestic supply	Liters per capita
1884–85	22,823	1,829	298	21,292	72
1886–90	31,364	1,835	46	29,575	98
1891–95	30,638	2,385	62	28,316	91
1896–1900	31,440	2,113	148	29,475	91
1901–6	38,177	1,337	85	36,921	111
1907–11	47,874	1,493	27	46,408	135

Source: Italy, MAIC (1914:42).

prices fell because of the recovery of French production, vineyards were returned to extensive cereals and laborers emigrated in huge numbers to, among other places, Mendoza, Argentina, where they helped establish a new industry (see chapter 11).

In the southern region of Puglia, the area of vines increased from 134,000 hectares in 1879–83 to 282,000 in 1913, and production was 5.2 million hectoliters, or 11 percent of the country's total. Like the Midi and La Mancha, Puglia benefited from low-cost land, which had previously been used for extensive cereals or grazing. The large estates were worked by sharecroppers, and if yields were low because of the dry climate, so too were the risks of vine disease. This low-cost but isolated region was linked by the railways to Italy's major urban markets after unification.

Yields were also low in central and northern Italy, but this was because the vine was just one of several crops produced on the same plot (table 2.7). This allowed farmers to vary labor inputs in relation to commodity prices but made it very difficult for contemporaries to estimate output or production costs.[50] In Tuscany, the large estates (*fattoria*) were divided up into small farms (*podere*) and let to sharecroppers, with the landowner marketing relatively large quantities of wine. The nobility and gentlemen farmers of Florence's Accademia dei Georgofili tried to improve the local wines, but much chianti remained poorly made from an excessively large number of grape varieties, fermented in musty wooden barrels and sold in the famous straw-covered *fiasco*.[51] By the late nineteenth century the loudest calls for change in viticulture and wine making came from northern Italy, and especially from Ottavio Ottavi. Experts called for the creation of regular vineyards, rather than having vines hanging from trees; the

[50] Marescalchi (1924).
[51] Loubère (1978:63–64). Giuseppe Acerbi in the early nineteenth century gives eighty-seven varieties used in Tuscany. Cited in Italy, Ministero dell' agricoltura e delle foreste (1932:55). For Tuscany, see especially Biagioli (2000).

TABLE 2.7
Wine Production by Region in Italy, 1870–74 and 1909–14

	1870–74 (millions of hectoliters)	1909–14 (millions of hectoliters)	Change (%)	Area of vines in polyculture (%)
North	9.8	17.3	77	92
Center	5.4	10.0	87	95
South	7.2	13.1	83	41
Islands	4.6	5.4	17	0
Total	27.0	46.0	70	80

Source: Pedrocco (1994:339); Italy, MAIC (1914:14–16).

specialization in a few tested and tried grape varieties; and the creation of large-scale, scientific wineries.[52] Changes eventually came, but well after 1914 because, as elsewhere in Europe, the vast majority of consumers were unwilling to pay a premium for better-quality wine.

WINE MAKING, ECONOMIES OF SCALE, AND THE SPREAD OF VITICULTURE TO HOT CLIMATES

The appearance of new vine diseases such as powdery mildew, phylloxera, and downy mildew was a major determining factor in growers' pursuit of technological change in viticulture. Indeed, it is a classic example of the "Red Queen" effect, namely, farmers being required to innovate simply to keep yields stable.[53] Other changes, such as the introduction of wire trellises or plows, had important labor-saving characteristics. While these changes increased capital requirements, however, they did not reduce the competitive advantage of the small, family-operated vineyard. This was not the case with technological change associated with some of the new wineries from the 1890s.

Pasteur's *Etudes sur le vin* was first published in 1866 and, according to Amerine and Singleton, "represents the application of the scientific revolution to the wine industry."[54] Pasteur identified the existence and activities of bacteria and yeast in wines and argued that the spoilage of wine was due to aerobic microorganisms producing acetic acid, which could be avoided by careful wine-making techniques. Pasteur showed that the grape's natural yeasts produced very different results, and he advocated instead the use of selected yeasts that had been scientifically produced in laboratories to achieve a better fermentation. These

[52] Loubère (1978:63–64).
[53] For the Red Queen effect, see Olmstead and Rhode (2002).
[54] Paul (1996:156).

allowed a quicker and more predictable fermentation, as dangerous microorganisms could be "swamped" by the addition of large number of wine-yeast cells.[55] The problem was especially acute in hot climates where the high temperatures in the vat ended fermentation prematurely and the unfermented sugar left in the wine allowed bacteria to appear that quickly ruined the wine, a development made more likely by the wine's lack of acidity. Excess heat was produced by three major factors: the initial temperature of the grapes, the heat generated by the rapid speed of the fermentation (which in turn was caused by the high quantity of sugar in the grapes); and the small amounts of heat lost through radiation and conduction in the vat. In theory, grapes could be collected in the early morning while they were still cool, and the wine fermented in small vats of less than 35 hectoliters, which reduced the heat loss because of the relatively large surface area. Yet neither was feasible even on small vineyards. For producers, especially the larger ones, the secret was to slow the rate of fermentation, which would limit the heat produced. Two very different methods were initially used: in parts of southern France, grapes with a very high level of acidity and low sugar content were grown; whereas in Algeria, where conditions were hotter, the fermenting must itself was cooled.

Viticulture excellence in regions of hot climates was limited to fortified dessert wines, such as port, sherry, and madeira, which had their fermentation interrupted prematurely by the addition of grape alcohol. However, there was little demand for dessert wines in Europe's wine-producing countries. Instead, the expansion of cultivation in the Midi after the railways was linked to the aramon grape variety, which was planted on a massive scale. This provided large quantities of thin, watery wines with a good acidity, but with an alcoholic content of only about 8 percent. The lack of sugar in aramon grapes considerably reduced the difficulties of wine making by limiting the heat generated during fermentation but produced a very unattractive wine for consumers. They were, however, ideal for blending with wines that were strong in alcohol content and color but lacked acidity. Initially these wines came from countries such as Spain, where the problems of high temperatures during fermentation were solved by their strengthening with alcohol before shipping. When the 1892 tariff effectively closed this market, the Midi merchants looked to Algeria, where growers chose the carignac and alicante bouschet grape varieties because their color, body, and alcohol complemented perfectly the acidity of the aramon. In Algeria the vintage was started early, when the grapes were capable of producing wine with 8 percent alcohol, but by the time the harvest had finished the figure had reached 12 or 14 percent. The higher acidity of the early wine helped complement the lower levels found toward the end of the season.[56] Early picking, however, carried an important economic loss, and Algerian wine in the 1880s and early 1890s

[55] Amerine and Singleton (1977:53).
[56] Bioletti (1905:14).

was notorious for its poor quality.[57] However, by 1914 new wine-making technologies allowed a stable, dry wine to be produced in large quantities and at a cheaper cost per hectoliter than could be achieved in small wineries.

The limited number of grape varieties found on the large estates in the Midi and Algeria produced huge harvests that had to be crushed in a short time period. New wineries were now designed to handle large quantities of grapes quickly, allowing animal-drawn carts to deliver the grapes directly to the hoppers prior to crushing. It was recommended that fermentation take place in well-ventilated buildings, "open freely to all winds, and constantly swept by draughts," to allow the heat to escape, but the wine was then best matured in the cool of cellars, which in the Midi were often separate buildings.[58] Wine was moved from the vat to the barrels by hand and later by electric pumps. One very large winery in Aude had twenty fermenting vats, each holding about 350 hectoliters and with a daily capacity of 1,400 hectoliters, which was fed from 215 hectares of vines and producing 30,000 hectoliters of wine.[59] In Algeria after 1895 the wooden vats were replaced by brick ones, and around the turn of the century by reinforced cement amphoras.[60] To crush the grapes aero-crushing turbines were deployed, which, instead of using rollers, applied a centrifugal force projecting the grapes against the vertical wall of the fixed cylinder of the turbine. When driven by steam, a turbine was able to crush from 180 to 200 tons a day, equivalent to about ten times the output of a medium-sized proprietor.[61]

The prolonged maceration at excessive temperatures gave Algerian wines a disagreeable earthy taste. This was the result of winemakers trying to ferment all the sugar out of the must, and devatting the wine fifteen or eighteen days after fermentation started. The problem was avoided if the must was cooled and a regular, short fermentation of five or six days carried out.[62] In 1887 Paul Brame successfully devised a system whereby the temperature of the must was reduced by pumping it through tubes that were immersed in water, although it was only after Algeria's "deplorable vinification" of 1893 that the system became widely adapted.[63] By the turn of the century it was noted that "there is probably not a single large cellar in Algeria, Oran, or Constantine which does not possess one or more of these machines, and by their use the production of a sound, completely fermented wine has been possible in all cases."[64] This interest soon faded,

[57] Isnard (1954:179–87).

[58] Roos (1900:130).

[59] The producer was Jouarres, at Minervois. Barbut, cited in Loubère (1978:199).

[60] Isnard (1954:204).

[61] Roos (1900:56–65). This machine was used by the Compagnie des Salins du Midi.

[62] Ibid., 209.

[63] Isnard (1954:189–90). The British consul general noted that "this remarkable progress in the history of Algerian viticulture is due, I understand, to the untiring efforts of M. Brame, of Fouka" (1898:3).

[64] Bioletti (1905:39). This author describes two differently types of machines: *attemperators*, which pumped water or other cooling liquids through a tube in vat; and *refrigerators*, which

however, given the relatively high cost of cooling large amounts of water in hot climates to produce cheap commodity wines. Other problems included the loss of color and extract in the wine, qualities the Midi merchants required.

The use of cultured yeasts became common in the large wineries after George Jacquemin established a commercial supply in 1891,[65] and they permitted a second method for controlling the temperature during fermentation, namely, sulfiting, or the pumping of sulfur anhydride through the must, which also became widely used in the 1890s. Sulfur dioxide was popular in wine making as it killed most undesirable organisms found in musts and wines. Wine yeasts can tolerate moderate concentrations, although it had long been known that, used in sufficiently high doses, it stopped fermentation completely.[66] In the South of France sulfur was often burned in casks before introducing the must, but by the late nineteenth century more efficient sulfurizer pumps were used so that the gas arrested the reproduction of the wine yeast and rendered it inactive for a certain time, but without killing it. By delaying fermentation, grapes could be transported to wineries situated in cooler areas, or begun at night. At the turn of the century a winery at Villeroy (Cette) produced 2,500 hectoliters of white wine annually this way,[67] but sulfiting was initially less successful with red wines because they were fermented with their skins, encouraging the leading winemakers to increase the output of white wine.[68] Some specialists suggested separating the process of maceration (extraction of tannins and coloring from the grape skins, seeds, and stem fragments) from that of fermentation (conversion of sugar into ethyl alcohol and carbon dioxide).[69] By the 1920s sulfiting—along with night fermentation, medium-sized fermenting vats (40–100 hectoliters), and cellars that were open to cool night air—was considered adequate in "not excessively hot" regions.[70] The process was usually much cheaper than cooling, and the result was that most wines were made this way in hot climates after the First World War, sometimes with excessive amounts of sulfur anhydride, leading to a disagreeable taste and aroma in the wine.

Pasteurization involved the heating of diseased wines to destroy all microorganisms and was carried out in wine merchants' cellars rather than in wineries.[71] In part this was because diseases often became apparent only after the wine had been sold, but also because of expense, with wine producers using sulfuric acid and other sulfides as cheaper substitutes.[72]

---

consisted of a spiral tube outside the vat, through which the wine was pumped, and which was cooled by a cold liquid on the outside.

[65] Pinney (1989:353).

[66] Jullien (1824:xv) defines *muet* as being "wine whose fermentation is stoped by sulphur."

[67] Bioletti (1905:50).

[68] Trianes (1908).

[69] A useful survey of the different methods is found in Castella (January 1922).

[70] Marcilla Arrazola (1922:105).

[71] Roos (1900:226).

[72] Gayon (March 17, 1904, p. 294).

The new wineries that appeared in the Midi and Algeria from the 1890s were efficient in a number of areas. The capital cost for constructing a small wine cellar was approximately double that of the largest cellars,[73] and according to Roos, the larger wineries produced a hectoliter of wine for about half the cost of the smaller ones.[74] In addition, 5–15 percent more wine was extracted from the grapes.[75] Better-quality wine was produced because skilled technicians were hired and scientific practices were used in wine making. These were indivisible inputs, implying that the total cost varied little whether a 100 or 10,000 hectoliters were produced. The new technology therefore resulted in both lower production costs and better-quality wines that could be sold for higher prices than those made in the old, family cellars. At times of scarcity and high prices, small growers had little trouble selling their wines straight from the fermenting vats, but when market conditions changed after 1900, they found themselves excluded from the market, as the wholesalers bought better-quality wines in bulk from the large producers (chapter 3). The major weakness of the large industrial wineries was that they often carried debt, which made them especially vulnerable in the early 1900s.

The new wineries offered greatest returns to producers in hot climates because of the nature of the technology and the need to have large supplies of grapes.[76] They contributed to a radical shift in the locus of production of cheap bulk wines from Europe's center to its southern periphery. In particular, the Midi and Algeria saw their output increase from the equivalent of less than 15 percent of domestic consumption in the 1820s to 50 percent by 1910 (table 2.8). The new wine-making technology was also a crucial factor in the expansion of viticulture in the New World. In theory, at least, the technology available in 1914 therefore made it possible to make a high-quality dry wine for popular consumption in most viticulture regions, something that was not true twenty years earlier. In re-

---

[73] In 1903 it was estimated that the cost of construction and equipping a wine cellar with a capacity of 200 hectoliters was 4,500–5,000 francs, compared to 160,000–200,000 francs for one of 20,000 hectoliters. Using a depreciation rate of 6 percent, this implied an annual charge of 1.35–1.50 francs per hectoliter for the smaller one, against 0.48 –0.60 franc for the larger cellar. The 200-hectoliter cellar required an additional 50 percent capacity (100 hectoliters) for fermenting vats, while the larger one only needed 20 percent more (4,000 hectoliters). The area of land was 50 m^2 and 2,000 m^2, respectively (0.25m^2 and 0.10m^2 per hectoliter), and the building costs were 6 and 4 francs the square meter. Roos, *Progrès agricole et viticole*, February 8, 1903, cited in Mandeville (1914:86–88).

[74] Mandeville (1914:93).

[75] Ibid., 75.

[76] In relative terms, they were many fewer in Europe than the New World. In the Midi there were 130 wineries by 1903 with a minimum capacity of 10,000 hectoliters each, although this was equivalent of only one for every 180,000 hectoliters of wine produced, implying that smaller wineries remained the predominant form of production. In Algeria there were 53 large wineries, equivalent to one for every 145,000 hectoliters, compared to the 34 wineries, or one for every 40,000, in Mendoza, Argentina. Gervais 1903; my calculations; and Barrio de Villanueva 2008b, cuadro 1). The Midi's harvest is taken as 23.4 million hectoliters in 1903.

TABLE 2.8
Growth in Midi and Algerian Production and French Wine Consumption (thousands of hectoliters)

	1852	1870–79	1900–9
French production	48,241	51,579	55,833
Net trade	–2,430	–2,440	+3,478
Wine available for consumption	45,811	49,139	59,311
Midi production	9,721	15,064	22,225
Midi production as % of national production	20.8%	29.2%	40.3%
Net imports from Algerian	0	–400	+7,516
Midi production and Algerian imports	9,721	14,664	29,741
Midi and Algerian wines as % of French consumption	21.2%	29.8%	50.1%

Sources: Midi—Lachiver (1988:606–9), and France, Annuaire statistique année 1933 (1934:179–80); Algeria—France, Ministère de l'Agriculture 1912 (1913:220–22); my calculations.

ality, wine producers chose their technology and production methods to maximize profits rather than quality, and the preference of most consumers was for cheap alcoholic wines rather than better-quality but more expensive ones.

LA VITICULTURE INDUSTRIELLE AND VERTICAL INTEGRATION: WINE PRODUCTION IN THE MIDI

Historians today often argue that small family farmers were more efficient than plantations in the production of commodities such as coffee, sugar, or cotton in the late nineteenth century because family workers had much greater incentives than wage laborers to properly care for the plants and to work quickly.[77] The same was usually true with grape production, and to this day small family vineyards remain highly competitive next to the large estates. This in part explains that, despite the appearance of scientific wine production, the integration of grape production and wine making remained the norm in Europe, and as late as 1934 some 86.5 percent of all French wine production outside the Midi was found in wineries of less than 400 hectoliters, an amount that could often be supplied from the family vineyard. By contrast, in the Midi about half of all wine was produced in small wineries, and 8 percent, or almost 2.5 million hectoliters, in those of more than 5,000 hectoliters.[78] The slow spread of large wineries outside areas of hot climates was caused primarily by the high transaction costs associated with obtaining grape supplies. The new wineries required grapes from

[77] Clarence-Smith (1995:157). For the efficiency of small-scale production of sugar cane, as opposed to the large economies of scale in its processing, see Dye (1998).
[78] Galet (2004); my calculations.

the equivalent of perhaps twenty or thirty times what a family vine-grower could produce, so the owners had to take the strategic decision as to whether to integrate backward into grape production themselves or purchase their needs from independent suppliers. In the New World, as shown later, the large wine producers sourced a large part of their needs from independent growers, but in Europe the number of grape varie-ties grown was much greater and quality was highly varied because of climatic conditions, making it difficult for growers and winemakers to create an efficient payment system for grapes. Fine wine producers in Bordeaux resolved the problem through direct cultivation and paying high wages, but this was not profitable for most cheap commodity wine producers, outside a few new areas of production.

By contrast, winemakers in the Midi and Algeria created large, integrated wineries on "green field" sites, and the problems associated with controlling work effort in the vineyards were reduced by a radical restructuring of production. There was significant inequality in landownership in the Midi, which probably increased during the second half of the nineteenth century. Rather than a source of conflict, however, the extremes in property ownership encouraged cooperation among growers. Even before phylloxera, the high demand for skilled labor on the large estates and the excess supply of labor on family farms helped to compensate for each other. This was particularly true of tasks such as pruning, with large owners being willing to let skilled vinedressers work a six-hour day, finishing at two or three o'clock each afternoon so that they could work their own vines.[79] The prospects of repeat work the following year provided incentives for the vinedressers to work diligently.

The mutual links between large and small properties were reinforced by phylloxera. The early appearance of the aphid in the region saw growers demanding state involvement to find a scientific solution to the disease. Local institutions, such as the University in Montpellier and the École nationale d'agriculture (La Gaillarde), played a major role in the introduction of American vines.[80] The Midi's large landowners were closely involved with these institutions, and through formal and informal labor contracts they provided a steady flow of information to the smaller growers. They lent equipment, money, the use of their wineries, information, and often the vines themselves in exchange for labor service.[81]

Augé-Laribé coined the term "industrial viticulture" for the large wine estates, first in the Midi and then later in Algeria. These were established on the fertile plains rather than the hills, and growers used large quantities of pesticides, fun-

---

[79] Smith (1975:365).
[80] La Gaillarde was the leading center in France in promoting the use of American vines. Already in 1889 it had 400 varieties, and this figure would grow to 3,500. From 1881 Gustave Foëx, author of the future classic of European viticulture *Cours complet de viticulture* (1886), became its director (Paul 1996:22–25).
[81] Frader (1991:36, 69).

TABLE 2.9
Production Costs in the Midi in the Late Nineteenth Century (in francs)

	Yields (hectoliters)	Average price (per hectoliter)	Gross income	Cultivation costs	Production cost (per hectoliter)	Net income	Value of land	Net return on capital (percent)
Plain	165	11	1,815	450	2.73	1,365	14,000	9.75
Lower slopes	90	14.5	1,305	550	6.11	755	9,000	8.40
Upper slopes	45	17	765	550	12.22	215	4,000	5.40

Source: Sempé (1898:45). Refers to the Hérault and Aude departments.

gicides, artificial fertilizers, and even irrigation to improve yields. When the black rot appeared in 1887, it "was so frightening that vignerons turned from vines grafted on *Vitis vinifera* to direct-producing hybrid vines, which scientists had singled out because of their resistance to diseases."[82] But disease was not the only factor. Rising production costs, the low opportunity cost on the old hillside vineyards, and the difficulties in obtaining sufficiently high prices to offset the lower yields associated with the better vines also drove many traditional growers to plant high-yielding hybrids on flat, fertile plains instead.[83] According to one contemporary, consumers demanded first and foremost wine, regardless of its quality, which encouraged growers to maximize yields: "it was a secondary detail whether the wine was good; during those years all wines were expensive, irrespective of their bouquet, color, or alcoholic strength. The trade paid more to the producers of quantity than those of quality."[84] Yields reached 59 hectoliters per hectare in the Midi in the period 1911–14 but 105 hectoliters on the huge estate of the Compagnie des Salins du Midi (CSM) when national average was only 33.5 hectoliters.[85] The industrial vineyards became considerably more profitable than those of the small hill farmers (table 2.9).

Growers increased yields by choosing new grape varieties such as the aramon, by carrying out only a light pruning, and by using significant quantities of artificial fertilizers. Being the first to suffer from phylloxera had the advantage that replanting took place at a time of wine shortages and rising prices, which attracted large quantities of outside capital to be invested in the region.[86] The need

[82] Paul (1996:14).
[83] See Gide (1901:226–27).
[84] Génieys (1905:38). Wines from the Midi were sold by alcoholic strength and color (Augé-Laribé (1907:192).
[85] Pech (1975:201–2). The CSM had 700 hectares of vines in 1900 and produced over 100,000 hectoliters. Its land was sandy and phylloxera free but used direct producers such as aramons and terrets, and massive amounts of fertilizers (ibid., 154).
[86] The Credit Foncier lent an estimated 20 million francs to growers between 1882 and 1902, equivalent of approximately 10 percent of the total cost of replanting in the department, assuming

to replant after phylloxera allowed landowners to redesign vineyards and grow vines on trellises in long, straight lines so that plows and horse-drawn sprays could move between them with ease, thereby cutting labor inputs and reducing monitoring costs associated with wage labor.[87] Augé-Laribé gives the example of a grower in Coursan (Aude), where plowing costs on the main vineyard of 50 hectares were four days, compared with the six days that were required on some of his other small plots.[88] Work skills were reduced by replacing pruning knives with secateurs from the late nineteenth century.[89] As Jules Guyot, perhaps the leading writer on viticulture at this time, noted, "skilful men certainly do more and better work with the knife, but when the proprietor is obliged to employ any ordinary vigneron or labourer to prune his vineyard, the secateur is preferable. It requires long practice to use the knife well and quickly."[90] As supervisors could easily walk between the rows to check an individual's work, they achieved greater control over the speed and quality of operations such as pruning, spraying, cultivation, and harvesting. Guyot writes: "A simple glance along the line of vines, permits the owner to spot the skill or the negligence of his vinedressers, just as the foreman can control with the same ease the quantity and quality of work of each of his workers."[91] The increasing labor scarcity in southern France attracted migrant labor from Spain (for vineyards in Pyrénées-Orientales, Aude, and Hérault) and Italy (for those in Var), and these worked in gangs (*colles*), which consisted of ten to fifteen skilled vinedressers who contracted for employment and benefited estate owners in that both the organization and monitoring were effectively subcontracted.[92]

According to one study at the turn of the twentieth century, economies of scale began to be important on vineyards of over 30 hectares and reached their maximum at between 60 and 80 hectares, with diseconomies appearing on estates of over 90 or 100 hectares.[93] This went a long way to resolve the problems associated with supplying the large new wineries with sufficient grapes. However, the estates had to be compact, as the potential economies of scale were quickly lost if the vineyard was fragmented into a number of small plots. This implied that there were often problems in establishing large vineyards in traditional areas of production as the land was already heavily fragmented into hundreds of plots.

a cost of 1,500 francs per hectare. The bank favored larger producers for economic and technical reasons (Postel-Vinay 1989:169).

[87] Génieys (1905:38) notes that "the period between 1890 and 1900 saw the triumph of the Aramon, planted on the old water meadows and trained on wire trellises and pruned according to the Guyot, Quarante, Royat, methods." See also Gide (1901:218–19).

[88] Augé-Laribé (1907:118).

[89] Smith (1975:371).

[90] Guyot (1865:37), English edition. See also Loubère (1978:83) and Frader (1991:31).

[91] Guyot (1861:19).

[92] Smith (1975) and Frader (1991:75).

[93] Cited in Augé-Laribé (1907:119–22).

In Algeria the area of vines increased from 17,614 hectares in 1878 to 174,490 hectares twenty-five years later.[94] Production was frequently large scale, with capital being provided by French banks, skilled labor by temporary Spanish migrants, and cheap, unskilled labor by local workers. In the first decade of the twentieth century, there were fifty-three wineries with a capacity of 10,000 hectoliters or more, and eighty-four vineyards with over 100 hectares.[95]

The appearance of powdery mildew in the 1850s acted as a major incentive for merchants to develop new sources of supply, integrating producers in the Midi with consumers in northern France and developing links between Spain's coastal regions and southern France. However, trade was still limited because of the high transport costs. By the time phylloxera started to devastate French vines in the late 1870s, European growers were much better placed to deal with the resulting shortfall. The increase in prices was more moderate than it had been with powdery mildew, as French merchants imported around a third of Spanish production. Phylloxera changed the nature of traditional viticulture, as growers needed both physical and human capital in what had been previously an occupation that required virtually no off-farm inputs, and skills were handed from father to son. Growing market specialization and new technologies also produced significant changes in wine making. Large, modern wineries not only lowered production costs but also produced better-quality wines. The problems of oversupply in the early 1900s were caused in part by fraud, but also by rising yields, resulting in part from the spread of hybrids and the better keeping quality of wine.

These changes also contributed to a major geographical reorganization of the industry within Europe and the Mediterranean basin. The reduction in transport costs allowed low-cost producers in areas such as the Midi, La Mancha, and Algeria to compete in both national and international markets. This specialization was based partly on low regional wages, but also on the capital intensiveness and greater skills found in the new vineyards and wineries. A distinction can also been made between Algeria, with its large, modern wineries and cooling equipment, and the Midi, which resorted, in part at least, to planting a highly acidic grape variety. By 1900 the Midi's most serious competition was no longer from wines produced from grapes in other French regions, but rather from the sale of adulterated wines.

[94] Isnard (1954:117).
[95] Gervais (1903) and Isnard (1954, 2:518). There were a further 144 vineyards of between 50 and 100 hectares, and 746 between 20 and 50 hectares.

CHAPTER 3

Surviving Success in the Midi: Growers, Merchants, and the State

A SERIES OF LARGE DEMONSTRATIONS took place in the Midi during the summer of 1907 protesting against low prices and the sale of artificial wines. At the same time, many of Bordeaux's leading quality wine producers were forced to look to their merchants for financial help, while growers of cheaper wines lobbied local and national governments to establish a regional Bordeaux appellation or brand. A few years later, in 1911, troops were needed to stop the destruction of large quantities of wines that had been brought from outside the Champagne region to Reims and Épernay for making into "champagne." Phylloxera and the market instability that followed in its wake were the economic origins of these very different events. Vine diseases, technological change, and market integration altered the distribution of power in the commodity chain, weakening the position of most growers while strengthening that of the merchants. This challenge was met by growers who, benefiting from changes in political organization that led to an increase in their political voice, obtained government support to create new institutions to protect their market interests and guarantee that the region would retain a reputation for militancy to this day.

Very different commodity chains were required to produce and sell French wines that were as diverse as vintage champagne, fine old claret, and cheap commodity wines. Chapter 2 showed that falling transport costs, urbanization, and rising real wages led to a significant increase in the domestic consumption of cheap wines and contributed, along with new wine-making technologies, to regional specialization. Merchants played a crucial role in organizing markets by blending wines from different areas to create a standardized product, and the major shortages caused by phylloxera after 1875 forced them to search for new sources of supply, often in foreign countries or through the production of artificial wines. As the recovery in domestic wine production was not accompanied by a significant reduction in these alternative supplies, prices and growers' profitability fell sharply, leading to demands that government intervene.

This chapter looks at the experience of the Midi, France's cheapest wine producer. After examining long-run changes in France's domestic wine supply, and in particular merchants' attempts to augment supply during the phylloxera epidemic by adulteration, it shows how the changes in political strength of small farmers and workers increased during the Third Republic, especially after the

1884 law permitting the formation of syndicates. Despite the presence of large vineyards in the Midi, the wine industry was relatively united in its attempt first to tackle phylloxera and replant, and then to demand state intervention to control fraud. Finally, the chapter considers how smaller growers started to establish cooperatives in response to another threat to their livelihood, namely, the increasing economies of scale and skills required for wine production and marketing.

PHYLLOXERA AND WINE ADULTERATION

The major wine shortages and high prices in France caused by phylloxera not only led to a rapid increase in production in countries such as Spain and Algeria but also encouraged both growers and merchants to look for domestic substitutes, which the authorities often tolerated because of the exceptional circumstances. One product that became popular from around 1879–80 was the manufacture of wine from raisins and currants, and in 1890 it was estimated that 300 liters of wine with an alcoholic strength of 8 percent could be produced from 100 kilos of raisins or currants, at the cost of just 0.15 franc a liter, considerably less than the price of real wine.[1] The first factories were established in the Midi, especially Hérault, but the fact that raisins were cheaper to transport than wine encouraged their location in, or near, major markets, and in 1890 there were reportedly twenty factories in Paris.[2]

The use of sugar in wine production also became very popular. Chaptal had shown that the addition of sugar to crushed grapes in years with poor summers, especially in northern Europe, improved quality. Output increased indirectly as viticulture was encouraged in otherwise marginal regions and the increase in the wine's alcohol content saved it from deteriorating. Sugar was also traditionally added with water to the remains of the grapes after their first pressing and repressed to produce second wines or *piquettes*. Coloring was sometimes added, with fuchsine being especially popular.[3] This practice was normally limited to wines for family consumption, "in theory by law and in fact by the abundance of good cheap wine and the relatively high price of sugar."[4] However, the wine shortages and desperate economic situation facing many growers led to the government's relaxing of the laws, and large quantities of *piquettes* were sold. By 1890 raisin and sugar wines together accounted officially for a sixth of total French consumption, although the real figure may have been considerably higher.

[1] Ordish (1972:149–50).
[2] *The Times*, August 9, 1900, p. 8, cited in ibid., 148–50. See also Sempé (1898:87–88, 96).
[3] Ordish (1972:144).
[4] Warner (1960:13).

Most wines kept badly prior to phylloxera, and any surplus wine at the end of the year was distilled. Growers might suffer from low prices after an exceptional large harvest, but there were few incentives for them or merchants to carry stocks from one year to the next. The improved wine-making technologies in areas such as Algeria or the Midi allowed wines to be kept longer, but chemicals were also used to preserve poor-quality ones. These changes, together with the high wine prices caused by phylloxera, led to a sharp decline in wine distilling, with production falling from an annual average of 8 million hectoliters in 1865–69 to 1 million in 1895–99.[5] At the same time technical developments in commercial distilling and the appearance of cheaper raw materials (grains, beets, and potatoes) produced a significant fall in the price of industrial alcohol. When French domestic wine supplies recovered and overproduction threatened in the late 1890s, the market to distill surplus wines had practically disappeared. Some of the industrial alcohol was used to manufacture artificial wines, but some also went to create new alcoholic beverages to be drunk in the *assommoir* or dram shops that competed directly with wine. The exact size of this competition is impossible to establish, but it was believed to have been extensive. For example, official wine consumption in Paris in 1903 was 185 liters per capita, half the 354 liters found in its suburbs. To reduce the incidence of taxation, wines were strengthened with alcohol to the legal maximum before being brought into the city and then watered down and adulterated with industrial alcohol once inside.[6] One report to the Chamber of Deputies in 1905 suggested that 20 million hectoliters of manufactured wine were circulating, while another gave a figure of between 10 and 12 million. In the same year the municipal laboratory in Paris randomly tested 617 wine samples and found that 500 had been doctored or adulterated.[7]

Finally, growers responded to higher prices by increasing the output of genuine wines. The area of vines in France peaked in 1874 at almost 2.5 million hectares and by 1914 had declined by almost a million hectares, or 40 percent. However, the new vineyards were more productive, and yields increased from 22 hectoliters per hectare in the 1870s to 33 in the 1900s. The 1893 harvest was the first in fourteen years to be above the long-term average of 1871–1913 (fig. 3.1). Faced with steeply falling prices, producers looked to the state for help to eliminate substitutes (imported and adulterated wines) and increase consumption of genuine ones (by reducing taxes and transport costs), rather than face the prospect of uprooting their vines and exiting the industry (table 3.1).

The tariff war of the late 1880s provided an excuse to raise duties on Italian wines, and from 1892 those from Spain were also increased, although the impact of this measure was initially limited by the use of free ports and the depreciation

---

[5] Degrully (1910:325).
[6] Ibid., 356.
[7] All cited in Warner (1960:37).

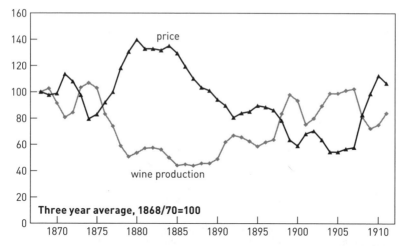

Figure 3.1. French wine production, consumption, and prices, 1851–1912. Source: France. *Annuaire statistique année 1938* (1939:62–63)

of the peseta.[8] Spanish exports to France in the 1890s were still 81 percent of what they had been in the 1880s but then fell sharply to 14 percent in 1900s. However, the decline from these markets was offset by the growth of Algerian imports, so that total French imports in the 1900s were still 5.6 million hectoliters, equivalent to 60 percent of the figure in the 1880s. In addition, with the recovery in domestic production, merchants found that many of their old export markets were now protected by tariffs, restricting demand for French wines, despite their lower prices.[9]

The production of raisin wines was a second area that legislators looked to control. The Griffe law of 1889 defined wine as being only made from fresh grapes but did not prohibit the manufacture or sale of raisin wines, so long as the consumer was aware of the drink's true content.[10] The Brousse Law of 1891 required that when substances such as raisin wine or gypsum were added to wines, this was marked on the casks and bottles. However, the real cause of the decline in raisin wine production was the succession of tax measures on imports, their manufacture, and the alcohol required to strengthen them. By 1897 Henri Sempé estimated that taxes alone worked out at 25 francs per hectoliter, equivalent to the farm gate price for wines, so the trade virtually disappeared.[11] Finally, in 1894 the watering down of wines was made illegal, regardless of whether the consumer was aware of the situation.[12]

---

[8] Sempé (1898:205). The peseta recovered after 1898. For the free port for Spanish wines, see chapter 5.

[9] For French exports, see Pinilla and Ayuda (2002:51–85); Simpson (2004:80–108).

[10] Stanziani (2003).

[11] Sempé (1898:102, table 3).

[12] Stanziani (2003).

TABLE 3.1
French Wine Supplies and Consumption, 1866–1909

	1866–75		1886–95		1900–1909	
	Thousands of hectoliters	% of total	Thousands of hectoliters	% of total	Thousands of hectoliters	% of total
Harvest	56,931	99.4	30,517	70.5	55,649	88.3
Imports	348	0.6	9,510	22.0	5,620	8.9
Sugar and raisin wines	?		3,249	7.5	1,840	2.9
Total	57,279	100.0	43,276	100.0	63,109	100.0
Exports	3,229		2,032	4.7	2,141	
Drunk by producers			9,859	22.8		
Sold to merchants			28,795	66.7		
Vinegar and distilling	5,000?		681	1.6	2,200?	
Total			41,367	95.8		
Waste (6% of merchants' wine)			1,831	4.2		
Difference between production and consumption			78			
Per capita consumption (liters)	c. 144		c. 110		c. 162	
Average wine yields	24.2		16.7		33.0	
Average wine price (Paris) franc/hl	28.5		32.2		18.2	

Sources: Sempé (1898:52); Degrully (1910:304, 428); Warner (1960:35); and Galet (2004:30–31).

The reduction of imports and piecemeal legislation on imitation wines did not resolve the problem of low prices. As on previous occasions when overproduction threatened, the long productive life of the vine, the low opportunity cost of much of the land on which it was cultivated, and the specialized skills that were required in its production made growers reluctant to uproot healthy vines. Markets had changed, however, as stocks could now be carried over from one harvest to the next with greater ease, restricting the ability of prices to recover after poor harvests. The postphylloxera vines had also required many growers to go deep into debt, so it was no longer sufficient to cover just their variable costs, as after a poor harvest the banks required interest and capital to be repaid, a problem that became especially acute on the Midi's and Algeria's large estates. Finally, the economic viability of small producers was threatened not just by competition from low-cost industrial wineries, but also from the market power of merchants who could source their wines from an increasingly greater geographical area.

POLITICS, PHYLLOXERA, AND THE VINEYARD DURING FRANCE'S THIRD REPUBLIC

The response of Europe's producers to problems such as adulteration, overproduction, and loss of competitiveness after 1900 varied significantly. Both political institutions and the ability of growers, winemakers, and merchants to organize collectively and establish new economic institutions differed across countries and over time. Continental Europe remained essentially rural long after the bases of an industrial society had been established. In France however, growers were able to influence government policy from a much earlier date than their counterparts elsewhere. As Tony Judt has written of this country, "the political organisations and doctrines which responded to the growth of an industrial and capitalist economy had a large and increasingly discontented rural population to whom they could also appeal (indeed, in conditions of a precociously established universal male suffrage, to whom they had to appeal)."[13] Although European peasants were attracted to a wide variety of different causes and supported parties across the spectrum, vine growers were more inclined to lean toward radical and socialist politics. A second factor was the extent and nature of vertical integration between grape growers and winemakers, and between winemakers and merchants, which could shape incentives to cooperate on some occasions or pursue strictly sectoral interests on others. The potential for conflict existed at a number of different levels: between small and large growers within a particular region, between growers within different regions, and between growers and merchants.

In 1907 the French economist Charles Gide wrote that "of all the industries in France, whether agricultural or manufacturing, the culture of the vine stands foremost."[14] Yet despite this claim, the exact numbers employed in the sector are difficult to establish. In 1868 Jules Guyot argued that the vine employed 1.5 million vigneron families, or six million inhabitants, and when auxiliary workers were included the figure reached two million.[15] By the end of the century the figure was almost identical, at a time when the active male population was still only thirteen million.[16] Even though many laborers and other professionals often owned only a very small plot of vines, it seems likely that these figures are inflated because some growers were also counted several times. Gide himself notes that the "number of vinegrowers is very great, is certainly over half a mil-

[13] Judt (1979:272). Emphasis in the original.
[14] Gide (1907:370).
[15] Guyot (1868, 1:i).
[16] Loubère (1978:167); Prestwich (1988:10).

lion for the whole of France," and gives a similar figure for the number of men who were employed full-time in the vineyards, which increased to two million during the harvest.[17] In addition, there were over 450,000 retail outlets (débitants) and 28,000 wholesalers.[18] Whatever the exact figures, the sector was a major source of employment, and of potential voters for politicians.

The rural nature of society was reflected in its political organization, and the electoral laws in France's Third Republic (1871–1940) implied that "at least half of the electoral districts for the Chamber of Deputies were predominantly rural, and many others contained a sizeable minority of peasant voters," while the Senate came to be known as *la grande assemblée des ruraux*.[19] Although the first mass-based political party was not established until 1901, politicians now had to take into consideration the interests of the small farmer, and to get elected they had to indulge, rather than penalize, the peasant producers. As Jules Ferry famously declared in 1884, "the republic will be a peasants' republic or it will cease to exist."[20] Even phylloxera was politicized, as Bonapartist and royalists whipped up antirepublican sentiment against republican governments that passed legislation giving officials the right to enter vineyards and uproot vines on the suspicion of infection. The bishop of Montpellier in 1876 "sanctioned devotional processions among the vineyards, not just to ward off the phylloxera but also against the 'republican menace.'"[21]

A decisive change for small producers and workers came with the 1884 Waldeck-Rousseau law on association in France, which removed the need for governmental consent for any association of more than twenty people. Originally designed to encourage moderate workers to join trade unions so that they would cease to be a vehicle for extremists, it had its biggest impact in the countryside among farmers and laborers. Agrarian syndicates were important because they allowed farmers to establish their own list of priorities rather than the large landowners or the church. Both the conservative and antirepublican Société des agriculteurs de la France (the umbrella organization of the syndicates was the Union centrale des syndicats agricole) and the republican Société nationale d'encouragement à l'agriculture looked to extend their political influence in the countryside, and by 1910 there were 5,146 agrarian syndicates with 777,066 members throughout France.[22] In 1881 Gambetta created a new ministry devoted solely to agricultural affairs in France and expanded the system of agricultural institutions and administration.[23] According to Sheingate, because French political

---

[17] Gide (1907:370–72).

[18] Figures refer to 1895. In Paris there were 27,000 retailers, one for every 87 inhabitants. Excluding Paris, there were 366,000 retailers in 1869 Sempé (1898:104–6).

[19] Wright (1964:14).

[20] Ibid., 13.

[21] Campbell (2004:168).

[22] International Institute of Agriculture (1911, 1:256–57).

[23] Sheingate (2001:54–55).

parties were still weak, "farm organizations enjoyed a competitive advantage vis-à-vis parties and became important players in national politics."[24] Winegrowers were especially quick to take advantage of the new legislation in their fight against phylloxera, a problem that growers could not solve individually. Syndicates collected and circulated information among members on the best way to deal with the disease and provided information and instructions on the use of new rootstock and grafting. A second area was the purchase of the vines, chemicals, and fertilizers, which benefited growers not just because bulk purchases were cheaper, but because the syndicate was able to control quality, especially important as fraud was a major problem in all countries until the 1920s. In France syndicates were also instrumental in checking another form of fraud, namely, the production and sale of artificial wines, which southern wine producers believed was the prime reason for the collapse in wine prices after 1900.

Conflicts within the wine community were present during the phylloxera epidemic and can often be explained by the timing of the disease's arrival in a locality. In particular, areas that were still free lobbied hard for a total ban on the entry of new vines (especially American rootstock) in their department, while those areas already infected wanted imports and government-funded nurseries to allow them to replant as quickly as possible. Once replanting was under way, the levels of cooperation between large and small grower, and winemaker and grower also varied, being relatively strong in the Midi but weak in Bordeaux. In the Midi, the fight against phylloxera provided a useful rehearsal for interest groups for when they faced low wine prices and fraud from the turn of the twentieth century.

THE MIDI: FROM SHORTAGE TO OVERPRODUCTION

High prices caused by powdery mildew and the major drop in transport costs produced by the railways acted as a catalyst for the Midi's viticulture, although while the area of vines virtually doubled between the late 1820s and early 1860s, yields changed little and the region continued to produce about a fifth of France's output (table 3.2).[25] Over the following two decades output increased to 30 percent of the national total, but this was now achieved by higher yields rather than extending the area planted. Phylloxera caused the area of vines to fall from its peak of 450,000 hectares in 1872 to 268,000 in 1886, but the exceptionally high wine prices and competitive nature of the region's viticulture led to recovery of

[24] Ibid., 64. According to this author, the institutional setting for agricultural policy revolved around three distinct variables: the pattern of political competition in each country, the structure of the agricultural bureaucracy, and the organizational strength of different farm groups (p. 39).

[25] In the ancien régime, the privileges enjoyed by Bordeaux also restricted markets for Midi wines (Dion 1977:393; Augé-Laribé 1907:34–36). The Midi's specialization in spirit production encouraged some growers to use high-yielding vines from an early date.

TABLE 3.2
Wine Production in the Midi and France, 1824–1909

	Midi*			Rest of France			Midi as % of France		
	Area (thousands of hectares)	Production (thousands of hectoliters)	Yield (hectoliters per hectare)	Area (thousands of hectares)	Production (thousands of hectoliters)	Yield (hectoliters per hectare)	Area	Production	Yield
1824–28	209	5,331	26	1527	32,948	22	12	14	+16
1862	394	9,713	25	1927	38,917	20	17	20	+18
1870–79	410	15,064	37	1968	36,515	19	17	29	+70
1880–89	295	8,946	30	1734	20,943	12	15	30	+106
1890–99	401	14,861	37	1361	21,354	16	23	41	+80
1900–1909	452	22,225	49	1247	33,608	27	27	40	+50

Sources: Calculated from Lachiver (1988:582–609, 616–18).

*Includes the departments of Aude, Gard, Hérault, and Pyrénées-Orientales.

TABLE 3.3
Movements in Factor and Wine Prices in the Midi, 1852–1907

	1852	1862	1882	1892	1907
Wine prices—francs per hectoliter[1]			33	17	9
Land prices—francs per hectare[2]	1,768	2,638	3,003	2,633	1,784
Day wages in francs[3]	1.7	2.3	2.6	2.4	2.9
Days work required to purchase one hectare of vines	1,040	1,147	1,155	1,097	615

Sources: Wine prices—Pech (1975:511–12); land prices—Chaffal (1908:66–68); wages—Sicsic (1991: 41–42).

[1] Average of three years centered on year given in table.

[2] "Second-class" land.

[3] Average from Aude, Gard, Hérault, and Pyrénées-Orientales. Weighted by summer wages (one-third) and winter wages (two-thirds).

the area planted to 462,000 hectares in 1900.[26] By the early twentieth century the Midi was responsible for two-fifths of French output and half of taxed wines sold, with the area under vines increasing by 15 percent compared with before phylloxera (1862); production growing by 130 percent; and yields being 50 percent more than the national average. The debate concerning overproduction from the turn of the century was linked specifically to the growth in output in Midi, as growers there were caught, more than elsewhere in France, between three adverse price movements: falling wine prices, declining land values, and rising labor costs (table 3.3). Small growers had problems to sell their wine even at the prevailing low prices.[27]

Land prices are notoriously difficult to measure, but the drop of 40 percent for vineyards between 1882 and 1907 given by Chaffal in table 3.3 is perhaps an underestimate. The Fédération des syndicats agricoles du Gard, which represented seventy-two syndicates, argued that between 1895 and 1907 the value of vineyards had fallen by two-thirds; the Chambre de Commerce de Béziers gave a fall of 80 percent between 1900 and 1907; and the Société centrale d'agriculture de l'Hérault a fall of 75 percent between 1895 and 1907.[28] The decline can also be seen by looking at individual properties. Thus the owner of one vineyard at Matelles (Hérault) of 144 hectares was offered 800,000 francs in 1896, but by 1900 it was valued at only 69 percent this figure. It was finally sold in 1905 at 28

[26] Pech (1975:496–97).

[27] The parliamentary commission established in 1907 to look into the causes of the national wine crisis noted that the Midi and Algeria were the worst-affected regions in France (Chambre des députés 1909:2307). It was headed by Cazeaux-Cazalet, a deputy from the Gironde.

[28] Quoted in Chaffal (1908:61–62).

percent of the 1896 figure, although the price had recovered to 47 percent when it came on the market again in 1912. Other vineyards tell a similar story.[29] Falling prices were of particular concern to the large vineyard owners who had borrowed heavily, as the interest and loans had to be repaid in full.[30]

By contrast, labor costs rose 20 percent from an average of 2.4 francs a day to 2.9 francs. Vineyard owners responded by cutting labor requirements to a minimum, and wage labor was employed only 230 days a year. When wine prices briefly recovered, the workers struck for higher pay, and the Office du Travail recorded close to 150 strikes between November 1903 and July 1904 in the Midi, involving almost fifty thousand workers.[31] The extent of the problems of higher wages and falling wine prices for estate owners can be clearly seen in table 3.3. While the wage:land price ratio remained relatively stable between 1852 and 1892, with slightly over a thousand days of labor needed to purchase a hectare of vines, this collapsed to little more than half the figure in the 1900s. In theory higher wages and lower wine prices should have been good for consumers. However, the importance of viticulture for employment and the depressed wine prices severely affected the rest of the local economy, and the number of vacant stalls at Béziers's market, for example, rose from twenty in 1904 to thirty-three in 1905, sixty-two in 1906, and sixty-eight by February 1907.[32] Despite this background of labor conflict, the large producers in the Midi successfully managed to identify their own economic problems with those of the whole region, making it easier to search for a political solution when prices collapsed in the first decade of the twentieth century.[33]

In the decade 1890–1900 wine in the Midi cost between 12 and 16 francs to produce and was sold for 18–20 francs per hectoliter.[34] Between 1900 and 1906 prices fell to as low as 5 francs, and growers sold at cost or below in five out of seven harvests.[35] Yet the causes of the crisis (*la mévente*) were not obvious. As some commentators noted, the net supply of wines in France in the early 1900s was not so different from the level immediately prior to phylloxera (table 3.1).

---

[29] For example, in Ginestas (Hérault) a vineyard was sold in 1901 for 37 percent of its estimated value in 1889, and in Bouches-du Rhône one was sold in 1905 for less than 20 percent of its value seven years earlier (Caziot 1952:226–28). Postel-Vinay (1989:184) suggests that creditors refrained from foreclosing properties in 1906 and 1907, preferring to wait until the wine market recovered.

[30] Postel-Vinay (1989:171–77) shows the high levels of debt facing the large vineyards in the Midi.

[31] For these conflicts, see Smith (1978: 106–11) and Frader (1991:75, 92–93, 121–25).

[32] Caupert (1921:22–28); Augé-Laribé (1907, chap. 10).

[33] Carmona and Simpson (1999:311); Simpson (2005).

[34] Gide (1907:370).

[35] Warner (1960:18). One calculation suggested that a local wine price of 10.7 francs per hectoliter was needed to cover variable costs, and 14.3 francs to cover fixed costs, but this second figure was reached only twice in the Midi during the 1900s (Atger 1907:23–27). For prices, see Pech (1975:512).

What many believed had changed was the increased sale of artificial wines. As already noted, the financial difficulties caused by phylloxera had led to the authorities tolerating, if not encouraging, the production of artificial wines. After the Brussels Sugar Agreement of 1902, which reduced taxes from 60 to 25 francs per 100 kilos, there followed what one writer has described as "an orgy of fraud" in the Midi.[36] Official wine production in thirty-five communities in Hérault totaled 1,004,915 hectoliters in 1903, but 2,284,848 hectoliters were actually sold, the difference being supposedly fraudulent wines. Nationally fraud was estimated at over 15 million hectoliters, equivalent to some 40 percent of the official harvest.[37] The high level of fraud in 1903–4 was unusual, however, and caused by a combination of a poor wine harvest, which led to prices reaching their highest since 1887, and the very low tax on sugar. Wine harvests quickly recovered, pushing prices down to very low levels once more, and the government restricted the amount of sugar that growers could use and increased taxes. Unless growers could obtain sugar illegally, the profitability of *sucrage* after 1904 was greatly reduced.

Yet even if this form of adulteration temporarily declined, problems remained. In particular, growers were aware that any increase in wine prices was now likely to lead to a return of widespread adulteration. A second complaint was the lack of statistical information on production and stocks that made it difficult to know accurately the supply of genuine wines. Finally, if low prices discouraged growers from carrying out fraud, others in the commodity chain still found it profitable to do so. Thus the 1907 commission looking into the causes of the crisis argued that it was not the result of overproduction, but of the fact that the price of genuine wines was driven down to those of the artificially manufactured ones, which had been made from poor wines that would have previously been distilled but since 1903 were being treated by *la chimie vinicole*.[38]

As the Midi growers competed on price rather than quality, they demanded an increase in national consumption by lowering taxes and transport costs. The reduction in rail tariffs in 1896 increased the region's competiveness against those from Spain and Algeria in the Bordeaux or Paris markets, which were transported by boat.[39] The *Loi des Boissons* in 1900 lowered taxes, and that of 1901 removed the *octroi*, halving the tax revenues from wine.[40] However, with per capita consumption at 168 liters in 1900–1904, there were obvious limits to a demand-side solution for low wine prices, especially as consumption of alterna-

[36] Warner (1960:14, 40).
[37] Degrully (1910:350, 353). Atger (1907:73) suggests a figure nearer 8 million hectoliters.
[38] France, Chambre des députés (1909:2307–8).
[39] Sempé (1898:175–77). Caupert (1921:36) notes that rail tariffs were cut by 50 percent in 1890.
[40] Warner (1960:32).

tives such as absinthe and beer had increased. For the 154,954 growers in the Midi, the continued sale of artificial wines was an obvious and visible explanation for low prices.[41]

Low prices hurt all producers, large and small, as well as those merchants who dealt with genuine wines. The major political influence in the Midi after 1901 was the Radicals, which defended the small producers, favored political democracy, and demanded greater social equality. Yet as Leo Loubère has noted, the wine defense group cut across party lines "and included deputies and senators who were moderate republicans, some who were royalists, and, after 1900, some socialists."[42] In particular, the wine crisis of 1907 showed relatively little of the class conscience that developed among industrial workers.[43] Modern forms of political organization, such as public demonstrations, accompanied more traditional ways of protest, including the mass resignation of local political officers and tax strikes. Starting on March 24 with a meeting of 300 people in Sallèles-d'Aude, Sunday protests in 1907 quickly grew in size, and by May 5 some 80,000 took to the streets in Narbonne. A week later the numbers were 120,000 in Béziers, followed by 170,000 in Perpignan. By May 26 the figures had reached 220,000 in Carcassone; almost 300,000 marched in Nimes seven days later; and, finally, over half a million in Montepellier on June 9.[44] Demonstrations of this scale were previously unknown in France and can be explained both by the large numbers of people in the Midi who depended directly or indirectly on viticulture and by the fact that the sector was united in its demands against the government in Paris.

The government had in fact already begun to respond, and the laws of August 1905 and June 1907 significantly reduced the amount of sugar that could legally be used in wine making, made it easier to prosecute fraudulent producers, and introduced measures to record growers' production.[45] The Confédération générale des vignerons du Midi (CGV) was created and became, in the words of Leo Loubère, "the wineman's bulldog in the fight against fraud" as in 1912 the state officially allowed its agents to sue for, and collect damages from, those convicted of fraud.[46] The broad base of its support within the wine communities was crucial to its success, and the CGV's thirty agents in 1911–12 carried out 3,042 investigations that led to 601 successful prosecutions for fraud. For every ten cases initiated and involving wine fraud, the public prosecutor was responsible for just two, and the CGV of the Midi eight.[47]

---

[41] Growers refer to 1900–1909 (Lachiver 1988:588–89). Some contemporaries also argued that high domestic tariffs restricted export markets (Warner 1960, chap. 3).

[42] Loubère (1974: xv, 177).

[43] Ibid., 3, 189.

[44] Frader (1991:141); Lachiver (1988:468).

[45] Warner (1960:41); and Frader (1991:145).

[46] Loubère (1978:355).

[47] Ibid.

FROM INFORMAL TO FORMAL COOPERATION:
LA CAVE COOPÉRATIVE VINICOLE

The CGV was also instrumental in encouraging growers to form wine-making cooperatives to take advantage of the cheap credit that the government made available to these groups after December 1906.[48] Family grape producers could compete with large producers because of the high transaction costs associated with using wage labor in the vineyards, and because if necessary they simply worked longer hours. However, this was not the case in the winery, and they became increasingly uncompetitive because of their lack of scientific wine-making skills and capital to purchase adequate equipment and cellar facilities. In addition, wholesale merchants faced much lower costs when buying from one or two big wineries rather than having to collect from large numbers of different growers scattered over a wide territory, and transportation costs were lower when wines was shipped in bulk.

The result was that large wineries sold their wine at higher prices. For example, the huge Compagnie des Salins du Midi, with facilities to produce 100,000 hectoliters, was paid an average of 19.25 francs per hectoliter of wine in the period 1893–1913, against a regional average of 16 francs. By contrast, one small producer of just 400 hectoliters (Gélly) received 27 percent less than the CSM during the period 1893–1906. This difference was even greater in years with the lowest prices, with Gélly being paid only 4.8 francs in 1904, against 11.5 francs received by the CSM.[49]

Lack of cellar space was another problem facing a number of small growers (the *non-logés*). In some cases this arose because the area planted after phylloxera did not coincide with the old area of vines and small growers failed to build cellars, preferring to sell their wines directly from the vat. In other cases, they lacked the necessary capital to increase the capacity of their wineries to cope with the larger yields from the new vines or failed to maintain capital investment in their cellars when they were producing little or no wine during phylloxera. In the village of Marsillargues (Hérault), for example, growers only had storage for around 85 percent of the production (300,000 hectoliters); in Saint-Laurent d'Aigouzes (Gard) it was 75 percent (of 160,000 hectoliters); and in Bompas (Pyrénées-Orientales) only 47 of the village's 227 winegrowers had cellar space to keep wines over the winter.[50] Consequently growers in Marsillargues in 1897 sold 55 percent of their harvest in September, 14 percent in October, and just 31 percent over the following ten months.[51]

The *crise de mévente* of the early 1900s led a number of influential French

---

[48] Caupert (1921:83, 112).
[49] Pech (1975:158).
[50] Mandeville (1914:43).
[51] Cot (1900), cited in Gervais (1913:50).

writers, including Charles Gide, Michel Augé-Laribé, and Adrien Berget, to encourage the creation of wine cooperatives to reduce small growers' production costs, and integrate vertically to absorb some of the marketing functions of merchants. By combining their scarce capital, several hundred family grape producers could build and equip a large winery to cut costs, pay for a skilled enologist to produce better-quality wines, and store these cheaply to be sold later in the year at higher prices. Wine cooperatives could also distill the remains of the grapes, and produce tartaric acid. Finally, they could act as banks, providing loans to their members.

The lower production costs associated with the new large, modern wineries were achieved by substituting capital for labor, an important factor even for family producers as labor requirements soared at the harvest time because of the need to rapidly collect the grapes, transport them to the winery, and crush them using labor-intensive methods. The state-of-the-art Flassans cooperative winery, for example, with a capacity of 5,000 hectoliters, required just nine workers employed for about twenty-five days to process 65,000–75,000 kilos of grapes daily, significantly less than the labor requirements for the 180 individual cooperative members if each had processed their grapes.[52]

The first attempt to establish a wine cooperative in France took place in Champagne in 1893 and ended in failure, but those from the turn of the century, especially in the South, were more successful.[53] Although the low prices between 1900 and 1907 encouraged the formation of wine cooperatives (referred to at the time as *filles de la misère*), there were still only thirteen in all of France in 1908. However, rapid growth followed the law of December 29, 1906, and decrees of May 30 and August 26, 1907, which allowed wine cooperatives access to long-term credit at the almost uniform rate of 2 percent interest over twenty-five years.[54] Capital was provided by the state but lent through regional credit banks, which were responsible for the loans. Transaction costs were greatly reduced because cooperatives were required to establish a specific legal structure to qualify. Between 1907 and 1914 around fifty cooperatives in southern France obtained loans covering an average of 47 percent of their capital costs, and the state gave subsidies of 815 thousand francs, or an additional 14 percent.[55] In 1914 the twenty-one cooperatives that had been created between 1903 and 1910 had an

---

[52] Mandeville (1914:95). The cooperative had a double crusher-stemmer that processed 24,000 kilos per hour; a pump that moved up to 20,000 kilos of crushed grapes per hour; a wine pump with capacity of 180 hectoliters per hour; a hydraulic press with a pressure of 120,000 kilos per square meter, etc. (Arnal, *Progrès agricole*, August 10, 1913), cited in Mandeville (1914:93).

[53] In France, societies for the collective manufacture of cheese (*fruitières*) date from the twelfth century and remained by far the most common of producer cooperatives, numbering 2,485 against only 39 wine cooperatives in about 1910 (International Institute of Agriculture, 1911, 1:280). In northern Italy a similar institution, the *turnario sociale*, dates from slightly later. For champagne, see Clique (1931:14). For the Champagne cooperative, see chapter 6.

[54] Caupert (1921:111); Gervais (1913); and Mandeville (1914:2, 11).

[55] Mandeville (1914:139). This author noted that some cooperatives were not finished and therefore expenditure would rise, reducing these figures slightly.

average of 160.6 members and a cellar capacity of 13,100 hectoliters, and in 1913 they produced 7.9 thousand hectoliters of wine, or about 50 hectoliters per member, equivalent to about 1 hectare of vines.[56]

Perhaps the most famous cooperative was that of Maraussan (Hérault) of the *vignerons libres*. This was originally established in December 1901 to sell wine to consumer cooperatives in Paris, although as early as 1906 the members had constructed and equipped a modern winery, with an initial capacity of 15,000 hectoliters, and at a cost of 175,000 francs. The Ministry of Agriculture contributed 30,000 francs, the local regional bank (under the 1906 Law) provided a long-term loan of 109,000 francs, and a further 30,000 francs was raised from consumer cooperatives in Paris. The subscription of the 120 members was just 25 francs each.[57] Therefore not only did growers in Maraussan and other cooperatives benefit from the economies of scale found in the larger wineries, but construction costs and the purchase of equipment were subsidized by the state.

There was widespread interest in Europe in the late nineteenth and early twentieth centuries in cooperatives, and although there were some spectacular successes, such as the Danish creameries, these were in general the exception. By 1914 there were still just seventy-nine wine cooperatives in France, of which fifty were found in the South, and their total production represented at most 1 percent of the nation's output.[58] Certainly winegrowers were "individualistic" and unwilling to trust some of their neighbors, as some contemporaries argued, but this was often for sound business reasons rather than any "cultural failing." Spacious, clean, and scientific wineries promised to improve wine quality, but cooperative members required them also to be profitable and to be operated "fairly." Two areas of potential conflict quickly appeared: the price to be paid for the grapes and the sale of the wine.

As discussed in chapter 1, many vineyards contained a number of different grape varieties as growers tried to reduce the risks of adverse climate conditions and disease. Grape quality also varied significantly from one year to the next, and winegrowers expected to receive a price that reflected these variables. However, cooperative workers could not check every basket of grapes used to make cheap wines and accordingly paid simply by their weight and sugar content. One solution was not to require members to sell all their grapes to the cooperative, but this risked them becoming depositaries for that part of the harvest that could not be sold elsewhere for higher prices. By contrast, when cooperatives forced their members to hand over all their harvest, the economic incentive for growers

---

[56] My calculations from ibid., tables 1 and 2. Using the full sample of fifty cooperatives, which includes some that were not finished, in 1913 there were an average of 135.2 growers, enjoying a cellar capacity of 11,200 hectoliters, with a production of 6,400 hectoliters, or 47 hectoliters per grower.

[57] Gide (1926:129–31).

[58] Simpson (2000, table 5). In Italy, a short-lived *cantina sociale* was established at Bagno a Ripoli near Florence in 1888, and by 1910 there were reportedly "slightly in excess of 150," and a further 40 cooperative distilleries were also active (International Institute of Agriculture 1915a, 2:152).

was to uproot their quality, shy-bearers, and plant high-bearers. In both cases, quality was driven down. Consequently, with the notable exception of Burgundy, wine cooperatives in France remained heavily concentrated in the South until at least the 1950s because in this region most producers already planted only a few varieties of high-yielding vines.[59]

The early cooperatives therefore failed to provide economic incentives for growers to improve quality, as Charles Gide in particular had hoped they would. They also had difficulties in selling directly to consumers. This was because members had limited information concerning market conditions to be able to monitor to their satisfaction the performance of cooperative managers, who on many occasions were simply fellow growers. Changes in wine quality and frequent, sometimes violent, fluctuations in market prices made it difficult for cooperative members to determine whether the prices they received were satisfactory and fair. The result was that while sometimes it was the cooperative that sold the wine, others preferred to leave it to the growers and avoid what might be considered a poor sale that might discredit the management.[60]

French cooperatives were also far too small to influence market prices.[61] By contrast, a trust controlled approximately four-fifths of the wine market in California at this time, and this served for a couple of ambitious attempts by the very large landowners in the Midi to corner the market.[62] Bartissol in 1905 proposed a marketing board that would sell annually 20 million hectoliters of branded wine in bottles directly to consumers. In case of overproduction, all growers would absorb the costs of distilling to reduce supply to support prices. However, many growers were reluctant to sign long-term contracts with an independent company. Even if a monopoly had increased wine prices, the capital requirements (300 million francs) and logistics of such a huge operation led to it remaining just a project. Palazy's proposal in 1907 was more modest and involved the direct participation of growers, who would sell wine to wholesalers and retail merchants. With a capital of 48 million francs, growers were required to enter agreements for five years, and the company hoped to sell a minimum of 12 mil-

[59] The growth in the cooperative movement in Burgundy from the late 1920s involved fine wines produced on miniscule vineyards often of less than half a hectare. The wine price covered the considerable monitoring costs associated with controlling vineyard activities. Growers were paid not just by the grapes' weight and sugar content, but also by the variety, and growers were obliged to sell all their production to the cooperative. Only certain shy-bearers were permitted, and the vendage was done collectively. See especially Clique (1931:21–60).

[60] This was the reason given for the Marsillargúes cooperative, the largest one in the Midi in terms of cellar capacity, at 60,000 hectoliters (Mandeville 1914:111–12). See also Cique (1931: 187).

[61] Hoffman and Libecap argue that cooperatives can raise prices only if the product is relatively homogenous, stocks difficult to accumulate, and a significant number of individual growers agree to output cuts that can be easily monitored (Hoffman 1991:397–411).

[62] Atger (1907:116–22); Degrully (1910:375–85); Postel-Vinay (1989:180–81). For California, see chapter 9.

lion hectoliters of wine. Although it claimed to have twenty thousand members and control 9 million liters of wine, it failed in 1906 as the Société civile de Producteurs de Vins naturels du Midi et de l'Algérie, and again in 1909 as the Société coopérative de Producteurs du Vins naturels du Midi, to negotiate discount privileges with the Bank of France and the Ministry of Finance.[63] Just as important was the question of growers' commitment, as there were strong incentives for them to remain outside the trust, as the potential benefits of higher prices would have been enjoyed by all, regardless of whether a grower was a member. When overproduction and low wine prices reappeared in the 1930s, growers looked directly to the state to resolve them, rather than create a private monopoly. Transaction costs associated with compliance were therefore reduced and absorbed by the state, with the resulting legislation (*Statut du Vin*) both regulating markets and helping the small producers.

The widespread protest in the Midi in 1907 marked a turning point in French wine history. Both powdery mildew and phylloxera had threatened the future of the nation's winegrowers in a most dramatic way, but in both cases they required a scientific solution rather than market intervention by the state. The 1900s were not the first time that growers had been faced with overproduction, low prices, and widespread financial losses, but markets now showed no signs of correcting themselves automatically. Biotechnological advances encouraged growers to plant high-yielding varieties, and imports from Algeria increased year by year. The situation facing winegrowers in the 1900s was not dissimilar to that of Europe's cereal producers in the 1870s and 1880s, when falling transport costs led to a growth in imports, and farmers could no longer benefit from high prices after a local harvest shortage, leading to the demand for tariffs to place a floor to domestic prices. A similar scenario of overproduction with wine was delayed by phylloxera. When prices did finally collapse after 1900, increasing tariffs on wines from low-cost producers such as Spain or Italy failed to resolve the problem because the cause of overproduction was found within France itself, and because cheap wines continued to be imported in increasing quantities from Algeria, part of France's single market. In addition, any recovery in prices because of a domestic harvest shortage such as in 1903 was limited, as growers and merchants were encouraged to turn once more to fraud. The 1905 law provided the possibility of prosecuting the sale of artificial wines, and the creation of the Confédération générale des vignerons du Midi in 1907 established an effective institution to police the sector. This resolved the problem of competition from artificial wines, but not the tendency for wine production to grow faster than consumption, and the sector looked to the government for financial help to distill excess wines from time to time. With the next major crisis in the 1930s, the government had to start restricting the use of high-yielding grape varieties.

[63] Caupert (1921:64–65).

By the First World War small producers were becoming less competitive because of the new wine-making technologies and the greater market power of merchants, but in France they were compensated by their ability to influence farm policy. The rapid growth in agrarian syndicates after 1884 was in part the result of the economic demands of farmers in an increasingly competitive market, but it was also encouraged by politicians looking for votes. Cheap credit and subsidies allowed growers to establish wine-making cooperatives that enjoyed economies of scale in production and marketing and allowed small family producers to compete with large commercial wineries. In time, the cooperative proved an efficient institution to disseminate the benefits of agricultural research to small growers, but also for the state to regulate markets by holding back surplus production after large harvests.[64] By 1952 cooperatives produced about a quarter of all French wine, and in the Midi about three quarters of all growers belonged to one of the 527 cooperatives.[65]

One possible solution to the weak domestic demand for French wines was to export. Charles Gide claimed that British and German consumers paid two or three times more than the French and continued:

Why is not the effort made? It is not clear whether the inertia of the French vinegrowers is to blame, or the indifference of English and German consumers, or, and this is probably the real cause, the exactions of the middlemen, of the English and German dealers, who play the part of importers, and who, snatching at bigger profits, try to sell the wines of South France at the same price as their clarets and ports, thus cutting them off from the average mass of buyers[66]

The following chapter looks at just this question.

[64] Knox (1998:15).
[65] Calculated from *Bulletin de l'Office International du Vin* 290 (1955): 40–41. There were fifty cooperatives in the Midi in 1924 (Augé-Laribé 1926:141).
[66] Gide (1907:374–75).

PART II

The Causes of Export Failure

THE POSSIBILITIES FOR FARMERS to sell in international markets increased enormously everywhere over the nineteenth century, as world trade grew annually by 4 percent, twice that of GDP among the world's leading sixteen nations. Rapidly falling freight rates produced by improved shipping and more efficient port infrastructure cut the cost of trade between countries. Tariffs on many products were reduced, at least until the 1870s. Inland transport was transformed by the railways and improved roads that linked producer regions to markets, and information costs declined with the development of the telegraph and rapid growth in the number and circulation of local newspapers. By 1914 there were few parts of the globe left untouched, but while both rich and poor regions exported goods, the demand for food and beverages was almost exclusively found in the rich markets of northern Europe and North America.[1] Britain was by far the country most dependent on international trade, importing the equivalent of almost 60 percent of its food and beverage needs on the eve of the First World War.[2] Over the century Britain's population increased from about eleven to thirty-seven million; GDP per capita tripled; and by 1900 the urban population was about 80 percent of the total. Before wine tariffs were slashed in the early 1860s, British consumers already drank virtually all the fine sherries and ports produced, and an important quantity of the best clarets and champagnes. Significantly lower taxes and other market reforms now offered a huge potential market for Europe's wine producers, as the British government tried to make wine competitive and a product of mass consumption.

The mechanics of collecting wine from many different vineyards, grading it, and delivering it in acceptable condition to consumers hundreds of kilometers away created plenty of problems for economic agents along the commodity chain. When consumers bought their wines from producers they did not know, reputable intermediaries were required to enforce contracts and reduce information costs. Fine wines were sold along the chain "by gentlemen to gentlemen," and problems of asymmetrical information concerning product quality between

---

[1] In 1913 Europe and North America exported 53.5 percent and imported 82.1 percent of all food and beverages traded internationally (Yates 1959:64, 66).

[2] Turner (2000:224–25).

producers and consumers were resolved by personal reputation and trust. The British wine trade in the early nineteenth century therefore consisted of a group of elite consumers paying exceptionally high prices for their wines. However, if merchants were going to exploit the lower taxes and transport costs to create a mass market for cheap wine among British consumers, they needed a different organizational structure to reduce marketing costs. In particular, retailing had to change from one that was information-intensive to one where consumers could easily and cheaply determine quality prior to the purchase of their bottles of wine from a retail outlet that was conveniently located near where they lived.[3]

The rise of the modern corporation and the major economies of scale and scope associated with the leading industries in the second half of the nineteenth century were matched by significant changes in distribution, as firms looked to sell their mass-produced goods to affluent urban consumer. In countries such as the United States and Britain, retailing was radically altered with the appearance of chain stores, consolidation of major brand names, and the introduction of mass-marketing techniques. Brands helped to reduce information costs for consumers by differentiating products, guaranteeing the good's purity, and creating producer reputations.

Many of these brands associated with food and beverages were linked to buyer- rather than producer-driven commodity chains. The governance structures of the two were very different and resided in the location of the barriers to entry. Agricultural production was often very competitive and globally decentralized, with millions of producers and few barriers to entry, so buyer-driven chains were created with the control at the point of consumption. With the growing dependence of British consumers on the world's food and drinks markets, a very small number of domestic wholesalers and retailers were able to control the chain. For example, the import of frozen meat reduced the need for skilled domestic butchers but required hygienic outlets for the "vast quantities of frozen and later chilled meat that were pouring into the United Kingdom."[4] By 1910 ten British firms between them had created a total of 3,684 retail branches,[5] and it was these firms that controlled meat quality, established brands, and enjoyed scale economies in marketing. Likewise with coffee, the economies of scale found in roasting and packaging (especially after vacuum sealing was invented in 1900) encouraged a concentration in the number of firms in consuming countries, where a growing share of the value was added.[6] By the end of the century the huge economies of scale found in processing, packaging, and distribution allowed firms such as Cadbury's, Heinz, Kraft, and Lipton's on both sides of the

[3] For a discussion on these two types of retailing, see Casson (1997:12–14).

[4] Imported meat increased from 10 percent of total consumption in 1870 to 37 percent in 1896 (Jeffreys 1954:182, 190).

[5] Ibid., 187.

[6] Over three-quarters of the retail price of coffee in the grocery trade was added in consuming countries by 1935 (Topik 1988:60).

Atlantic to spend heavily on branding and advertising, thereby constructing high entry barriers to potential competitors. By contrast, producer-driven networks are controlled by core firms at the point of production and today include such products as aircrafts and computers. High entry barriers make it difficult for new firms to start production, and the large corporations can exert control through backward linkages with suppliers and forward control into retailing.[7] Producer-driven chains are generally rare with regard to agricultural commodities, but the limited supply of top-quality land implied that a number of such chains existed among wine producers, including such well-known brands as Château Margaux and Moët & Chandon, as will be discussed in part 3.

Despite the growing opportunities to trade, both buyer- and producer-driven chains failed to increase wine sales to nonproducer countries such as Britain, so that although approximately 12 percent of the world's production was exported in 1909–13, France accounted for nearly half of all imports. The next chapter looks at why exports to nonproducer countries were not greater, and why bulk importers were unable to brand wines as was being done with breakfast cereals, soups, and beer.

[7] Gereffi (1994:104).

CHAPTER 4

Selling to Reluctant Drinkers: The British Market and the International Wine Trade

> The Prince of Wales had a small quantity of remarkably fine wine; and his household chose to drink it out. The prince ordered some for the table, and none could be got; there were only two bottles left. The man who had the management of the wine went to the City, to a merchant, and stated what he wanted. The dealer said, "Send me a bottle of what remains, and what I send must be drunk immediately, I can imitate it." The trick was successful, and was repeated three or four times.
>
> —*Report from the Select Committee on Import Duties on Wines*, 1852:663

THE DIFFICULTIES IN CREATING a mass market for wine in nonproducer countries can be examined by considering in detail the nature and organization of the British market. Britain's growing, comparatively wealthy urban population offered significant potential for wine producers everywhere: if it had consumed just one-tenth of the French figure in the 1890s, this would have created a demand for wine equivalent to 18 percent of Spain's total output, 14 percent of Italy's, or 105 percent of Portugal's.[1] Yet although imports almost tripled between the late 1850s and the mid-1870s, there was no permanent change in drinking habits, and wine consumption then declined, so that on the eve of the First World War per capita consumption at a bottle and a half was no higher than it had been a century earlier, and the city of Paris consumed about seven times as much wine as the whole country.[2]

The major cause of the failure to develop new export markets was the inability to solve problems of asymmetric information that existed between seller and consumer, resulting in consumers facing three different but interrelated forms of confusion and deception. In the first instance there were the problems of artificial or adulterated wines, which were manufactured with substances other than just grape juice and sold to consumers as being the genuine article. While fraud had always been present in most food and beverage markets, its nature changed significantly over the second half of the nineteenth century because of the growing physical separation between producers and consumers, the shortages of gen-

---

[1] Average French consumption, calculated by adding imports to production and subtracting exports, was 45.9 million hectoliters between 1891 and 1900 (Direction Générale 1934:177).

[2] W. & A. Gilbey, letter to *The Times*, October 14, 1891, p. 14.

uine wines caused by phylloxera and other vine diseases, and the development of new preservatives, such as borax, benzoic acid, and salicylates, that allowed manufacturers to mask food deterioration and to lower costs, often making food adulteration imperceptible to consumers.[3] Wine quality varied significantly, which provided greater opportunities for fraud and adulteration than were present with other imported goods, such as cereals, butter, coffee, or frozen meat. To put it bluntly, cheating in the wine trade was very easy, and an increasingly belligerent local and national press presented consumers with large numbers of real and fictional food scares, so that by 1900 it was widely believed in Britain that most wines were adulterated in some form or another.

A second problem involved the mislabeling of wines. The commercial success of an individual (Château Lafite) or collective (claret) brand encouraged producers from other regions and countries to label their own wine in such a way that they could also benefit from the name. While companies could protect their private brands in the courts, this was not possible with claret and sherry, and these became generic terms for all wines enjoying certain vague characteristics. Confusion for British consumers was increased by the fact that it was legal to sell in that market "British claret," "Hamburg sherry," or "Spanish port." The result was that even strong private brands, such as Château Margaux, Moët & Chandon, and Gonzalez Byass, saw their sales threatened by the decline in the collective reputation of claret, champagne, and sherry, respectively.

Finally, attempts by importers to create buyer-driven commodity chains by selling large quantities of wine under their own brands enjoyed only limited success. Unlike breakfast cereals, canned soup, or bottled beer, the fact that wine quality varied significantly from one small producer to the next, and deteriorated quickly if not properly stored, implied that there were few economies of scale to be achieved through bulk purchases and retail chains.

This chapter shows that wine was traditionally a luxury because of the high and discriminatory import duties, which benefited Portuguese and Spanish producers at the expense of the French. With the reforms of the early 1860s there was a temporary increase in consumption and a switch in preference away from Iberian fortified wines toward French table wines. Merchants blended cheap commodity wines from different locations to minimize quality fluctuations, but although retail prices remained remarkably stable during the phylloxera-induced period of shortages, this was achieved only by significantly reducing product quality. Poor wines and numerous press reports concerning their adulteration led to falling consumption. The failure of buyer-driven commodity chains such as the Victoria Wine Company or Gilbeys to significantly cut marketing costs implied that the small family retailer remained competitive, but neither could simultaneously cut prices and guarantee product quality for consumers.

---

[3] Law (2003:1116).

The Political Economy of the Wine Trade in Britain prior to 1860

The importance of wine in international trade, its status as a luxury item, and the comparative ease for most governments to tax imports encouraged restrictive mercantilist trade policies from the seventeenth to the mid-nineteenth centuries. The British had hoped that the profusion of vines they found in their North American colonies would free them from European producers, but the indigenous American wild vines produced little or no wine that was drinkable.[4] Policy instead turned to discriminate between European producers and to limit trade with countries with which Britain was at war, especially France.[5] The mercantilist nature of policy was made clear in the preamble to a late seventeenth-century law: "It hath been found by long experience that the importing of French wines, vinegar, brandy, and other commodities of the growth, produce, or manufacture of France . . . hath much exhausted the treasure of our nation, lessen the value of the native commodities and manufactures thereof, and greatly impoverished the English artificers and handicrafts, and caused great detriment to this kingdom in general."[6]

A growing tendency to drink Iberian rather than French wines was reinforced by the Methuen Treaty of 1703, which limited duties on the imports of Portuguese wines to a maximum of two-thirds of what French wines paid; in exchange the Portuguese repealed the restrictions on the entry of certain types of English cloth.[7] The Methuen Treaty made Portuguese wines more competitive than French, but wine remained very expensive for British consumers. From 1831 French wines paid the same duty as those from Portugal and Spain, but while Iberian wines were responsible for around three-quarters of imports during the 1850s, French wines accounted for just 6 percent of the market.[8] Some commentators attributed the failure to increase market share after 1831 to the unsuitability of French table wines for the English market, while others believed that they continued to be discriminated against, as the fortified Iberian wines (ports, sherries, etc.) paid less duty per unit of alcohol. Sherry and port were also much more expensive than ordinary French wines, so the duty

---

[4] Pinney (1989:5–10).

[5] Unwin (1991:242).

[6] 6 William and Mary, c. 34., cited in Briggs (1985:24).

[7] Francis (1972:106, 130). Wine duties therefore were based on volume, rather than being an ad valorem tariff or one based on alcohol content. A similar duty might have been signed with Spain, but it was considered at this time to have no textile industry that threatened the sale of English cloth.

[8] From an average of 21,000 hectoliters in the eight years between 1823 and 1830, to 15,000 between 1831 and 1838. Calculated from Redding (1851:384).

represented a smaller share of the retail price,[9] and French exports were limited to expensive claret and champagne.[10]

There were few British wine drinkers before 1860. According to George Porter, author of *The Progress of Nations*, consumption was limited to "the finer kinds of wine," and these were only within reach of "the easy classes."[11] In a similar vein, W. B. James believed that most wine was drunk by the half a million heads of families that paid income tax, or less than 10 percent of the population, which would imply an average per capita consumption of fifteen bottles a year for this select group.[12] Import duties averaged 6s. 7d. a gallon during the first half of the nineteenth century, or just over a shilling a bottle, at a time when the average weekly wages for male workers was about 16 shillings.[13] Taxes on wine imports contributed around 3 percent of government revenues, or £1.75 million over the period 1820–60, but taxes on all forms of alcohol and public places for alcohol consumption accounted for a massive 30 percent during the nineteenth century.[14] The trend toward free trade in Britain led to widespread debate on the wine tariff, and, as John Nye has recently noted, "the problem for nineteenth-century leaders was how to reform the protectionist side of the ledger while preserving the revenues that the state had come to rely upon."[15]

A parliamentary select committee was created in 1852 to consider the implications of tariff reform on the wine market. Sir James Tennent argued that as wine was a luxury, the rate of duty was less important in determining demand than other factors such as taste and fashion,[16] so that cutting duties would not increase consumption, but simply reduce government revenue. He noted that while in France annual per capita wine consumption was about 90 liters, the British each drank 95 liters of beer and 4.5 liters of wine and spirits, and he believed that these beer drinkers would be unlikely to consume more wine.[17] In a similar vein, John McCulloch argued in his *Commercial Dictionary* that lower duties would be irrelevant for consumers of the "finer" wines, while "inferior and

---

[9] Nye (1991:37) suggests, using French sources, that Portuguese wines were valued about five times more than French ones. However, wines were especially difficult to value, which explains why suggestions to introduce an ad valorem tariff were rejected in the nineteenth century.

[10] In 1789 England and Ireland accounted for just 1 percent production in the Gironde but 6 percent of the value (*Dictionnaire du Commerce*, cited in Cocks and Féret 1908:86).

[11] Porter (1847:570).

[12] United Kingdom. Parliamentary Papers (1852), James, p. 380. My calculations.

[13] The figure for duty comes from Simpson (2010, table 2); for wages, Bowley (1900, appendix 1).

[14] Taxes from alcohol were collected by duties on spirits, wines, and beer, and from selling licenses for alcoholic drinks, although the greatest burden of taxation fell on the production or import of drinks by brewers, distillers, and wine merchants, rather than the retailers (Wilson 1940, chap. 18).

[15] Nye (2007:100).

[16] Tennent (1855:28).

[17] United Kingdom. Parliamentary Papers (1852), Tennent, p. 350.

low priced wines" would find little demand among "the bulk of the population" even if all duties were totally abolished, so free trade would simply imply a loss of revenue for the Treasury.[18]

By contrast, a number of other influential commentators, including Cyrus Redding and T. G. Shaw, both authors of popular wine books, and the port-wine shipper J. J. Forrester, believed that if duties were cut sufficiently, new social groups would be attracted to wine, and government revenue maintained.[19] Forrester demanded radical changes and a reduction in duty to a shilling per gallon to convert wine from being a luxury to a necessity, allowing "everyone" to be able to afford to drink it.[20] Forrester, as many others in the debate, was not a disinterested observer. According to one retailer, there were "three classes of traders who prefer high duties: those who supply the richer classes; those who . . . concoct solely for fraud; and those who ship improper mixtures for the drawback: these two latter could not do it, except under a high duty."[21] The Iberian traders of fortified wines were among the first category, and, as Shaw noted after the tariff reforms of the early 1860s, they tried to keep their dominant position in the industry by opposing change.[22] Rivalries among interest groups within the drinks trade continued throughout the period.[23]

Many of those who gave evidence before the Select Committee argued that high wine duties were responsible for the widespread adulteration and production of artificial wines. These included "British wine," which, according to William Gladstone, the chancellor of the exchequer who was responsible for the liberalization of trade in the early 1860s, in the hands of "respectable merchants" was made of raisins, sugar, and brandy. British wine was much cheaper than the genuine article, as "the duty paid on these materials is reckoned as amounting to 1s. 2d. a gallon. Therefore you have a duty on foreign wine of 5s. 10d. the gallon—on colonial wine of 2s. 11d., and on British wine of 1s. 2d. the gallon. The result is the consumption of foreign wine diminishes, the consumption of colo-

---

[18] McCulloch (1845), *Wine*, p. 1416, cited in Tennent (1855:148).

[19] United Kingdom. Parliamentary Papers (1852); Shaw (1864:182–83); and Redding (1851: 659–60).

[20] United Kingdom. Parliamentary Papers (1852), Forrester, p. 448.

[21] Ibid., Dover, p. 648. The drawback involved the mixing of Cape wines (which paid half duty) with port, for example, and then exporting it as port and reclaiming the full duty.

[22] Shaw (1864:20).

[23] As the shipper A. G. Sandeman noted:

You have the merchants connected with Spain and Portugal who wish to have their strong wines introduced on the most favourable terms. You have the dealers in light wines who wish to have a monopoly, as it were, and have the duty kept high on the strong wines. . . . Then you have those connected with the spirit trade who look with great jealousy upon a reduction of the duty upon strong wines as likely to affect their interests; so that really it is most difficult to reconcile every interest in the trade (United Kingdom. Parliamentary Papers 1878/79, Sandeman, pp. 142–43).

nial wine has increased, and the consumption of British wine has doubled within the last ten years."[24]

Gladstone believed that lower duties would allow cheap, "good and whole-some" wines to be sold instead of adulterated ones, and if duties were reduced sufficiently, the increase in trade would be "immense."[25] The comparison with tea was made explicitly:

> Here especially we are met by the cry that wine is the rich man's luxury. It is the rich man's luxury. . . . Is tea the rich man's luxury? No. It is the poor man's, and above all, the poor woman's luxury. But I speak in the year 1860. In 1760, tea too was the rich man's luxury. In 1760 there was no more tea consumed per head of the population than there is wine now. In 1760 there were 4,000,000 lb. of tea consumed; now the annual con-sumption is 76,000,000 lb. The price of tea which is now sold at 3s. per lb. was some-where about that time advertised by the cheap houses at £1 per lb. Wine is the rich man's luxury, and you may make tea or sugar, or any other article of consumption, the rich man's luxury if you put on it a sufficient duty.[26]

A radical reduction in duty from five shillings and nine pence to one shilling per gallon, such as debated by the 1852 select committee, would have required annual consumption to grow from 275,000 to over 1.36 million hectoliters just to maintain tax revenues. Some contemporaries questioned whether the supply of wine was sufficiently elastic, arguing that if output failed to increase, the ben-eficiaries of the lower duties would be foreign growers and exporters who would enjoy higher prices, and there would be no increase in imports for British con-sumers. Although British imports in the early 1850s were significantly less than 1 percent of Europe's total wine production, possible supply shortages were taken very seriously.[27] Approximately three-fourths of all imports were port and sherry, and genuine wines came from specific geographical areas where the po-tential to increase output by extending the area of cultivation or increasing yields was strictly limited.[28] In 1851, for example, the total production of port was 94,123 pipes (equivalent to about 500,000 hectoliters), of which 41,403 pipes were classified as first quality and 20,000 set aside for export to Europe (essen-tially Britain and Ireland).[29] With sherry, William Tuke, a London wine broker, did not "believe that *fine* Sherry could be produced in larger quantities than it is

---

[24] Hansard's Parliamentary Debates (1860:clvi), February 10, p. 847.

[25] Ibid., 849–50.

[26] Ibid., 845. The quote continues, "By that means you will not only effectually bar access of the poor man to it, but will reserve to yourself the proud satisfaction of saying with literal truth, 'Our indirect taxes are paid by the rich; none are levied upon articles consumed by the poor.'"

[27] Tennent (1855:81–83).

[28] As one wine broker noted, "take John Bull in the aggregate, and he only knows those two wines, Port and Sherry" (United Kingdom. Parliamentary Papers 1852, Tuke, p. 165).

[29] Ibid., Forrester, p. 18. Harvest size and annual shipments differed significantly, as most port was not exported for several years.

now," but he admitted that Spain could produce 100,000 pipes of white wine "not Sherries."[30] Possible supplies of fortified wines from other regions were not promising, as the "repeated exertions ... made by enterprising importers, to introduce into this market cheap but wholesome wines, the lower growths of France, Spain, Portugal, Germany, Italy, and other countries, failed, with the '*solitary exception*' of Marsala."[31]

Others believed, however, that the supply was more elastic, especially as, because of adulteration, many people already consumed wines that were not genuine port or sherry. The huge potential supply of French and Spanish light table wines was recognized, but this would require consumers to drink a very different sort of wine from that which traditionally had been imported.

GLADSTONE AND THE RISE AND DECLINE IN CONSUMPTION IN THE LATE NINETEENTH CENTURY

The debate was ended with the Anglo-French (Cobden-Chevalier) Treaty in 1860, when lower duties on wines were used as a bargaining tool to encourage a French reduction in duties on a wide range of British industrial goods. By 1862 duties on wines with an alcoholic strength of under 14.8 degrees Gay-Lussac (equivalent to 26 degree proof Sykes) had been cut from 5s. 9d. a gallon to 1s., with those between 14.8 and 24.5 degrees paying 2s. 6d.[32] The new legislation also introduced major changes in the marketing of wines, especially the Single Bottle Act of 1861, which allowed general retailers, on the payment of a relatively small license fee, to sell bottles of wine for consumption off their premises, while another license fee permitted alcohol to be consumed with food in refreshment houses "of good repute."[33]

---

[30] Ibid., Tuke, p. 146. The supposedly inelastic supply of port and sherry, especially the better qualities, led Sir James Tennent (1855:22) to argue that "any great augmentation of the demand even for the lowest class of wines—the supply of which is assumed unlimited—is pretty certain to be followed by a considerable enhancement of their costs at the places of their growth and shipment."

[31] Tennent (1855:67). Emphasis in the original. One importer noted that "the public certainly want a cheap wine, but it must be Port, and my opinion is, that neither Masdeu, Figueira, which is a Portugal wine also, nor red Sicilian wines, however low the price may be, will ever come into competition with genuine Port wine" (United Kingdom. Parliamentary Papers 1852, J. C. Gassiot, p. 127).

[32] Rates changed in both 1860 and 1861. Those of 1862 lasted until 1886, when the lower rates were extended to wines of up to 17 degrees. In 1899 the duty was increased to 1s. 3d. for wines of less than 17 degrees, and 3s. for those above it. For sparkling wines an additional duty was imposed in 1888 of 2s. 6d. per imperial gallon, or 1s. 3d. if the value of the wine did not exceed 15s. This was reduced to 2s. irrespective of value in 1893, and 2s. 6d. from 1899. Finally, ordinary bottled wines paid an additional 1s. a gallon after 1899.

[33] The license fee was between £2 10s. and £10, depending on the value of the premises. By 1880,

TABLE 4.1
Impact of Duties on Retail Wine Prices in the United Kingdom, 1850s and 1860s

	London import price	Duty	50% retailer markup	Total	Price per bottle	Duty as % of retail price
Tariff levels 5s. 9d. a gallon; in use 1840–1859						
Good sherry	£28	£31 3s.	£24 11s.	£89 14s.	2s. 9d.	35
Tavern sherry	£14	£31 3s.	£22 11s.	£67 14s.	2s 1d.	46
Table wine	£6	£31 3s.	£18 11s.	£55 14s.	1s. 9d.	56
Tariffs: Less than 14.8 degrees, 1s per gallon; 14.8–24.5 degrees, 2s. 6d. per gallon; in use 1862–1886						
Good sherry	£28	£13 10s.	£20 15s.	£62 5s.	1s. 11d.	22
Tavern sherry	£14	£13 10s.	£13 15s.	£41 5s.	1s. 3d.	33
Table wine	£6	£5 4s.	£5 12s.	£16 16s.	6d.	31

Source: Based on United Kingdom. Parliamentary Papers (1852) Dover, p. 653.
Notes: Calculations are made using a sherry butt of 108 gallons, with one gallon taken as six bottles. Sherry is assumed to have an alcoholic strength of 14.8–24.5 degrees; table wines less than 14.8 degrees. Figures are rounded.

Table 4.1 gives an idea of the theoretical impact of these changes on wine prices. Before the tariff reforms in the early 1860s, duty accounted for half the retail price of ordinary table wines and about a third on "good" fortified sherry, so it was the cheap, low-alcohol wines that benefited most from the cuts. If the supply of all types of wine had been perfectly elastic, allowing wine producers to increase output to meet demand without raising their prices, the new tariffs would have involved price cuts at the retail level of up to 70 percent.[34]

These changes caused wine consumption in Britain to almost triple between the late 1850s and mid-1870s. It was not maintained, however, but fell 30 percent between 1871–75 and 1909–13 to just over half a million hectoliters on the eve of the First World War, a figure equivalent to little more than 1 percent of French output.[35] Per capita consumption, which had reached about three bottles in 1873 and 1876, declined by half by 1909–13 (fig. 4.1).[36]

These movements in aggregate consumption were accompanied by two other important changes. First, the decline in duties encouraged British merchants to

3,895 licenses had been granted to sell wine. Wine dealers, who could not sell wine in quantities of less then 2 gallons (twelve bottles), paid £5 5s. In 1866 the distinctive rates of tax on wine imported in barrels and bottles was abolished (Wilson (1940:322–23 and table 25; Briggs 1985:37–38).

[34] For fortified wines this was an average tax reduction of 57 percent, and for the cheapest unfortified ones it represented 83 percent.

[35] Revenue for the Treasury consequently fell from its peak in 1857, when it represented about 10 percent of total revenue from liquor. It then declined during the rest of the century, although the fall was partly offset by growing revenue from licenses.

[36] From 0.56 to 0.26 gallon (Wilson 1940:332–33).

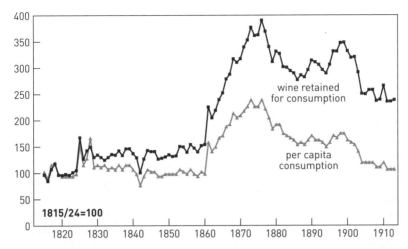

Figure 4.1. Wine consumed in the United Kingdom, 1815–1914. Source: Wilson (1940:332–33 and 364–65)

import wines with a lower alcoholic content, so that while in the late 1850s almost 85 percent of all wines were fortified to over 14.8 degrees (essentially port, sherry, and cape), by 1875 this figure had fallen to 75 percent, and it continued to decline to 56 percent by 1882.[37] The changes in 1886, which raised to 17 degrees the ceiling for those paying the lowest rate of one shilling duty per gallon, resulted in little more than a third of all imported wines paying the higher rates.[38]

A second, related change was the origin of the wines. Port and sherry shippers now argued that the new tax regime discriminated against them, as their strong wines had to pay more, and made them less competitive, leading to French wines increasing their market share from about 5 percent in the 1820s to 40 percent in the 1880s (table 4.2).[39] The switch to light French wines was initially slow, as an English merchant (in a letter quoted by Pasteur) noted in 1863: "at first these Wines met with a ready sale, but importers soon found that this branch of Trade was far from lucrative, as it entailed endless trouble and frequent loss, in consequence of the difficulties in keeping these Wines in marketable condition."[40] By

[37] United Kingdom. Parliamentary Papers (1878/79), pp. 315–17; and *Ridley's*, June 1894, pp. 361, 363).

[38] For example, 38 percent in 1893 and 36 percent in 1914 (*Ridley's*, June 1894, p. 362; June 1915, p. 418). As wines were often subsequently blended, this does not in itself imply a switch in consumption habits.

[39] The new duties in 1886 implied that by 1893 only 37 percent of Spanish wines paid the higher rates compared to 79 percent in 1882. However, the total consumption of Spanish wines declined by 160,000 hectoliters gallons (22 percent) over the 1882–93 period, and this decline can be attributed almost entirely to the drop in imports of the higher-strength wines, as imports of the lower-strength wines remained constant in absolute terms.

[40] Quoted in *Ridley's*, March 8, 1905, p. 210. The *Parfait Vigneron* was less polite: "It should also

TABLE 4.2
Source of Wine Imports to the United Kingdom

	Total (millions of liters)	Total imports (%)			
		France	Portugal	Spain	Others
1816–20	21.85	3.8	55.2	19.2	21.8
1821–25	24.71	4.5	51.2	22.3	22.0
1826–30	29.72	5.4	45.6	29.6	19.4
1831–35	28.44	4.0	43.1	34.9	18.0
1836–40	30.67	5.7	41.3	36.3	16.7
1841–45	27.86	6.4	38.4	39.2	16.0
1846–50	28.74	5.8	40.9	39.0	14.3
1851–55	30.58	7.7	37.7	39.8	14.8
1856–60	32.47	10.2	29.6	39.8	20.4
1861–65	49.62	20.1	24.6	41.7	13.6
1866–70	65.66	27.2	20.0	41.7	11.0
1871–75	77.97	29.2	20.4	39.9	10.5
1876–80	75.78	37.5	19.4	33.4	9.6
1881–85	65.81	40.3	19.5	29.7	10.6
1886–90	63.28	39.3	22.7	25.8	12.2
1891–95	65.56	38.7	23.8	23.1	14.4
1896–1900	73.52	37.1	22.5	23.8	16.6
1901–5	62.28	32.7	23.3	24.6	19.4
1906–10	54.71	30.9	25.6	23.6	19.9
1911–16	49.85	24.7	29.7	25.7	19.9

Source: Wilson (1940:361–63).

the 1870s this was no longer true, and the leading British trade journal, *Ridley & Co.'s Wine and Spirit Trade Circular*, noted on a number of occasions how wine merchants had become more professional in their handling of these wines.

Beyond these aggregate changes, most individual wines experienced a period of rapid growth followed by a slump. Sherry, or more correctly "Spanish white," saw a boom from the 1840s to 1875, when a massive 42.7 million liters was imported, or 43 percent of all wines (fig. 4.2). Imports then declined virtually every year and in 1896 fell below 9 million liters. By contrast, "Spanish red," which was produced mainly in Tarragona (Catalonia) and considered as a cheap alternative to port, grew from less than 4.5 million liters in 1871 to peak at 12.7 million liters in 1900 before also declining rapidly.[41] Exports of red wine from the Gironde (Bordeaux) increased dramatically after 1860 but peaked in the early 1880 and then fell sharply for the rest of our period (fig. 4.3). Champagne continued

---

be added that, as few, if any, English cellarmen have knowledge or experience of the management of any but strong brandied kinds, which scarcely anything can hurt, and only of our ripe well-fermented qualities, merchants who desire to purchase new wines, and to rear them themselves, will find it absolutely necessary to engage experienced French cellarmen." Quoted in Shaw (1864:101).

[41] *Ridley's*, various years.

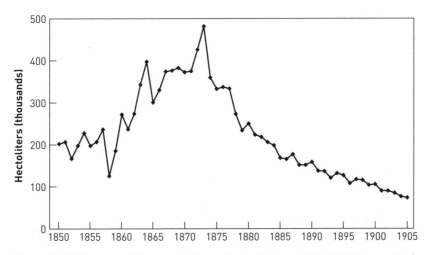

Figure 4.2. UK imports of sherry and "Spanish white" wines, 1850–1905. Sources: Wilson (1940:362–63) and *Ridley's* (various years)

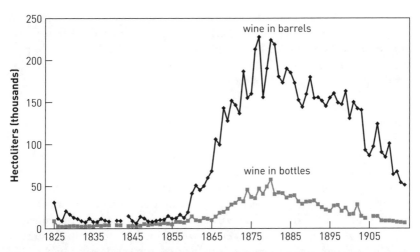

Figure 4.3. Wine exports from Bordeaux to the United Kingdom, 1825–1911. Source: France. Direction Générale des douanes (various years)

to be a success story despite the general fall in wine imports after 1875 (fig. 4.4) but also saw a decline in the years prior to the First World War.[42] Finally, although port imports avoided the sharp downturn prior to the First World War, there were also significant fluctuations in demand over the nineteenth century (fig. 4.5).

[42] There is no distinction between champagne and French sparkling wines in the UK statistics before 1894.

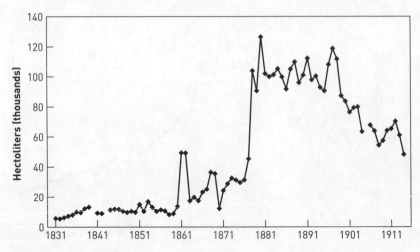

Figure 4.4. Exports of champagne to the United Kingdom, 1831–1911. Source: France. Direction Générale des douanes (various years)

Some contemporaries attributed these shifts in consumer preferences to changes in fashion. While there is clearly some truth in this, the literature also suggests that once a wine had become popular, it was frequently imitated by producers elsewhere, and the subsequent criticism in the local and national press led to a decline in the popularity of the genuine article.

THE RETAIL MARKET AND PRODUCT ADULTERATION

The British retail trade for foods was still highly specialized in 1850, and in the grocery trade, for example, "the blending of tea, the mixing of herbs and spices, the curing and cutting of bacon, the cleaning and washing of dried fruits, the cutting and millgrinding of sugar, and the roasting and grinding of coffee were essential functions of the retailer, apart entirely from the tasks of weighing out and bagging."[43] Wine merchants were equally skilled, especially in the area of blending different wines to meet a particular consumer's demand, and to "refresh" old ones.[44] At times, as noted at the start of this chapter, merchants could literally create new wines. In general, contemporaries believed that blending was acceptable if it was not carried out in order to deceive, either by mislabeling a

[43] Jeffreys (1954:2).
[44] The blending of wines held in bond in British ports was strictly controlled in the 1850s, and it was illegal, for example, to blend French wines with Portuguese wines for home consumption, although this was allowed if the wine was then reexported. These measures were not an attempt to protect consumers, but rather to avoid the mixing of wines that paid different rates of duty (United Kingdom. Parliamentary Papers 1852, Reay, p. 485).

Figure 4.5. UK imports of port wine, 1814–1910. Source: Wilson (1940:362–63)

wine's origin or through adulterating wines with other substances, especially those that were dangerous to health. One wine merchant, Christopher Bushell, argued that "if you have a pure wine from the south-east of France, and you blend it with a Port wine, it is true that it is not a Port wine, but equally true it is not adulterated."[45] Another, J. W. Dover, noted quite simply that "mixing two good wines together is not fraudulent; if they mix cider with wine, it is fraudulent."[46] Cyrus Redding agreed, writing in his classic work, *A History and Description of Modern Wines*:

> By the adulteration of wine is not to be understood the mixture of two genuine growths for the sake of improvement . . . but, in the first place, a clandestine amalgamation of an inferior kind of wine with one which is superior, to cheat the purchaser, by passing it off for what it is not; and secondly, what may be denominated with more propriety the product of fictitious operations passed off as genuine growths, having little or no grape juice in its composition. The first of these heads may be divided into adulterations of wines before and after they are imported.[47]

The problem facing the trade was that although blending by skilled merchants was legitimate and often improved quality and reduced prices for consumers, any unskilled or unscrupulous merchant could mix two wines and mislabel them for selling. The problem for consumers was to know how to identify the honest merchant and avoid the dishonest. Klein and Leffer suggest that a firm will maintain a long-term relationship with consumers and not cheat on quality if the present value of expected profits from future, repeat sales exceeds any short-

[45] Ibid., Bushell, p. 764.
[46] Ibid., Dover, p. 649.
[47] Redding (1833:321; 1851:345).

term, temporary increases due to shirking on quality.[48] In other words, a merchant who had invested heavily in reputation was much less likely to cheat than one that had recently only started business, especially if it was in an area of rapid growth, such as the wine trade after 1860. For expensive, fine wines a limited number of private buyers and repeat purchases was expected, so many merchants were willing to maintain quality standards. Because consumers often came from similar geographical and social backgrounds, short-term, opportunistic behavior by a wine merchant selling poor wine at inflated prices to one consumer was likely to quickly be known to others. This had always been true of the best French wines, but from the late eighteenth century it also included expensive vintage ports and fine sherries.[49]

While economic incentives for well-established wine merchants selling fine wines to cheat their consumers were limited, this was less true for retail merchants and tavern owners who sold cheap wines. Indeed, consumers might prefer adulterated wines if the result was that they could consume goods that otherwise would be inaccessible to them.[50] In this respect, just as today there is a demand for false Gucci handbags, nineteenth-century consumers were happy to drink "port" that had been produced a considerable distance from the Douro valley. In the 1850s South African cape paid only half the duty of other wines, and considerable amounts were imported. Yet according to H. Lancaster, a London wine merchant, "it comes in as Cape, and pays Cape duty; and I hear of no Cape selling as Cape, except some little Constantia that may be asked for now and then."[51] Most was sold as port. From the 1870s Tarragona red wine from Spain was imported for a similar use (chapter 7). Blending wines from different locations therefore allowed retailers to create new products for their customers. However, not only did this increase the problems of classifying wines, but in an age with few public health regulations and limited ability of consumers to detect fraud, it

[48] Klein and Leffler (1981).

[49] Francis (1972:300). He continues, "Good wines could only be brought from high-class merchants and they were apt to confine their dealings to a distinguished and discerning clientele" (p. 306). It did not imply that wine merchants were averse to using modern sales techniques, as Shaw (1864:23) noted:

more means and instruments are used to make sales then is generally supposed; for it is not seldom that a finely-booted and spurred independent-looking gentleman in the hunting-field is, in reality, *sub rosâ*, a wine merchant's "help"; and, by "incidental" hints, there and at table, about Château this and Château that, he earns money for his employer and commissions for himself. However good a stock may be, there is so much competition that the merchant's sales will be very slow indeed, if he imagines that the excellence of his cellar will absolve him from practising solicitations; with much, besides, very repugnant to the feelings of a gentleman. Not least of these is the payment to servants of money which, properly, should be given to their masters, and favours of various kinds to persons of influence in clubs, &c.

[50] Alsberg (1931).
[51] United Kingdom. Parliamentary Papers (1852), Lancaster, p. 108.

encouraged cheating through mislabeling and the addition of substances other than wine.

High import duties had previously encouraged merchants to adulterate wines. Cyrus Redding, writing in the early 1830s, noted that sherry was sometimes mixed with cheap wines to create "inferior sherries" and then strengthened with brandy before being exported, but they were never adulterated in Spain.[52] By contrast, in England "sherry of the brown kind, and of low price, when imported is mingled with Cape wine and cheap brandy, the washings of brandy casks, sugar candy, bitter almonds, and similar preparations, while the colour, if too great for pale sherries, is taken out by the addition of a small quantities of lamb's blood, and then passed off for the best sherry by one class of wine sellers and advertisers."[53]

Not surprisingly, British retail merchants were not popular among foreign wine producers, and in 1825 one port shipper referred to them as the "most rotten set in London," while a couple of decades later Cyrus Redding noted that "no branch of trade is open to the practice of more chicanery and fraud than that of wine dealing."[54]

The publication of Dr. Arthur Hassall's scientific work in the *Lancet* between 1851 and 1854 showed both the extent of food adulteration at this time, and its implications for public health.[55] Whether this was actually greater than at earlier periods is impossible to determine, but the increasing importance of national and local newspapers resulted in wider public awareness and the demand for regulatory action by local and national authorities. Henry Bartlett, a fellow of the Chemical Society, believed in 1872, as Cyrus Redding had earlier, that most adulteration took place in England rather than in Spain, and that the major problem was mixing wine with "bad spirits."[56] Until the 1870s most packaged alcohol was bottled by retailers, and these often sold wines under their own brand. This was encouraged by the domestic wine trade press, and the *Wine Trade Review* noted that "it is to the real merchant that whatever value which may attach to a brand should belong, as it is he who should be responsible to the consumer for the quality of the article supplied him."[57] However, as Paul Duguid has reminded us, wine shippers had used long iron brands to mark their names on the casks.[58] This, he argues, represented not so much a fight between

[52] Redding (1833:191).

[53] Ibid., 322.

[54] Cited in Duguid (2003:413).

[55] Burnett (1999) and United Kingdom. Parliamentary Papers (1854–55), First Report, pp. 1–45. This report claimed that wine adulteration in Britain was equivalent to 20 percent of wine imports (ibid., Second Report, p. 35).

[56] United Kingdom. Parliamentary Papers (1872), pp. 206, 213. However, the British consul and others in Jerez argued that local producers were also producing artificial wines.

[57] *Wine Trade Review*, January 15, 1864, p. 2, cited in Duguid (2003:425).

[58] Duguid (2003:431). Hence the origin of the word "brand."

rival producers, but an attempt to control "what was done in (or with) their names by suppliers or clients" along the supply chain.[59] Development of an effective national framework to combat fraud began in Britain with the Sale of Food and Drugs Act of 1875, which defined adulteration in terms of risk to the consumer's health, or deception concerning the description of the product.[60] This placed the regulatory emphasis on the retailer rather than the producer and allowed the mixing of wines ("compound foods"), provided that these were not "injurious ingredients" to health and were adequately labeled.[61] It proved to be insufficient to control the wine trade.

The heterogeneous nature of wine made it a difficult product to sell, but the problems of marketing in the nineteenth century were also caused by extreme market volatility. Mark Casson has suggested a number of types of volatility that organizational structures have to overcome.[62] In the first instance, supply and demand fluctuated within established markets caused, for example, by a harvest failure in Bordeaux or a business depression in the manufacturing districts of northern England. Although the timing of these events could not be predicted accurately, they were not entirely unexpected, and short-term movements in prices were usually sufficient to balance supply and demand. However, the fact that wine quality also changes with each vintage significantly increased the problem of classification.

A second type of volatility involved major structural changes in the market. Gladstone's innovations of 1860–62, by reducing tariffs and creating new forms of distribution, significantly increased the size of the potential wine market. These changes, together with improvements in transport, also encouraged new merchants in both the traditional and new regions to trade with Britain. However, volatility was also produced by breakdowns and interruption of supply and these, as we have seen, were frequent during the second half of the nineteenth century (table 4.3). The first important one was oidium or powdery mildew, which reduced harvests and wine quality, especially between 1853 and 1856, and caused prices to soar briefly. Phylloxera led to French production slumping and prices rising by a third between the early 1870s and the early 1880s before declining once more in the face of massive imports, adulteration, and a recovery in domestic production. Finally, the appearance of downy mildew in the 1880s not only reduced the size of harvests but, in Bordeaux especially, ruined wine quality (see chapter 5).

---

[59] Ibid., 419.

[60] French and Phillips (2000:36–37). The act allowed local authorities to inspect retail outlets, but not manufacturers. The three acts of Parliament between 1860 and 1872 had failed to provide an efficient legal basis to combat adulteration of food and drink.

[61] Ibid., 4. The major problem appears to have been the addition of potato spirit. United Kingdom. Parliamentary Papers (1872), p. 207.

[62] Casson (1997:10–11).

TABLE 4.3
Vine Disease, Harvest Size, and Prices in France

	Years of maximum intensity	Impact on harvest size[1]	Change in farm price[1]	Change in Paris retail price[1]
Powdery mildew	1853–56	−57%	na	+56%
Phylloxera	1879–92	−50%	+39%	+3%
Downy mildew	1882–86[2]	−20%	+15%	+7%

Sources: France, Annuaire statistique année 1938 (1939:62–63); Singer-Kérel (1961, table 11).
[1] Harvest and price movements compared to previous five years
[2] Mildew was especially virulent in Bordeaux in these years.

The wine trade in producer countries tended to respond to these supply disruptions in two very different ways: by raising prices of fine wines because of the shortage of good-quality stocks, or by reducing quality by blending with cheaper, often inferior wines from other regions to compensate for the smaller local harvests. For example, powdery mildew caused the price of fine old sherries to triple in Britain between 1850–53 and 1860–63, but at the same time the reduction in import duties and the ability of shippers to export "sherry" from other regions led to a fall in prices of the cheaper wines.[63] In fact, London prices for a wide range of cheap commodity wines from Iberia stagnated over several decades, despite significant fluctuations in both the size and quality of the vintages, as well as the tendency for wine prices in producer countries to increase from the late 1870s (fig. 4.6). The same was true of other wines. Thus it was noted in 1907 that the price of "vintage" champagne "may fluctuate considerably," but for nonvintage ones it was "pretty constant,"[64] while in Bordeaux Ridley's reported in 1881 that, as a consequence of the significant decrease in harvest size caused by phylloxera, "no pure Claret can now be put on board under £8, and then of less satisfactory quality than was a few years since easily obtainable at £5; whilst blends with Spanish Red, South of France, and other Wines are sold, occasionally under their true designation, but generally under the usurped title of 'Claret,' at from £5 to £7 per hhd."[65]

In conclusion, although merchants maintained stable retail prices for cheap wines by blending (or adulteration), quality in the late nineteenth century dropped, leading to a decline in consumption. The problem was not limited to these wines. As Akerlof notes in a different context, buyers with insufficient knowledge concerning quality prior to purchase were discouraged from buying all types of wines because of the continual negative reports in local and national

[63] Shaw (1864:235); Simpson (2005).
[64] Ridley's, July 1907, p. 514.
[65] Ibid., January 1881, p. 5. A hogshead (hhd) was equivalent to between 220 and 225 liters.

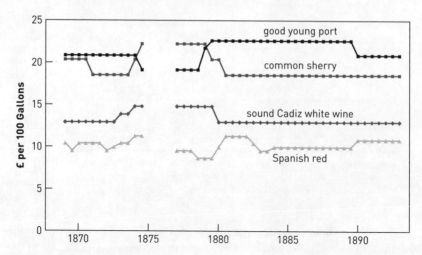

Figure 4.6. Prices of cheap Iberian wines in London, 1869–94. Source: *Ridley's* (first issue in January and July of each year)

newspapers.[66] In particular, the markets for fine sherries and clarets collapsed in part, as we shall see in later chapters, because large quantities of "villainous trash" were being sold under these names, which eventually undermined the reputation of even the best brands.[67]

WHO CONTROLS THE CHAIN? EXPERIMENTS AT "BUYER-LED" COMMODITY CHAINS

In the early part of the nineteenth century British producers and exporters in places such as Porto or Jerez established the commodity chain and developed wines specifically for their home markets. These wines were bought by the cask from the retail merchant who then bottled it, often in the consumer's house. The lower duties and the Single Bottle Act of the early 1860s provided opportunities to package and market wines in new ways, such as creating national brand names, and developing new forms of advertising. Retail merchants jealously tried to defend their right to bottle and saw the growing sale of bottled wines, whether expensive vintage champagne or cheap ports and sherries in multiple chain stores as a threat to the traditional retail business, believing that it would reduce them to mere "penny-in-the-slot machines deprived of all judgment in their buying

---

[66] Akerlof (1970).

[67] *Ridley's* used this expression as early as January 1870 (p. 5) with respect to claret. It was not the only factor, however, as the false rumors over the supposed health safety of sherry and the quality of claret from the leading châteaux declined substantially in the late nineteenth century. See chapters 5 and 8.

and selling."[68] *Ridley's* complained as early as 1869 that the foreign shippers rather than the merchants enjoyed "the lion's share of the plunder," and the consequence of this change was not lost on the British trade journal, which continued: "It is daily becoming a more interesting question both in our own and other branches of commerce—"how is the middle man to exist?," the only rejoinder from the producer and consumer, between whom he was formerly the medium of communication, being apparently the somewhat cynical one,—"there is no necessity *that he should exist!*"[69]

These changes encouraged the appearance of buyer-driven commodity chains. In theory importers, by blending wines regardless of geographic origin, could improve quality and lower costs by avoiding poor local vintages, and make large purchases after good ones. In other words they could perform, on a larger geographical scale and in Britain, the sort of operations for which the Bordeaux négociant, for example, had traditionally been responsible. Wines could then be bottled centrally and sold under the importers' brand via select retail outlets. The Victoria Wine Company, which was established in 1865 in Mark Lane in the City of London, had by 1886 some ninety-eight retail stores throughout the country.[70] The company bottled its own wines and placed the fact that they were "unadulterated" prominently in its advertisements.[71] Gilbey's was even more successful, being responsible in 1875 for 5 percent of all wine, six times the share of its nearest rival.[72] Taking advantage of the lower duties and especially the Single Bottle Act, it sold all its wine in sealed and labeled bottles.[73] The firm's success was achieved by importing wines in bulk directly from the country of origin, and by the use of its "Castle" brand and some 2,000 carefully monitored agents throughout the country. These agents were often already well-established grocers and were stocked with a selection from Gilbey's list of two hundred different drinks. Finally, the importer Peter Burgoyne claimed to have spent £300,000 to sell his brand of Australian table wines by the turn of the twentieth century (see chapter 10).

Yet these companies were unable to halt the decline in wine sales from the mid-1870s. A number of problems can be identified. First, large-scale retailers had few advantages over smaller ones unless they could achieve buying-and-selling economies and introduce standardization and stock control.[74] The fact

---

[68] *Ridley's*, July 1907, p. 514.

[69] Ibid., August 1869, p. 2, emphasis in the original. *Ridley's* noted that merchants would have to accept the lower status of being a shopkeeper or give up, and "already, Wholesale Dealers, with certain pretensions, are opening retail branches in prominent thoroughfares."

[70] Briggs (1985:9).

[71] The *Illustrated London News*, December 13, 1873, cited in ibid., 48. By 1880 the company was listing sixteen sherries, together with fourteen ports, eight clarets, and "sundry wines," including a Hambro sherry (ibid., 53).

[72] Faith (1983:12).

[73] United Kingdom. Parliamentary Papers (1878/79), p. 157.

[74] Jeffreys (1954).

that wine quality varied greatly not just from one harvest to the next, but even on an individual vineyard in the same year, made it very difficult to create standardized products to brand. Gilbey's, for example, depended on leading shippers in Jerez and Porto to select its wines to be sold under its own brand. When this firm integrated backward and purchased Château Loudenne in the Médoc in 1875, it was done as much to reduce information costs in its search to buy suitable wines from local growers as to cut production costs.[75] As an investment, the purchase of Château Loudenne occurred at perhaps the worst moment possible because of phylloxera and was a drain on company resources, but it did allow Gilbey to market itself as a wine producer. Retailing was more complicated than with most other foods and beverages, even when the retailers did not have to bottle the wine themselves, as *Ridley's* noted:

> One has only to go about Town and Country with one's eyes open, to see what is offered in the shops of certain small Grocers and other Dealers, and the manner of so offering Wines which, perhaps may once upon a time have been sound and good, are found standing upright in the window in the full glare of the sun, till they must have entered on the last stage of degeneration, but nevertheless ready to be supplied to the guileless customer whose knowledge in buying is about on a par with that of the Vendor in selling. It is when the purchaser gets his bargain home, however, that the dénouement takes place. He may not be much of a judge, but his palate is sufficiently sensitive to know sound Wine from bad, and it tells him that it is not the former which he has got. He, therefore, not only eschews that particular Wine for the future, but in many cases tars the entire *genus* with the same brush, discarding the juice of the grape for Spirits and Beer.[76]

Yet even *Ridley's* had to agree that when chain stores used retailers who were both responsible and knowledgeable, the consumer was likely to be better served than by the "small and inexperienced" dealers who bottled their own wines. In particular *Ridley's* noted that the Civil Service and the Army and Navy Cooperative Societies, which at their commencement were "extremely unpopular with the distributive Trade, owing to the severe competition which they brought about," conducted their business professionally and the reputation of wine did not suffer, unlike "other stores."[77]

The problems in the wine market contrast strongly with the beer and spirits industries, where growing economies of scale in the production and the producers' ability to create a standardized product made it much more susceptible to branding. According to Gourvish and Wilson, with beer "dilution and adulteration, the old resort of distressed publicans, seem to have been increasingly stamped out after the mid-1880s, as analysis became common, and brewers

---

[75] Gilbey first bought wines directly in France in 1863 (Maxwell 1907:18). For Château Loudenne, see especially Faith (1983).

[76] *Ridley's*, September 1899, p. 620.

[77] Ibid., 621–22.

fought long and hard to remove the practice of the 'long pull.'"[78] Markets were guaranteed by the forward integration of the leading breweries into distribution, through their purchase of retail outlets (public houses). *Ridley's* contrasted these improvements with the problems of wine in 1901:

> It cannot be denied that the quality of Beer—if we mean by quality, condition, attractive appearance and taste—has greatly improved in the last thirty years. When one recalls the turbid slop which was called fourpenny Ale, and compares it with the bright, brisk and light beverages now sold in most places at the same figure, we shall realise, apart from the low price, how, if in the lowest grade of Beer such an improvement has taken place, the palatability and attractiveness to all classes in the higher grades of Beer have increased. *What improvement has taken place in the manner of selling or the style and quality in Wine as an article of consumption? The answer must be none whatever.*[79]

Yet the decline in consumption from the 1870s cannot be explained by a switch to other alcoholic beverages, as the movement in demand for both spirits and beer was similar to that of wine. Rather than the temperance movement, which before 1900, according to one historian "had promised much and delivered little," the shift away from alcohol needs to be explained by the appearance of new forms of consumer goods and leisure activities.[80] Although Gladstone's "wine revolution" failed to take off, the retail trade in Britain was changed. Traditionally, considerable amounts of port, sherry and claret had been drunk in rural areas, and contemporaries noted a weakening in this trade, especially with the onset of the agrarian crisis and the fall in land rents after 1873.[81] Rural consumers had often purchased their port and sherry in barrels and bottled them on their own premises, but urban consumers had limited space for a wine cellar and preferred to buy their wines already bottled.[82] Public as opposed to private drinking grew in importance, and brewers from the turn of the century entered into the wine trade at the cheap end of the market, not just supplying their publicans, but also purchasing wine merchants' businesses.[83]

There was also a growing demand from hotel and restaurants as eating out

---

[78] Beer prices remained stable in the second half of the nineteenth century at between 3.5d. and 6d. a quart (Gourvish and Wilson 1994:207–8).

[79] *Ridley's*, April 1901, p. 258. My emphasis.

[80] Weir (1984:96); Simpson (2004:84). For the temperance movement, see especially Harrison (1971).

[81] Sherry imports peaked in 1873 in part because of the rumors that duties were to be raised (Pan-Montojo 1994:106). Thudichum's infamous letter to *The Times* was also published that year. See chapter 8 .

[82] *Ridley's*, April 1903, p. 241; September 1904, p. 639; and September 1899, p. 621.

[83] See, e.g., ibid., September 1899, p. 622; January 1900, p. 8). According to Samuel (1919:170), "A very large proportion of wine—it has been estimated that it may be, perhaps, 70 percent.—is consumed at wine bars and in public-houses and cheap restaurants, though this would not represent 70 percent. of the taxation, owing to the higher duties on sparking wines and better quality port."

became fashionable, which again encouraged the growth in proprietary brands.[84] While as late as 1875 John Cordy Jeafferson noted that "there is no Continental capital so poorly provided as London with establishments were the stranger may obtain a fairly good dinner for a small sum, or an excellent dinner for a great price," by the turn of the century a complete "revolution" had taken place, and now it was claimed that there was a greater variety of hotels, restaurants, and eating houses to suit "all tastes and all purses" than anywhere.[85] A bottle of Pommery 1893 in 1905 might cost "a gentleman" 8s. a bottle at "his club," but the price at a fashionable hotel for the same wine was 25s. or 27s. 6d., although this had the consolation in that it helped to "keep the vulgar out."[86] Smaller and more modest hotels sold much cheaper wines, but markups were often of a similar magnitude, and the wines often "execrably bad," adding just one more perceived injustice to the retail merchant.[87]

The evidence suggests that Gladstone's desire to extend wine drinking to new social classes had had only limited success by 1914. In the early 1870s a very rough estimate by Leone Levi claimed that members of the working class purchased 75 percent of all beer and spirits, but they consumed just 10 percent of wine.[88] Taking the average annual per capita consumption of wine in the early 1880s as two and a half bottles, this implies that 70 percent of the population drank on average just a third of a bottle, compared to nine bottles for the rest of society.[89] A more detailed attempt at measuring wine consumption for the years 1913–14 (table 4.4) found that those enjoying annual incomes of £150–200 consumed four times more wine than those with £50–£100.[90] Feinstein gives the average earnings in 1911 for Britain's 15.88 million wage earners as £58.6, with the police top earners at £96 followed by coal miners with £85.6.[91] A further 4.5 million, or 22 percent of the total, were considered as "nonmanual." As about a third of imports were still fortified wines in 1913, which paid the highest rate of duty, and assuming an average family size of five, the nature of per capita consumption was probably similar to that of the early 1880s. *Ridley's* had come to a similar conclusion a few years earlier:

> Wine worth drinking in this Country does not come within the means of the masses. It may not be generally recognised that the number of persons with incomes of over £200 a year are, judging from the Income Tax Returns, considerably under one million, whilst not half that number have £500 per annum, and it is to these latter that the

[84] *Ridley's*, October 1915, p. 661.
[85] Simon (1905:154, 156).
[86] Ibid., 157.
[87] *Ridley's*, August 1869, p. 16.
[88] Levi (1872).
[89] The working class is taken as 70 percent. See Rowntree and Sherwell (1900:9).
[90] United Kingdom. Parliamentary Papers (1927), xi, p. 92, cited in Burnett (1999:142).
[91] Feinstein (1990:603).

TABLE 4.4
Estimates of Duty Paid on Alcoholic Beverages by Income Levels, 1913–14

Income	Duty (consumption of husband and wife)			Total duty (£ s. d.)	Equivalent in bottles of table wine*	Equivalent in bottles of fortified wine*
	Spirits (% of total)	Beer (% of total)	Wine (% of total)			
£50	44	52	3	2. 0. 0	6	3
£100	53	39	5	2. 15. 0	13	5
£150	54	38	8	3. 5. 0	26	11
£200	54	31	13	4. 0. 0	48	20
£500	59	7	33	5. 0. 0	160	67
£1,000	65	2	33	7. 10. 0	240	100
£2,000	51	1	48	13. 10. 0	624	260
£5000+	36	—	64	27. 10. 0	1,680	700

Source: United Kingdom. Parliamentary Papers (1927), p. 92.
*Per capita consumption of either table wine with an alcoholic strength of below 30 degrees or fortified wines between 30 and 42 percent.

Wine Trade has to look for its customers. With a clientèle limited to about a million individuals, it is useless to anticipate an unlimited extension of the Trade.[92]

The fine wines imported into Britain prior to 1860 were expensive because of high production costs (caused by the labor intensive nature of viticulture, the low yields, and the capital needed for making and maturing wines for the British taste); the high marketing costs (long credit, slow turnover, and a limited number of consumers); and elevated import duties. The reduction in duties brought about by the Cobden-Chevalier Treaty saw wine imports double from slightly less than 350,000 hectoliters in 1860 to 700,000 in 1870. Over the next three decades they stagnated before dropping by almost a third, even though wine duties remained virtually unchanged. By 1913 per capita consumption was no greater than it had been in 1815, despite the retail price of the cheapest wines being one-fifth of that in Wellington's time, and, as *Ridley's* noted in 1901, stagnant demand for wines after the 1870s occurred during a period of rapid economic change.[93]

This chapter has argued that import duties were not a major obstacle to selling wine in the British market after 1860. This is suggested by the experiences of different types of wine, which all saw rapid increases in consumption at one moment or the other during the mid-nineteenth century, but this was followed by a steep drop in their popularity. For sherry, this decline can be dated from the early 1870s; fine claret, from the early 1880s; ordinary claret, from the late 1880s; and

[92] *Ridley's*, 1908, p. 1081.
[93] Ibid., April 1901, p. 258.

TABLE 4.5
Production and Trade in Wines in Selected Countries, 1909–13 (millions of hectoliters)

	Production	Exports	Imports
France	46.4	2.0	8.2
Italy	46.0	1.6	0.0
Spain	14.9	3.1	0.0
Austria-Hungary	7.9	1.4	0.4
Portugal	4.8	1.1	0.0
Greece	3.2	0.6	0.0
Germany	1.8	0.2	1.1
Great Britain	0.0	0.0	0.6
Belgium	0.0	0.0	0.4
Netherlands	0.0	0.7	0.9
Switzerland	0.0	0.0	1.6
Europe total			
Algeria	7.9	−6.7	0.0
Argentina	4.4	0.0	0.6
Chile	2.0	0.0	0.0
United States	1.9	0.0	0.3
Brazil		0.0	0.7

Sources: Production—table 1.2; trade—International Institute of Agriculture (1927:292–93).

champagne, from about 1900. Although imports of port avoided the sharp downturn prior to the First World War, the wine also experienced major fluctuations in demand over the century. A significant decline in the quality and reputation of individual wines precipitated these changes in consumer behavior. The heterogeneity of wines made it much harder for department stores or retail chains to sell wines as they did other foods, such as breakfast cereals or canned soup. Despite this, virtually all the fine sherry and port continued to be sold to the British market, and it also remained the largest for bottled claret.[94] Only with champagne did others of any size exist.

As shown in previous chapters, the production of artificial wines and the practice of adulteration were as widespread in producer countries as in Britain. Annual per capita wine consumption also declined in France, from an average of 145 liters per person in the decade 1869–78 to 98 liters in 1883–92, because of high prices and adulteration, but this then recovered from the 1890s, and by the 1900s annual consumption had reached over 150 liters per person. Crucially, while wine-producing countries legislated against the making and marketing of artificial wines using raisins and other products, this did not happen in Britain, and in 1912 "British" wines were equivalent to 8 percent of the total market.[95]

[94] Britain accounted for about a fifth of Bordeaux's exports in bottles in 1909–11.
[95] Prest (1954:83).

TABLE 4.6
Tariffs on Commodity Wine Imports in Selected Countries, 1898 (francs per hectoliter)

Country	Tariff	Country	Tariff
Argentina	40	France	20
New South Wales	138	Italy	20
Victoria	165	Spain	50
South Australia	165	United Kingdom	28
Chile	125	French wine price (1896–99)	26
United States	55		

Sources: Cocks and Féret (1898:775–84); for France—Sempré (1898:196) and France, *Annuaire statistique 1938* (1939:62–63).

Note: Figures are for ordinary red wine. The French figure ranges from 12 francs for wines of up to 10 degrees alcohol to 20 francs for those of 15 degrees.

Finally, the poor export performance of the sector was not limited to Britain, as only 12 percent of the world's wine production was exported in 1909–13, and if Algeria—which sold virtually all its wines to France—is excluded, the figure declines to 8 percent (table 4.5). Even more striking was the fact that imports to countries other than France were equivalent to just 6 percent of world production. Although the flow of new ideas and technology was increasing across national frontiers after 1900, the international trade in cheap commodity wines was limited, especially as tariffs were often higher than domestic prices. This was because taxes on alcohol were an important source of national revenue, and most countries protected producers from foreign competition, in exchange for high excise taxes on domestic output of beer, spirits, and wine (table 4.6).

PART III

Institutional Innovation: Regional Appellations

CHAPTER 4 ARGUED that buyer-driven commodity chains in Britain failed to establish strong brand names as they had been able to do with many other foods and beverages. This was caused by the major volatility in wine quality, which was accentuated by vine disease; concerns of widespread fraud and adulteration; and the difficulties in establishing cheap impersonal exchange mechanisms to allow high sales volume, rather than relying on the personal reputation of wine merchants and the associated low volume and high margins. This section examines the response of traditional producer-driven commodity chains to the growth in market opportunities after 1860, and the demands by growers to establish collective regional brands for claret, champagne, port, and sherry. Three major characteristics were present in each of these regions: a limited supply of high-quality land; the concentration of production or exports in the hands of a few houses; and the presence of thousands of small producers of grapes and wines of very different qualities. It was the response of these distinct groups to the opportunities opened by trade and the problems associated with adulteration and fraud, together with the nature of political institutions, that would determine how the industry was reorganized and rents distributed from the turn of the twentieth century.

The areas where fine wines could be produced in each of the regions were strictly limited, but they were surrounded by much larger areas of vines that were used to produce lesser-quality wines. Claret, for example, was a term that applied equally to the exclusive wines of Château Lafite and to those of the Palus region, although prices differed by a factor of at least ten. Furthermore, as production costs and prices in Bordeaux for cheap clarets were double those of the Midi, there was a major incentive for local growers and merchants to blend their own wines with those from outside the region and sell them under the Bordeaux brand. A similar situation existed with the other wines described in this section. The declining collective reputation of claret, champagne, port, and sherry in the British market and elsewhere directly threatened the local producers of the cheaper wines, and indirectly even the brands of the leading firms.

Trade was highly concentrated. Many of the major houses were already established before 1840, and their names continue to be important today, although not necessarily as independent houses. In Jerez, the leading three houses were responsible for 23 percent of all exports to Britain in 1848–52, a figure that reached 36 percent with the top five houses. Market concentration was even greater with port, at 32 and 46, percent respectively.[1] In the Champagne region, a few large houses, such as Krug and Moët & Chandon, also dominated, and even in Bordeaux before 1860 the leading estates often preferred to use the major shippers, such as Johnston, Schröder, and Schyler, to add an additional guarantee to their brand.[2] The market dominance of these leading producer-shippers was achieved by the combination of their control of the wine produced from the limited area of high-quality land; their ability to draw on large quantities of capital to stock and mature fine wines; and the marketing and commercial connections that they enjoyed in the key British market. In all cases it was recognized that brands of a respectable producer-shipper added value to the wine.[3]

Both sherry and port were fortified dessert wines, and this allowed exporters not just the possibility of creating a product that would withstand transportation and poor handling by retail merchants, but also of producing a homogenous product in sufficient quantities each year that was relatively easily branded. The leading champagne firms also carried significant stocks of "house" champagne, and in this case the nature of the drink implied that it could be exported only in bottles, usually under the brand of the manufacturer. In Bordeaux the leading châteaux were relatively small, producing a fraction of the wine sold by the champagne houses, for example. However, the 1855 classification institutional-

---

[1] Calculated from Shaw (1864:192, 235, tables 12, 13) Quantity did not imply quality, however, and *Ridley's*, February 1878, p. 49; March 1878, p. 89, ran a campaign that succeeded in 1878 in stopping the publication of these annual shipping lists.

[2] Salavert (1912:66). See chapter 5.

[3] Shaw (1864:17) suggested that brands could add up to 50 percent to the value of the wine. *Ridley's*, January 1884, p. 3, argued that consumers were willing to pay higher prices for branded items because of the greater security that they were not adulterated. A reputation could also be ruined, as noted in 1882:

> Too much of this cheap Wine has unfortunately been shipped and put into bottle as the veritable article, and it is to be feared will not in the results tend to advance the name of Port in popular estimation. During the past year we have had occasion particularly to notice a large parcel forwarded to London by one of the best know and oldest established firms at Oporto, the shipping price being, we believe, £18 free on board, and the importers an equally well known firm in this city. As a specimen of "Port Wine" it was perhaps as bad as anything ever so misdescribed, but for all that, it has figured, in lots of ten, fifteen, or twenty pipes at a time, month after month, in catalogues of Public Sales, very prettily marked, boldly put forward as "Port," of —— & Co's. shipping and sold, on the faith of the name, at a small profit. We humbly venture to caution the celebrated shippers in question that no surer course could be adopted to damage the unquestionably high reputation of their brand, to say nothing of the harm likely to result to the Port Wine trade generally from the distribution of such stuff amongst the consuming public. We are glad to notice that some of the leading shippers, including the firm alluded to, have lately notified their intention of selling these common Red Wines under their true designation

(*Ridley's*, January 12, 1882, pp. 6–7).

ized a marketing structure that had developed over the previous half century or more, allowing consumers to select their wine by the geographic location of a specific estate or village. Fine claret also required large quantities of capital, as a producer's reputation was linked to the spectacular quality of a particular vintage achieved only once every four or five years. Therefore Bordeaux merchants encouraged the sale of fine claret under the proprietary brand of the wine estate, not just because it created a luxury item in short supply that they were best placed to sell, but also because they themselves were very often the owners of the leading estates or benefited from exclusive contracts with them.

The economic incentives for the established export houses to cheat on the quality of fine wines were small, as any short-term profits would not compensate for the long-term decline in value of the brand.[4] However, after Britain reduced tariffs in the early 1860s, the established export houses could both compete in the market for expensive fine wines as well as use their distribution channels to sell in the potentially much larger market for commodity wines, where price was the decisive factor rather than quality.

The manipulations of wine by merchants that led to a sharp drop in quality and the hostile press comments on the supposed widespread adulterations created a critical situation for the small producers in these regions. Local growers had long claimed that claret, port, champagne, and sherry constituted collective trademarks for a wine made only from grapes produced in these specific regions, and from 1900 onward they were vociferous in their demand for regional appellations. Appellations, they argued, would improve consumer information about wine quality and guarantee that only local wines would be sold. Critics, by contrast, claimed that appellations would simply create local wine monopolies and shift rents to growers, and one contemporary compared this to a return to the provincial monopolies of the ancien régime.[5]

While growers were united in their opposition to outside producers appropriating the local name, they faced formidable organizational problems in creating a regional brand. The sixty thousand or so growers in Bordeaux in 1900, for example, might agree not to make wines from raisins or to sell cheap Midi wines as claret to maintain the region's collective reputation, but few would have been willing to make the sacrifice if they believed others would continue to cheat.[6] Only when growers believed that a system could adequately identify and punish

[4] Casson (1994:47) notes that "a firm that has sunk a large amount of non-recoverable capital into the product is far more vulnerable to loss of custom than a 'hit and run' entrant with versatile equipment and negligible sunk costs. This is one reason for long-lived firms enjoying a reputation for quality and integrity that start-up firms do not." Gary Libecap (1992:245) argues that small firms would be more likely to cheat on quality than larger ones, as "producers with large market shares absorb more of the industry-wide losses of a demand shift to substitutes if consumers cannot isolate the violating firms."

[5] See the debate on Bordeaux in chapter 5.

[6] Olson (1965:10).

cheats were they likely to respect the rules themselves. This required government legislation and was opposed by those growers living outside the appellation, and by the merchants who looked to strengthen their private brands and demanded the right to purchase and blend wines from wherever they considered necessary to sell cheaply in the highly competitive international markets.

Defining the characteristics of wines such as claret or champagne also caused problems. If claret came from within certain boundaries in the Gironde, did that mean that all wines produced within the geographical area could be considered as claret, regardless of their quality? And how were wines to be classified that had been grown in one village, then crushed and fermented into wine in another? Could champagne be made with grapes grown in the Marne but crushed in Germany? Or with grapes grown in the Midi and mixed with wine from Épernay to make champagne? These questions led to bitter conflicts, as merchants argued that a wine's reputation derived from their own wine making and blending skills. By contrast, growers believed that the reputation of a wine was located in the terroir and not the winery, and therefore only wines made from grapes produced in certain designated areas should carry the collective regional brand. Finally, establishing an effective regional appellation might allow legal acceptance at a national level, but the process of international recognition in some cases was not achieved until the end of the twentieth century.

The following chapters consider in detail the nature of cooperation and conflicts in each of the four wine regions—in particular the response first to phylloxera and later to the collapse in wine prices after 1900—and explain why some regional groups (growers, merchants) were more successful than others in establishing new market-supporting institutions. As noted, the costs and benefits from regulation were unequally distributed, and this created incentives for individual groups to capture "rents" at the expense of others. However, while the thousands of small growers in Bordeaux and Champagne were able to use their political influence to create advantageous institutions for themselves, such as regional appellations, the situation was often less favorable in those countries with "elite democracies." In Spain, for example, authorities routinely dismissed the demands by small sherry growers in Jerez for a regional appellation or local bank, which would have made them less dependent on a handful of powerful Spanish shipping families. By contrast, in Porto, the small vine-growers found a considerably more sympathetic state, and a regional appellation was reestablished. The foreign nationality of most of the leading port shippers placed them at a distinct disadvantage when negotiating with the Portuguese state, although it did allow them to negotiate successfully with the British government the Anglo-Portuguese Commercial Treaty Acts of 1914 and 1916.

CHAPTER 5

Bordeaux

> The Gironde has been for a long time France's leading wine *département*, less for the extension of its vineyards (around 150,000 hectares) than for the variety and care in their cultivation; for its winemaking skills, and the character and remarkable qualities of its products; for the low prices of the inferior ones and the astronomical price for the best; and finally for the vast size of the domestic and foreign wine trade.
>
> —Jules Guyot, 1868, 1:429

THE ENGLISH OCCUPATION of Bordeaux between 1152 and 1453 led to a major growth in the wine trade, and imports peaked at 102,724 tons, or about 900,000 hectoliters, in 1308–9.[1] Trade declined once more after the loss of the city and fluctuated over the centuries according to the political situation between the two countries. From the late seventeenth century, duties on French wines were set at higher levels than on those from Portugal and Spain and resulted in British consumers, in the words of David Hume in 1752, being obliged to "buy much worse liquor at a higher price."[2] Although French wines of all descriptions accounted for less than 10 percent of the British market as late as the 1850s, a relatively high proportion of Bordeaux's best wines were exported to Britain, to be consumed by a very small group of wealthy consumers.

The long history of commercial relations between many British ports and Bordeaux, together with the Gironde department's large and varied production of wines, made the region a potential source of fine and commodity wines after the reduction in duties in 1860. This chapter begins by examining the long-run changes in wine production and trade during the nineteenth century and the organization of wine production. After a period of prosperity that lasted from the mid-1850s to the early 1880s, there followed three decades of depression caused by the appearance of vine diseases, the decline in reputation of both fine and beverage wines, and overproduction in the French market. Fine wine production in the Gironde was concentrated in the region of the Médoc, and the production on the large châteaux contrasted with that of the thousands of small, family vineyards found in the region. Information problems for consumers of

---

[1] Trade was especially strong after the French recaptured La Rochelle in 1224 (Penning-Rowsell 1973:80–81).

[2] Hume (1752:88).

TABLE 5.1

Wine Production by Quality in Bordeaux, 1870s (thousands of hectoliters)

	Red	White
Ordinary and great ordinary wines	2,088	984
"Fine wines"	147	20
"High-classed wines"	50	10

Source: Cocks and Féret (1883:90).

fine wines were reduced by the 1855 classification, but the growth in market power and economic independence of the leading estates was checked in the late nineteenth century, partly because of the expenses associated with vine diseases, and partly by the decline in wine quality and collapse in the reputation of their wines among consumers. Finally, small growers successfully used their political voice to achieve legislation to establish a regional appellation, which limited to wines of the Gironde the right to carry the Bordeaux brand.

CLARET, TRADE, AND THE ORGANIZATION OF PRODUCTION

Bordeaux, unlike Jerez or Porto, was itself a major production center for both fine and commodity wines. Red wine represented almost 70 percent of the Gironde's wine output in the early 1870s, and 86 percent of fine wines (table 5.1). Fine red Bordeaux wines were blended from a variety of grapes. According to one wine specialist, even today "this is only partly because Merlot and Cabernet are complementary, the flesh of the former filling in the frame of the latter. It is also an insurance policy on the part of growers in an unpredictable climate."[3] The exact proportions of cabernet sauvignon, cabernet franc, and merlot used varied according to the local terroir and the nature of the vintage. There were also long-term changes in the grape varieties used, especially among commodity producers, and the historian Philippe Roudié argues that the appearance of powdery mildew in the 1850s reinforced the differences between the *grand crus* and *petits vins*, as the growers in the less favorable areas for wine production uprooted the merlot and planted malbec (red), enrageat, and sémillon (white), all varieties that were more disease resistant and produced higher yields.[4]

French trade statistics provide figures for wine exports from Bordeaux both in bottles and in barrels (the Bordeaux tun contains four hogsheads of 220–250 liters each). The export figures for bottles serve only as a very rough indicator for fine wines, as an unknown quantity of high-quality young wine was exported in casks to be matured and bottled at its destination. There could also be significant price variations between the different wines that were exported in bottles. While

[3] Robinson (2006:90).
[4] Roudié (1994:69–71).

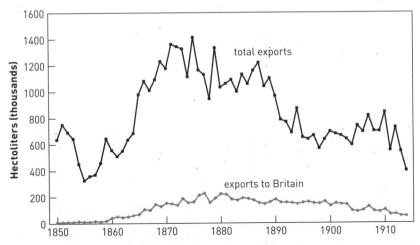

Figure 5.1. Bordeaux wine exports in barrels, 1850–1914. Source: France. Direction Générale des douanes (various years)

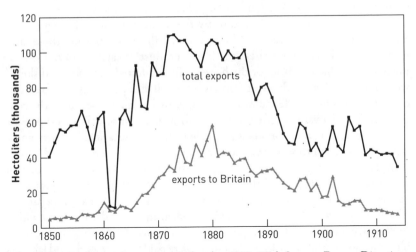

Figure 5.2. Bordeaux wine exports in bottles, 1850–1914. Source: France. Direction Générale des douanes (various years)

recognizing these limitations, figures 5.1 and 5.2 provide a reasonably accurate picture of the long-run changes in the exports of fine and ordinary table wines. In particular, figure 5.2 shows that Britain was the major market for fine wines in the second half of the nineteenth century. Even at the end of the period, when exports had declined significantly, *Ridley's* believed that consumption of the "better class Wines" per capita was as great in England as it was in France.[5] Ex-

[5] *Ridley's*, June 1907, p. 445.

ports of commodity wines in barrels to the United Kingdom increased from less than 20,000 hectoliters in the 1850s to ten times this quantity in the late 1870s. Exports of both categories to the Britain peaked in 1880 and then started to decline, the fall being much greater with bottled wines.

Wine exports to other markets peaked earlier, with those in barrels doing so in 1875, and those in bottles in 1866 and 1873. Wine shortages and higher prices in France encouraged the development of viticulture in some of Bordeaux's traditional markets in the New World, and this continued when prices started falling in the 1890s, as foreign producers sought refuge behind protective tariffs.

The growth in demand for wine in the mid-nineteenth century led to the area of vines in the Gironde increasing from 133,000 hectares in the early 1840s to 188,000 by the 1870s, when phylloxera and lower prices caused it to decline once more to 138,000 hectares by the early twentieth century.[6] The basis of Bordeaux's reputation rested on the production of fine wines from less than one hundred growers that represented only a small part of the Gironde's total production. As late as 1924, the 285 growers with more than 30 hectares (some of which produced commodity wines) were responsible for only 11 percent of the region's output (table 5.2). On the famous large estates, the vines were almost never leased and the owners managed directly the production of their own grapes. Suitable incentives were required for laborers to diligently carry out those tasks in the vineyard that might affect the future productive capacity of the vine or influence the characteristics of the harvest. The preferred contract used skilled workers (*prix-faiteurs*) who were responsible for all the operations on a fixed area of vines, which in the Médoc was usually slightly less than 3 hectares. The contract stipulated a variety of rewards, both formal and informal, to compensate workers, including accommodation, heating, cheap wine, a small garden, and a salary.[7] In addition to secure employment, the *prix-faiteurs* organized the work themselves, and they were given enough free time to tend their own vines or earn extra wages by doing piecework. The mutual benefits of this type of contact can be seen in the long, continuous employment of workers, with many *prix-faiteurs* on Château Latour, for example, having worked there for decades, and with a son following his father by being employed as a *vigneron*.[8] By contrast, the more physical and nonskilled tasks on the estates were paid by piecework, and laborers could earn relatively high wages. The need for piecework increased significantly in the second half of the nineteenth century on account of vine disease, and this provided much-needed employment for local growers who had insufficient vines to keep themselves fully occupied.

[6] Lafforgue (1954:293); Cocks and Féret (1908:6).

[7] Féret (1874–89:461) gives a salary of 225 francs, while Bowring suggests 150 francs. The operations included pruning, collection of cuttings, putting down vine props and laths, fastening the vines, and twice drawing the cavaillons (United Kingdom. Parliamentary Papers 1835:150).

[8] Higounet (1993:101). By contrast, on the eve of the Second World War there were reports concerning the work quality of the *prix-faiteurs* (France, Ministère de l'Agriculture 1937:163).

TABLE 5.2
Farm Structure and Output in the Gironde, 1924

	Number of vineyards	Area (hectares)			Production (hectoliters)		
		Average size	Total area	% of total	Average per hectare	Total output	% of total
Less than 3 hectares	50,301	1.13	56,842	42	23.0	1,284,806	28.5
3–10 hectares	8,909	5.05	44,975	33	45.0	1,994,117	44.5
10–30 hectares	1,096	16.00	17,536	13	40.0	711,583	15.8
30+ hectares	285	53.80	15,133	12	33.5	504,138	11.2
	60,951	2.22	134,486	100	34.4	4,494,644	100

Source: France, Ministère de l'Agriculture (1937:157).

Small producers of commodity wines faced different problems from those of the large estates, namely, the need to reduce their exposure to a poor harvest or low prices. One possibility in the less prestigious wine areas of the Gironde was intercropping (*joalles*). There were two main types, *les grandes*, with rows of vines planted every 6–12 meters and cereals grown in the space between, and *les petites*, with vines planted closer together, every 3–6 meters and with legumes and forage crops sown.[9] By the late nineteenth century, with phylloxera and mildew devastating their vines, some growers also turned to fruit trees.[10] Wine yields were higher than those of fine wines, as the vines were found on relatively fertile soils and intercropping required the use of fertilizers.

THE 1855 CLASSIFICATION AND THE BRANDING OF CLARET

The most important vineyards, such as Château Margaux and Château Lafite, were well-known among British consumers of fine wines in the eighteenth century and enjoyed a considerable premium over lesser wines. Growers hoped to sell their wines about six or eight months after the harvest to the négociants in the Chartrons area of Bordeaux, who then matured and exported them. Growers therefore saved themselves the cost of racking the wine from its lees and losses caused by evaporation. The fine wines shipped in the early nineteenth century were very different from those later in the century. The fact that the grapes were not removed from their stalks during fermentation made the wines very hard and full of tannin. To overcome this they were blended, as André Jullien noted

[9] Jouannet (1839:229–30) suggests that the *joalles* were found on the hills, but Cocks and Féret (1883:52) also give the Palus.
[10] *Revue Agricole, Viticole et Horticole Illustrée*, May 15, 1906, pp. 154–55.

in 1816: "The wines of the first growths of Bordeaux as drunk in France do not resemble those sent to London; the latter, in which is put a certain quantity of Spanish and French Midi wine, undergo some preparations which give them a taste and qualities, without which they would not be found good in England."[11]

This same author noted elsewhere that

> The English houses at Bordeaux, immediately after the vintage, purchase a large quantity of the wines of all the best vineyards, in order that they may undergo la travaille a l'Anglaise. This operation consists in putting into fermentation part of the wines during the following summer, by mixing in each barrel, from thirteen to eighteen pots of Alicant or Benicarlo, or the wines of the Hermitage, Cahors, Languedoc, and others; one pot of white wine, called Muet (wine whose fermentation has been stopped by the fumes of sulphur) and one bottle of spirits of wine. The wine is drawn off in December, and then laid up in the chais (cellars) for some years. By this operation the wines are rendered more spirituous and very strong, they acquire a good flavour, but are intoxicating. The price likewise is increased.[12]

When this custom ended is difficult to establish, as contemporaries often simply repeated in their accounts what earlier authors had written. Penning-Rowsell notes that the seventh edition of the *Encyclopedia Britannica* (1842) suggests it was still common, but it seems likely that tastes did alter around the mid-nineteenth century.

Producers and consumers were well aware of the importance of growth and vintage in determining the price of a wine. In the early 1830s a first growth—a wine from one of the region's best vineyards—could fetch 3,000 francs per tun in a good year, fourteen times more than a Côtes or Palus wine did after a poor harvest (table 5.3). However, in the early 1830s many of the lesser growths after a good harvest actually sold for a higher price than wines from a top château after a poor one. Other French and foreign wines were also sold in Bordeaux, usually for blending purposes. This large range of wines made it crucial that they were adequately classified and that accurate information concerning their quality was provided to potential purchasers. By the late eighteenth century the use of both the grower's and the shipper's name (brand) was becoming frequent with fine clarets in the important British market, although the addition of the vintage was still rare.[13]

The many different wines produced and sold in the region led to the appearance of informal classifications for those involved in the trade from an early date.

---

[11] Jullien (1826), cited in Penning-Rowsell (1973:156). Henderson (1824:183) appears to cite earlier editions of Jullien. Cavoleau (1827:128) speaks of the *premier crus* of the Médoc being mixed in this form to produce an inferior article, but one that met consumer demand.

[12] Jullien (1824:113).

[13] In 1797 Christie's sold "vintage" claret—six hogsheads of 'first-growth claret' from 1791 (Penning-Rowsell 1973:109).

TABLE 5.3
Prices and Output of Different Bordeaux Wines in the Early 1830s

	Price per tun, according to harvest quality			Output (tuns)	Average output (thousands of francs)[4]
	Good	Middling	Bad		
Red wines					
Region: Médoc					
First growth[1]	3,000	1,750	400	300	525
Second growth	2,700	1,400	350	620	868
Third growth	2,400	1,200	325	650	780
Fourth growth	1,800	1,000	300	750	750
Fifth growth	1,500	900	300	1,100	990
Best villages, a[2]	1,000	600	280	6,000	3,600
Best villages, b[2]	600	400	250	20,000	8,000
Common wines, lower Médoc	450	300	220	20,000	6,000
Region: Graves					
Ch. Haut Brion	2,700	1600	350	60	96
First class	1,500	700	300	200	140
Second class	800	500	280	3,000	1,500
Common	500	300	250	10,000	3,000
Region: St. Emilion					
First class	700	400	225	10,000	400
Second class	450	280	200	3,000	840
Regions: Palus, Côtes, etc.					
1	400	250	180	30,000	7,500
2	300	200	180	100,000	20,000
Total red				196,680	33,476
White wines[3]					
Best	1,100	700	300	450	315
Common	150	120	90	12,000	1,440
Other				15,677	3,930
Total white				28,127	5,685

Source: Calculated from United Kingdom. Parliamentary Papers (1835), p. 138.
[1] Included are Ch. Lafite, Ch. Latour, and Ch. Margaux.
[2] St. Estèphe, Pauillac, Soussan, Margaux, Cantenac, etc.; two qualities are given.
[3] Categories have been reduced here to two.
[4] Based on the "middling" harvest quality.

A hierarchy of regions, such as that shown in table 5.3, was the first step, which included both the quality of the vintage and the region of production (Médoc, Graves, Saucere). Subdivisions included the important wine-producing villages (St. Emilion, St. Julien). Finally, fine wines were identified by growths (first, second), and the best by their name (Château Haut-Brion).

In 1816 André Jullien published the first classification of Bordeaux's leading vineyards in his *Topographie de tous les vignobles connus,* which listed five distinctive classes, although specific properties were found in only the first two.[14] Other lists soon followed, published in both French and English.[15] However, it was the Bordeaux classification of 1855, compiled by wine brokers (*courtiers*) for the Universal Exhibition of that year, that became and remains the reference for the wine trade and consumers alike.[16] This listed fifty-seven red wine producers in five different growths and twenty-two white wine producers in a further three. All the red producers, with the exception of Château Haut-Brion (Graves), were found in the Médoc. The success over time of this classification was due to three factors. First, rather than a subjective study based on taste, it used prices that had been paid for different wines over many years. Second, it was compiled by the relatively impartial brokers, not the growers themselves. Finally, wine merchants considered it as only a rough guide and were quite willing to pay higher or lower prices when they thought a wine warranted it. This was important because the quality of a château's wine varied not only according to vintage and growth, but also over time on account of changes in the level of investment and the quality of management.[17] The 1855 classification became widely known in the second half of the nineteenth century, but consumers still often lacked good information on the quality of the different vintages, or how individual wines would develop over time.

The 1855 classification, however, ignored the great majority of vineyards in the region. As these included many good wines, the increasing popularity of claret encouraged growers to establish their own brands by adopting impressive names for their vineyards. The leading wine guide of the region, Charles Cocks

[14] Jullien (1816). Some of the *courtiers* had their own private lists in the eighteenth century. See Markham (1998) for a history of the classification of Bordeaux wines.

[15] For example, in French, Franck, *Traité sur les vins; Le guide* (1825); Paguierre, *Classification* (1829); Cocks, *Bordeaux* (1850); and in English, Jullien, *The Topography* (1824); Henderson, *History* (1824); Paguierre, *Classification* (1828); Redding, *History and Description* (1833); and Cocks, *Bordeaux* (1846).

[16] The 1855 classification was the response of the Bordeaux Commodities Market to a request from the local chamber of commerce. It took only two weeks to compile, although, as noted above, there was a tradition of classifying the estates. It was not supposed to be a definitive guide (and there would have been bitter fighting among interest groups if they thought this was going to be the case), but, with just two exceptions—in particular the promotion of Château Mouton-Rothschild to a first growth in 1973—the list has remained unaltered to this day. Brokers provided exporters with the knowledge of where suitable wines could be bought and their price.

[17] Only Châteaux Mouton-Rothschild and Léoville-Barton have remained with the same family since 1855 (Robinson 2006:175).

and Édouard Féret's *Bordeaux et ses vins*, listed 318 properties with the label "château" in its 1868 edition, but this increased to 800 in 1881, and 1,600 by 1900.[18] Previously obscure vineyards, which may have produced excellent wines to sell under a shipper's name, now gained a distinctive identity for themselves.

Having established a recognizable brand name, growers had to protect it. To avoid merchants mixing their wines with others, estate bottling was introduced, and with it the use of distinctive labels and branded corks. A leading shipper, Finke, noted for the 1868 vintage: "from several of our purchasers, namely Latour, Lafite and Larose, we obtained the right to keep and bottle the wines at their respective Châteaux, and it is our intention for the future to give as much extension as possible to this feature, as owing to the increased trade in Bordeaux wines, there is a greater demand for pure growths under their proper name, instead of, as formerly, for so-called 1st, 2nd, 3rd classes."[19]

Only those wines considered of sufficient quality were bottled in this way. The high prices paid after a good vintage easily compensated for the lower prices received for poor ones, and, at least in theory, wines from poor vintages were sold as *vin ordinaire* rather than estate-bottled wines.[20] The growing reputation of the leading châteaux also encouraged some of their owners to purchase other vineyards and to sell the poorer-quality wines using the estate's brand name, but labeled as "second wine." British wholesale merchants, however, criticized the trend of château bottling, claiming it was "a guarantee of origin, not of quality,"[21] and feared that its widespread use would give producers considerably greater control of the market and a greater share of the profits, as was happening with champagne.

By contrast, for the cheaper Bordeaux wines, merchants were faced with having to purchase large quantities of very different wines from a considerable number of growers, and these required blending to achieve a product of consistent quality and stable price regardless of the nature of the local harvest. The Bordeaux wine trade would appear to have been in a position to establish and maintain the reputation of their wines better than most other regions. The 1855 classification of the leading growers provided a sufficiently accurate guide for fine wines, and by 1873 the well-established export houses were able to draw on the produce of a large and highly diverse wine district, reaching almost 200,000 hectares for cheap wines.[22] However, exports for both fine and ordinary clarets peaked in the early 1880s, and producers faced serious eco-

---

[18] See Roudié (1994:142).

[19] *Ridley's*, March 1869, p. 5.

[20] "It is said that £37 has been offered for the Mouton Rothschild 1887, with the right to have it bottled at the Château; but the proprietor is reported to have refused to allow the bottling clause" (*Ridley's*, April 1888, p. 164).

[21] Army & Navy Co-operative Society, cited in *Ridley's*, June 12, 1889, p. 298.

[22] This was significantly more than it had been earlier in the century. Franck in 1824, for example, gives only 130,000 hectares. *Traité sur les vins*, cited in Roudié (1994:31).

nomic problems by the turn of the century, although for very different reasons. Consequently we need to look at both segments of the market, starting with fine wines.

SUPPLY VOLATILITY, VINE DISEASE, AND THE DECLINE IN REPUTATION OF FINE CLARET

Phylloxera was officially noted in the Gironde in 1869, reaching the high-quality wine region of the Médoc in 1875. Its subsequent spread was much slower than in areas such as the Midi, as growers were worried that the new American vines would ruin wine quality and therefore spent heavily on chemicals to protect the traditional French stock. The négociants were just as skeptical of the new vines as the vineyard owners. One of the clauses of the 1906–10 *abonnement* (see below) that négociants agreed upon with Château Latour was that "the vineyard can in no way be increased during the period of the contract, and grafted American vines must be excluded, save those that are already there."[23]

From 1879 a number of the smaller growers started injecting carbon disulfide around the vines. Growers with less than 5 hectares could request state aid at the rate of 25 francs a hectare, although this represented only about half the cost of chemicals required. By the mid-1880s there were eighty-eight local syndicates of growers established to protect vines in the Gironde, and they treated almost 3,838 hectares.[24] Another strategy was the spraying of vines with sulfocarbonates, but because this required expensive pumping equipment, it was limited to the major growers, while other large producers installed pumps to flood their vineyards. The massive use of chemicals, the need for new equipment, and the heavy labor requirements made all these measures very expensive. As table 5.4 suggests, the high value of many of the Bordeaux vineyards led to a much higher percentage of the land being treated than elsewhere in France.[25] The fight against phylloxera increased demand for labor, and this, together with the clearing of old vineyards and their replanting with the new American ones in areas of cheap wines, created much-needed employment for smaller growers whose own harvests were reduced by disease.

This heavy expenditure was successful in delaying phylloxera until growers and négociants were convinced that the new vines would not produce inferior

[23] Higounet (1993:276–77).

[24] Government aid was 95,967 francs in total. The eighty-eight syndicats had 2,535 members with 5,652 hectares of vines in total. In Libourne there were forty-six syndicats, 1,565 members, and 3,830 hectares, compared with Lesparre (representing most of the Médoc), which had only thirty-nine members and 145 hectares (*Feuille vinicole*, September 15, 1887).

[25] This was also true within the Gironde itself—for example, in 1898 the villages of Cussac, Cantenac, Margaux, St. Estèphe, and St. Julien, which accounted for 2.5 percent of the vines but 4 percent of the area flooded, 37 percent of the vines treated with sulfocarbonates, and 12 percent of those treated with carbon disulfide (Arch. Gironde 7 M 219).

Table 5.4
The Response to Vine Disease in the Gironde, 1882, 1894, and 1905

	1882	1894	Gironde, as % of France	1905
Free of disease	32,151*	35,833		25,940
Infected but not treated	138,100	41,040		36,738
Vines being treated by:				
Flooding	2,639	9,675	27.1	7,621
Carbon disulfide	600	6,997	14.1	4,767
Sulfocarbonates	2,378	3,945	44.9	3,053
Subtotal:		20,617		
Replanted with American vines	132	38,520	6.1	72,543
Replanted with direct producers	—	2,154		2,859
Total area of vines	176,000	138,165	7.6	138,080
Total area destroyed by phylloxera	21,800			75,402

Sources: Feuille vinicole, June 15, 1882, Arch. Gironde 7 M 219, and Pouget (1990:98–97); and Feuille vinicole, November 22, 1906.
*My estimate from official sources.

wine. According to the historian René Pijassou, wine from the leading Médoc growers was still being produced from the old French vines until about 1900, and only after 1920 did it come predominantly from the new grafted ones.[26] Therefore the large Bordeaux growers did not help small growers by experimenting with the new American vines to the same extent as large growers had done in the Midi or would do in Champagne. Pijassou has also argued that the string of poor harvests between 1882 and 1892, and the decline in popularity of fine clarets, cannot be attributed to phylloxera itself. Strictly speaking this was true, but the successful attempts to delay phylloxera by using chemicals resulted in an increase in the quantity of manure also being used. Traditionally vines had been manured once every twenty years in the Médoc, with Château Latour using 200–250 m³ each year before 1880. By 1884, after new antiphylloxera methods had been introduced, the quantities had multiplied five or six times because of the weakened vines,[27] increasing yields on the leading Médoc estates and reducing the wine's quality.

Today scientific advances in vineyards and wineries have significantly reduced the number of poor vintages, and good-quality wine is produced in most years.

[26] Pijassou (1980:763).
[27] Ibid., 779–80.

In the nineteenth century quality was much more varied, and in the first two-thirds of the century, except for the years 1808–10 and 1835–39, there was at least one good harvest every three years, and each decade, with the exception of 1820s, enjoyed a minimum of four good harvests.[28] The difference in the quality of the harvests was reflected in the price of the wine, with the price of Château Latour in the 1860s, for example, fluctuating from a minimum of 550 to a maximum of 6,250 francs a tun. For producers of fine wines, good years were those of high grape quality, and these were expected to be dispersed fairly equally over time. By contrast, the size of the harvest was much less important in determining revenue.

This pattern was interrupted by downy mildew, a disease that reduced the wine's alcoholic strength and its keeping quality and devastated five consecutive harvests between 1882 and 1886. Between 1877 and 1893 no harvests were considered excellent, and six of the crops gathered between 1879 and 1886 were classified as poor, one as average, and just one as good.[29] Even worse for the trade, the poor keeping quality of these wines was not at first appreciated, and wines believed to be good were bottled and sold by reputable merchants.[30]

The most famous case was that of Château Lafite, which after the harvest of 1884 sold its wine for £14 per hogshead, with "the right to bottle at the Château with the brand and label."[31] Several years later, when part of the consignment had been already sold by the shipper, it was discovered that the wine had turned bad. The legal dispute over who was responsible, together with the bad publicity that it generated, resulted in estate bottling losing its popularity among wine producers until the 1920s.[32] The high prices paid for fine clarets depended on reputation, and *Ridley's* noted a decade later that confidence had still not been restored:

> In the ordinary course of event it might have been expected that the vintage of 1887 to 1893 would repair the damaged caused by those of 1882 to 1886, but this is evidently not the case, and Médoc Wines are still suffering from the discredit which the mildewed years brought them. As with Sherry, we here find how difficult it is to restore to popular favour an article upon which a stigma has once been placed.[33]

[28] Petit-Lafitte (1868, table C). In 1820–29 there were only three good harvests. The 1811 vintage was considered exceptional.

[29] Cocks (1969:137). The measure of wine quality is not an exact science, especially as an initial judgment of a wine's quality can change by the time it is drunk. Lafforgue (1954:299) provides slightly different results.

[30] In 1891 it was noted that the "instances in which Wines had gone wrong after being put into bottle were by no means few and far between, and the trouble caused both to shipper and importer became at one time a very serious matter. Of late, however, complaints have been far less frequent" (*Ridley's*, January 1891, p. 5).

[31] Ibid., January 1887, p. 35.

[32] According to Nicholas Faith (1978:96), poor-quality wines, together with the increase in duty on bottled wine in Britain after 1888, "set back the cause of bottling near or at the grower's for a generation." See also *Feuille vinicole*, January 27, 1887.

[33] *Ridley's*, July 1894, p. 400.

TABLE 5.5
Wine Output in the Médoc and the Gironde

	Gironde	Grand crus, Médoc
1864–78	100	100
1879–87	43	67
1888–97	78	121
1898–1907	123	138
1907–21	119	72

Sources: Lafforgue (1954:301); and Pijassou (1980:776–77) (1855 classification).

The considerable drop in revenue caused by downy mildew occurred just as phylloxera was driving up costs, which encouraged growers to increase output, a trend that had already begun because of the heavy use of fertilizers. As table 5.5 shows, not only did the fine wine producers in the Médoc suffer less than other growers in the Gironde in the period 1879–87, but output in 1888–97 had more than recovered to the 1864–78 level. Château Latour, for example, increased output by 252 percent, but prices fell by 58 percent between 1879–87 and 1898–1907, while Château Margaux's output increased from 450 hogsheads of supposedly "premier wine" to 1,200–1,400 hogsheads in 1903 of "indifferent" or "bad" wine.[34] The leading négociants also suffered because although some had large stocks of vintage wines maturing in their cellars, most of their costs were fixed, and the decline in wine prices implied that their production costs rose from the equivalent of about 14 percent of the price of wine between 1875 and 1885 to 24 percent by 1905.[35]

The very success of names such as Château Lafite, Château Margaux, and others associated with the 1855 classification now contributed to this decline, as less reputable merchants in both Bordeaux and Britain exploited these brands by selling the large quantities of wines from the poor vintages that the desperate growers had in stock. Thus *Ridley's* argued that

> The Public, who unfortunately know more about Growths, than Vintages, receive Circulars offering Château this or Château that at apparently low rates, and on the strength of the name, purchase Wines which can but prove intensely disappointing. They then are apt to argue that, if Wines bearing the names of the best Estates of the Médoc be so inferior, those of lower grades must be bad indeed. Thus their faith in Claret, instead of in the Merchant who has sold it them, is shaken, and an inducement is at hand to try Wine from some other district.[36]

Exports of bottled claret to Britain, by far the leading market, slumped from 58,030 hectoliters in 1880 to 7,165 hectoliters in 1913. This was often explained by the growing popularity of smoking and drinking coffee instead of fine claret

[34] Higounet (1993:297); *Ridley's*, April 1903, p. 675.
[35] Béchade (1910:102).
[36] *Ridley's*, September 1897, p. 576.

after dinner, the absence of good vintages from the early 1880s, and the decline in the reputation of the leading brands.[37] In fact between 1899 and 1914 there were three excellent vintages, seven good ones, three average, and only three poor.[38] Quality wines were being produced again, but the merchants had difficulties convincing skeptical consumers. The problems of declining reputation, together with rising tariffs, led to consumption slumping everywhere, with sales of bottled claret to Sweden falling by 86 percent, to Germany by 43 percent, to the Netherlands by 42 percent, and to North America by 26 percent, with only Belgium seeing an increase, by 38 percent.[39] As a result, the British market for bottled claret in 1900–1904 was still larger than the rest of Europe and North America combined (table 5.6). The decline in sales to Latin America can be attributed to a combination of growing national production and rising tariffs.[40] Tariffs were obviously not a factor in France's colonial markets, and by 1900–1904 Senegal had established itself as the third largest market for bottled claret, after Britain and the United States.[41]

Table 5.6 shows that while the leading growers almost doubled output in the five-year period from immediately before the outbreak of downy mildew and to the turn of the twentieth century, exports of bottled wines more than halved. The economic depression for the leading Médoc growers was both long and deep. Many were forced to sell their properties, and land prices in the Médoc fell by up to 80 percent in the thirty or forty years prior to the First World War.[42] Thus Château Malescot-Saint Exupéry (a third growth) was sold in April 1901 for 155,000 francs instead of the 1,076,000 francs it had reached in 1869, and Château Monrose (a second growth) was sold for 800,000 francs in 1896, down from 1.5 million francs in 1889.[43] If the plight of Bordeaux's leading growers in the early 1900s was as desperate as that in the Midi, the nature of the problem, and consequently its solution, was very different. In particular, the Midi's growers and merchants competed on price and not quality, and the homogenous na-

---

[37] "English habits ... have undergone a considerable change during the past 30 years, and the after-dinner half-hour is now monopolized by coffee and tobacco, while Britons have not yet accustomed themselves to serve fine claret or burgundy with roast meat or game" (letter by Gilbey to *The Times*, September 29, 1896).

[38] Cocks (1969:137).

[39] I include those years when no figures are given (Germany 1881; Canada 1880, 1901, 1902; Sweden 1904), even though some exports are probably included in "all other countries" for these years. By excluding these years, the decline in imports is in Germany –55 percent; in North America, –23 percent; and in Sweden, –82 percent.

[40] Salavert (1912:187–88). For tariffs, Cocks and Féret (1908:1051–64).

[41] Export to Senegal averaged 3.7 million hectoliters in 1900–1904, against 17.7 million to Britain and 4.1 million to the United States.

[42] France, Ministère de l'Agriculture (1937:159).

[43] Land prices fell in the thirty to forty years prior to 1914 by 80 percent (Pijassou 1980:815–16). Farm wages doubled in the Gironde between 1826 and 1870, but from this date until the First World War wages in the Médoc increased from 2.75 to 3 francs, considerably less than the national average (France, Ministère de l'Agriculture 1937:173, 459).

TABLE 5.6
Production and Exports of Fine Bordeaux Wine (hectoliters)

	1877–81	1900–1904	Change (%)
Production[1]	39,375	76,932	+95
Exports of wine in bottles[2]			
Britain	47,210	17,724	−62
All other markets	53,651	26,882	−50
Total exports	100,861	44,606	−56
Other European markets	13,486	10,230	−24
Non-European markets	40,165	16,652	−59
Latin America	21,478	2,695	−87
North America	6,254	4,528	−28
French colonies	–	6,229	
Dutch Indies	4,551	–	
British India	1,630	–	
Other countries	6,252	3,200	

Sources: Cocks and Féret (1898:104–5) and (1908:90–91); France, Direction Générale des douanes, various years.

[1] Red wine from the leading growers (*grands crus*) of the Médoc. Château Haut-Brion is excluded owing to the absence of output figures for 1877–81.

[2] Includes all wines in bottles.

ture of their wines encouraged a group response. By contrast, the fine wine producers in Bordeaux needed to restore their own personal reputation for quality, and this could only be achieved if they limited output once more and invested heavily in their vineyards. Virtually all growers lacked both the capital and the means to improve their wines and change the negative perception of them in the market. The shift in market power to the producers during the quarter century or so following the 1855 classification and the introduction of estate bottling was now rapidly reversed, and the Bordeaux merchants once more dominated the commodity chain. In 1906 and 1907 a number of merchants entered into agreements (*abonnements*) with more than sixty growers, including half of the classed growths, to buy all their production at fixed prices over the following five or more harvests. Clauses were inserted in the contracts to limit output in an attempt to improve quality.[44] Backward integration also took place, with a number of the leading properties being bought by merchants or their families.

However, the decline of claret sales required more than just uncoordinated, individual actions to improve wine quality. In 1909 a proposed "Trust" between growers and merchants to raise money to promote Bordeaux's wines was debated but came to nothing, although the much more modest *Fête des vendanges*, which

[44] Cocks and Féret (1908:xvii–xxii); Higounet (1974:335). The négociants insisted for Château Latour that "the vineyard can in no way be increased during the period of the contract, and grafted American vines must be excluded, save those that are already there" (Higounet 1874:276–77).

helped promote the Bordeaux "marquee," was established.[45] The British wine trade journal *Ridley's* criticized the Bordeaux merchants' marketing efforts:

> The competition of Australian and Wines other than those of Bordeaux, and even more that of Whiskey, has undoubtedly diverted the public taste from Claret, but perhaps the most vital reason of all is a change in fashion. . . . It is, we believe, in a large measure because the Bordeaux trade do so little to attract the attention of the Public. . . . Think of the huge amount of money that has been spent on advertising Whisky, but no one would suggest that a similar amount should be spent on advertising Wine, the natural production of which is limited. Whisky did not appear to either the palates or inclinations of the Public at first, and it was nothing but the continual and also judicious advertising that has brought it into such World-wide repute. Similarly, on a smaller scale, with Australian Wines. Where would they be to-day, we would ask, had it not been for advertising? On the other hand, during the time those concerned in the sale of Whisky, Champagne and Australian Wines have been embarking on so vigorous a crusade of advertising, the Claret folk have done practically nothing, and unless they wake up and take steps by way of an outside appeal to the Public, they will incur the danger of getting still further behind. They have an object lesson before them in the case of Sherry, which from 1873 went continuously down, until the Jerez Shipping and their Agents woke up to the conclusion that something must be done, and this "something" was the formation of the Sherry Shippers Association, and the expenditure of a moderate amount of money on advertising. By this means they have not only stayed the decline, but turned the corner, and their Wines are slowly gaining in the estimation of the Public.[46]

RESPONSE TO OVERPRODUCTION: A REGIONAL APPELLATION

On the eve of phylloxera's appearance, the port of Bordeaux was also France's leading export center for cheap commodity wines, and a significant part of the growth in British imports came from here. Phylloxera caused a shortage of local wines, and French wine prices increased from twenty-five to forty francs per hectoliter between 1874–78 and 1880–84, despite imports increasing from the equivalent of less than 1 percent of production in the early 1870s to a third in the 1880s. In 1891 French wine production was 30 million hectoliters, imports 12 million, and exports just 2 million.[47] By the mid-1880s Bordeaux was itself importing more wine that it was exporting, leading the British consul to note that "it is probable that about 50 per cent of all wines shipped from here last year to British ports in wood were 'vins de cargaison,'" namely, local wines mixed with

[45] Roudié (1994:227–29).
[46] *Ridley's*, January 8, 1914, p. 5.
[47] *Annuaire statistique* (1938:63, 179–80).

imported ones.[48] The British press presented an even more dramatic story. Thus in the *Telegraph* we read: "An immense proportion of the wine sold in England as Claret has nothing to do with the banks of the Garonne, save that harsh heavy vintages have been brought from Spain and Italy, dried currants from Greece, there to be manipulated and re-shipped to England and the rest of the World as Lafite, Larose, St. Julien and St. Estephe."[49]

Although it seems unlikely that wines of the 1855 classification actually suffered this fate, the frequent newspaper references to the supposed mixing with foreign wines and adulterations undermined claret's reputation. Even after the French wine market switched from one of shortage to overproduction in the early twentieth century, British imports continued to fall, with competition now from the so-called basis wines, concoctions manufactured in Britain from imported grape juice and other substances and then mixed with French wines. According to *Ridley's*, "people drink so called 'Claret,' composed of one-third of the genuine article and two-thirds of the British imposition, and condemn, not the latter, but Claret."[50]

Bordeaux's merchants argued that the high price and poor quality of local wines in the 1880s and 1890s required them to look elsewhere for supplies for blending. Nevertheless, this was not how local growers viewed the problem, especially as many were suffering from the heavy costs associated with replanting their vines after phylloxera and treating other vine diseases. As domestic harvests began to recover after phylloxera, pressure from the growers resulted in an increase in French import duties from two to seven francs per hectoliter in 1892. To protect the part of Bordeaux's trade based on cheap *vins de cargaison*, merchants were allowed to import foreign wines duty free for the sole purpose of mixing with local ones for export.[51] However, not only did Bordeaux's export of these wines amount to little more than 200,000 hectoliters a year, but they were often sold in countries that were also major markets for fine wines, once more increasing the bad publicity.[52] The free port was closed in 1899 and this, together with the growth of Cette as a distribution center for Midi and Algerian wines, and Hamburg's highly competitive free port, contributed to Bordeaux's decline as the international center for cheap commodity wines.[53] As figure 5.3 suggests,

[48] British Parliamentary Papers, *Consular Report*, Bordeaux, 1889, no. 501, p. 9. For trade, see Roudié (1994:180).

[49] *Ridley's*, July 1887, p. 315.

[50] Ibid., April 1906, p. 338.

[51] Roudié (1994:212–13); Gallinato-Contino (2001:171–82).

[52] Audebert (1916:15), in Arch. Gironde, 8 M 13.

[53] Farou, *La Crise Viticole et le Commerce d'Exportation* (1909), in Arch. Gironde, 7 M 169, 7–12. Phylloxera in Catalonia and Navarra also created difficulties in obtaining wines from Spain. By contrast, Hamburg merchants bought wines from the cheapest producers, whether in Portugal, Greece, Turkey, or Hungary, were more efficient at creating wines, and even enjoyed lower freight rates to Buenos Aires than did Bordeaux.

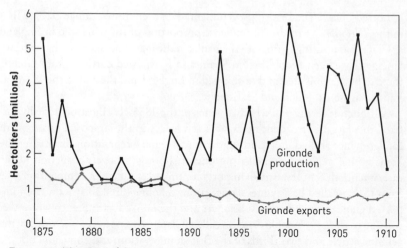

Figure 5.3. The Gironde's wine production and exports, 1875–1910. Sources: Lafforgue (1954:301) and Direction Générale des douanes. Tableau Général du Commerce de la France

exports drifted downward from the late 1880s, just when the Gironde's wine production was recovering.

By the late nineteenth century Bordeaux's growers (who numbered almost half as many as those in the Midi) were also losing their competitive edge in the domestic market for ordinary beverage wines, as lower rail freight rates made those from the Midi cheaper, not just in Paris, but also in Bordeaux itself. By the 1900s yields in the Midi exceeded those in the Gironde by more than 70 percent, and production costs were considerably lower. The 1907 parliamentary commission, which was chaired by Cazeaux-Cazalet, a deputy from the Gironde, noted that the wine crisis in the Gironde was caused by low-cost competitors (*fraude par substitution*) rather than overproduction (*fraude par multiplication*).[54] The Midi sold in Bordeaux the equivalent of 28 and 32 percent of the Gironde's harvest in 1902 and 1903, respectively, and wines from elsewhere accounted for a further 36 and 52 percent. However, when the average for the longer period between 1902 and 1906 is taken, the figures fall to 22 and 27 percent, respectively, implying that two-thirds of all the wines sold in the Gironde were still produced locally (table 5.7). Yet the Gironde's growers were correct to identify the threat posed by the Midi. Local harvests in 1902 and 1903 were just 82 and 60 percent of the average between 1902 and 1906, while French wine prices were 5 and 47 percent higher. The Midi merchants sold less wine in the Gironde in 1904, 1905, and 1906 only because prices in those years were too low. Therefore, by the turn of the twentieth century competition from the Midi had effectively placed a price ceiling on Bordeaux's *vin ordinaire*, and the region's higher production

[54] *Revue Agricole, Viticole et Horticole Illustrée*, June 15, 1907, no. 183.

TABLE 5.7
Wine Supplies in the Gironde, 1902–6 (thousands of hectoliters)

	1902–3	1904–6	1902–6	1902–6 (% of total)
Local harvests	2,478	4,144	3,478	67.6
Wines purchased from:				
Midi	746	750	747	14.5
Other areas of France	358	277	309	6.0
Foreign imports	700	557	614	11.9
Total	4,282	5,728	5,148	100

Source: France, Chambre des députés (1909:2352).

costs made cultivation unprofitable in many years. Rather than exit the industry, the Gironde's growers looked to regulate the market by restricting the use of the "Bordeaux" brand to their own wines and excluding all competitors, regardless of the quality of their wines.

The 1905 law provided the legal framework to establish a regional appellation. Local growers argued that their wines were better than those from the Midi and elsewhere and pointed out that consumers required more information and a guarantee of quality if they were to pay higher prices. The most vocal opposition to the idea of an appellation came from the merchants within Bordeaux itself. They were opposed because they believed it would be harder to maintain stable prices and quality after a poor local harvest, as they could no longer use wines from other regions for blending and still sell them as "Bordeaux." Indeed, they argued that if they were not able to mix local wines with those from elsewhere, they might not even buy from local growers after poor harvests.[55] Many of the wines from Entre-Deux-Mers, Palus, and Réole, it was argued, required blending with the stronger Roussillon and Dordogne wines if they were to be transported.[56] The merchants also claimed that a regional appellation guaranteed a wine's origin but not its quality. Another complaint was that the new appellation would increase merchants' operating costs precisely at a time of low prices. Merchants who brought wines from outside the Gironde were now obliged to keep two sets of books, and the government took the opportunity to levy new taxes on the required labels that showed the origin of the wine (five francs per hundred bottles). There were also concerns about implementation. To reduce fraud, the law of June 1907 required growers to sell no more wine than what they declared they had produced. But according to the Syndicat du commerce en gros des vins et spiriteux de la Gironde, growers ex-

[55] *Feuille Vinicole*, May 12, 1910.
[56] Vitu (1912:70).

aggerated the size of their harvests in 1907 and 1908, presumably to enable them to buy cheap, non-Bordeaux wines to blend with their own.[57] Another fear for the merchants was that if the regional appellation were successful in raising local wine prices, growers would respond by increasing output through planting on less suitable soils and using high-yielding vines. Finally, merchants believed that foreign governments might be tempted to impose higher duties on a supposed "luxury" wine, as many were already doing with champagne.[58]

Fine wine producers also considered a regional appellation largely irrelevant, as their consumers were supposed to be both rich and well informed. Nevertheless the leading growers and their merchants signed a joint agreement in July 1908 to find a solution to the sale of fraudulent wines. Although merchants questioned the need for outside controls of their stocks, they agreed to support the measures so long as they were not "inconvenienced," and providing the measures were accompanied by a strict monitoring of growers' harvests. The Ligue des Viticulteurs, which represented Bordeaux's small growers, strongly criticized this joint agreement, claiming that the 1905 and 1907 legislation required merchants to control their stocks rather than apply voluntary controls.[59] These producers viewed the influx of cheap wines brought by the merchants as the major cause of low prices.

Contemporaries interpreted the creation of a geographic appellation in two different ways: either as giving local growers the privilege of using the Bordeaux name regardless of the quality of their wine, or as an attempt to maintain quality by excluding inferior wines from outside the region. Opposition to the regional appellation was especially strong outside the Gironde itself, and growers and merchants claimed that it was an attempt to restrict competitive markets and represented a return to the privileges of the ancien régime. According to one writer, "with this system France will no longer be a country of free trade, such as was achieved with the Revolution, but a cluster of provincial monopolies protected by excise officers. We shall return slowly to the Middle Ages."[60]

Historically a number of different wines had been transported down the Garonne and Dordogne rivers in the old administrative area of Guyenne and sold in Bordeaux. In particular, the producers from Marmande (Lot-et-Garonne) and Bergerac (Dordogne) argued that their wines were crucial for making up the wine deficit of the Gironde and contributed to the characteristic flavor of many of Bordeaux's wines. The commission dismissed these arguments. In the first instance, the department's archivist, Brutails, provided historical evidence that the time of year when wines from outside Bordeaux could be sold was strictly controlled prior to the French Revolution, and that they were required to be sold in

[57] Arch. Gironde, 7 M 190.
[58] *Feuille Vinicole*, May 12, 1910.
[59] Arch. Gironde, 7 M 169.
[60] Cited in Vitu (1912:55–56).

different-sized casks from those of the Bordelais.[61] This suggests that these wines had not been considered crucial for blending with those from Bordeaux for the export market. Furthermore, Brutails noted that numerous nineteenth century authors, including Édouard Féret, had clearly identified Bordeaux wines with the Gironde, rather than with these other regions of the old province of Guyenne.

Statistical evidence suggests that whatever the importance of these regions in the past in supplying Bordeaux with wines, it had declined significantly by the early twentieth century. The Gironde's average harvest between 1902 and 1906 was 3.5 million hectoliters, and it exported and sold to other regions in France a total of 4.3 million. Of the balance, together with wine used for local consumption (1.7 million hectoliters), just 11 percent came from Lot-et-Garonne, Lot, and the Dordogne, compared with the 37 percent imported from other countries (including Algeria), and 45 percent from the Midi (table 5.7). The commission in charge of establishing the appellation concluded that the neighboring departments provided "relatively negligible" quantities of wine, and therefore they were unlikely to be important for blending with Bordeaux's wines.[62] Yet for many of Bordeaux's growers of ordinary wines, the regional appellation was a cheap and easy way to restrict competition from growers in the Midi and elsewhere who enjoyed lower production and marketing costs. As a measure for guaranteeing quality, however, consumers would have to wait until after the Second World War, when the *appellations contrôlées* included restrictions on grape varieties and production methods that could be used.[63]

In retrospect, Gladstone's attempt to allow British consumers to drink cheap claret was badly timed, as shortages in France soon drove up prices. In their attempt to sell large quantities of cheap commodity wines through the new commercial outlets in Britain after 1860, merchants found it difficult to maintain the reputation of claret because of the absence of an easily defined product, or the security that wines would not be adulterated after leaving their cellars. When these problems coincided with a shortage and high price of wines because of disease in the late nineteenth century, even the traditional marketing arrangements based on the individual reputation of Bordeaux's négociants and the local retail wine merchant were threatened. Once reputation had been lost, growers and merchants found it very difficult to restore consumer confidence even when quality improved later. As Penning-Rowsell noted, the failure by the wine trade to recognize initially the quality of the 1906 vintage exemplifies the state of the market at this time, as it was later considered the best between 1900 and 1920.[64]

---

[61] Brutails (1909), in Arch. Gironde 7 M 187.

[62] Cazeux-Cazalet (1909:4–10), in Arch. Gironde 7 M 187.

[63] Cooperative wineries in the region did not appear until the 1930s, in part because the growers cultivated a greater variety of grapes, which presented difficulties in measuring wine quality.

[64] Cited in Higounet (1993:330).

CHAPTER 6

Champagne

> In the eyes of the ordinary consumers, Champagne, if dear, has at least the merit of being what it is represented to be, which is more than can always be said of other wines figuring on retailers' lists, and the public, as a body, prefer certainty, with high charges, to uncertainty, even when baited with low prices.
>
> —*Ridley's*, January 1883:3

CHAMPAGNE PRODUCERS were the most successful of all producers in establishing brand names, informing consumers of wine quality, and associating the drink with the needs of the rapidly changing lifestyles of the middle and upper classes in rich urban societies during the nineteenth century. Output increased from just 300,000 bottles, equivalent to the production of about 150 hectares of vines on the eve of the French Revolution,[1] to almost 40 million bottles by 1910. The major champagne *maisons*, or houses, successfully exploited three interconnected features to achieve this growth. First, they developed new wine-producing methods that allowed quality champagne to be produced with the desired level of pressure and benefited from the high degree of specialization in production.[2] Second, the prosperity of the champagne industry was closely linked to the growth of the global economy, with exports accounting for over 85 percent of all sales in the 1870s and 1880s and two-thirds in the 1900s. Finally, the industry was a pioneer in modern marketing methods, which allowed the *maisons* to exploit major economies of scale in marketing.

The production of sparkling wine is not unique to the Champagne region. In France the sparkling wines from Gaillac (southwestern France) and Limoux (Languedoc) date from the Middle Ages, while those from Saumur (Loire valley) competed with champagne in both the domestic and international markets in the nineteenth century. Italian *spumante* and Spanish *cava* also enjoyed local popularity before 1914.[3] Yet none of these wines came anywhere near to providing serious competition with those produced in Reims and Épernay. The grape growers and champagne houses inevitably claimed that a major part of their success was the terroir, which precluded the possibility of imitating the

---

[1] René Crozet (1933), cited in Forbes (1967:150). My calculation—assuming a yield of 20 hectoliters per hectare with no wastage.

[2] Paul (1996:198).

[3] Loubère (1978); Pan-Montojo (1994).

drink elsewhere. Yet the comparative strength of the champagne houses lay, and continues to do so, in their ability to blend a wide variety of grapes and wines to obtain the required characteristics of the final product. The region was relatively close to many of Europe's leading courts, which in the seventeenth and eighteenth centuries were the markets for the small quantities of champagne consumed, suggesting that the association of champagne with luxury was not just a clever marketing invention of the nineteenth century. As with port, sherry, and claret, trade was concentrated in the hands of a small number of houses, and André Simon, writing in 1934, noted that "there are barely a score of Champagne shippers of world-wide repute, and of these a dozen at the most could be described as being in a large way of business, but not less than seven of them date back to the eighteenth century, and all the others are over or else nearly a hundred years old."[4]

Most of the large champagne houses purchased the greater part of their grapes from the thousands of small growers and were successful because they enjoyed the economic resources and technical skills to organize the highly specialized tasks involved in the conversion of grapes into vintage champagne. The fact that fine champagne was made from the limited supply of quality grapes, and that grape and wine production were highly specialized, labor-intensive activities, led to production costs being significantly greater than in most other regions. However, by marketing the drink as a luxury, and by successfully defending their reputation for quality, shippers were able to charge high prices and enjoy good profits. The 1880s and 1890s were golden years for both growers and producers. The growers' prosperity was ended by the appearance of phylloxera and subsequent decline in local harvests, so that the rapidly developing demand for cheap champagne from the 1890s in the domestic French market encouraged the houses to look elsewhere for their grapes. The disastrous harvests in the Marne between 1907 and 1911, which were only a third the size of those in the prephylloxera years of 1892 to 1896, led to thousands of local growers demanding the creation of a regional appellation and an end to the use of outside wines being used to make "champagne." Protest were bitter because disputes occurred over where the borders of the appellation should be drawn.

This chapter looks briefly at the early history of champagne and the dramatic increase in production in the late nineteenth century, the organization of the commodity chain favoring the champagne houses over British retailers, the response of the champagne houses and small growers to the phylloxera crisis, and the collapse of local production and importation of large quantities of outside wines after 1906. In the end, despite the crisis, the champagne producers were more successful than those in other wine regions in controlling the quality of their product.

[4] Simon (1934a:116–17).

The Myth of Dom Pérignon and the Development of Champagne

The Champagne region was famous for its table wines long before the production of sparkling wines.[5] Situated on the crossing of important trade routes, the medieval trade fairs of Champagne (including Provins and Troyes) were among Europe's most famous in the Middle Ages. Reims was the principal gateway for the export of wines from Champagne and Burgundy from at least the fourteenth century, while cheaper beverage wines were transported from Épernay along the Marne and Seine rivers to Paris.[6] In the second half of the seventeenth century, the region's most famous wine was Sillery, which, according to Nicholas Faith, "was a brand name for a blend of wines, only part of which came from the village of the same name."[7] About the same time the Marquis de Sillery, the exiled Marquis de St-Évremond sold a sparkling wine in England that soon became fashionable in the court of Charles II.[8]

Champagne is a blended wine produced on the very edge of Europe's northern limits of wine production. The major grape varieties used are the pinot noir, followed at some distance by the pinot meunier and the chardonnay. Grapes are grown in three major regions of the Marne department: the Mountains de Reims, the Marne valley, and the Côte d'Avize (Côte des blancs). Sparkling wines develop naturally when the cold winter weather halts fermentation prematurely, allowing the remaining sugar to begin a secondary fermentation when the temperature warms once more in the spring; if the wine has already been bottled, it retains the carbon dioxide and allows the bottle to "pop" when it is opened, with copious amounts of bubbles produced when the wine is poured. It seems likely that the first sparkling wines (*vins mousseux*) were produced by accident and considered undesirable, not just because there was little demand for this type of wine, but because large numbers of bottles were broken and the wine lost, as producers had little idea of how to control the process.

Although historians have dismissed the popular myth that attributes the discovery of champagne to Dom Pérignon, a blind monk who was cellarer at the Abbey of Hautvillers in Épernay between 1668 and 1715, they continue to give him a key role in its early development. In Harry W. Paul's historical study of the science of wines, Dom Pérignon is a "revolutionary oenologist ... worthy of joining that happy band enjoying mythical status such as Pasteur, Ribéreau-Gayon, and Peynaud, because of his ruthless emphasis on quality both in vines

---

[5] The old, prerevolutionary province of Champagne was split into five departments: Aisne, Aube, Haute-Marne, Marne, and Seine-et-Marne. The major champagne towns are Reims and Épernay.

[6] Brennan (1997:44, 53).

[7] Faith (1988:25).

[8] Simon (1934a:21–22).

planted and those kept in production and on the importance of the quality and maturity of the grapes pressed."[9] Rather than an inventor as such, he was a keen observer and experimenter, whose contemporary fame rested on the fact that he consistently produced excellent wines in a region where this was notoriously difficult. Pérignon used black grapes, especially the pinot noir, for white wine production (*blanc de noirs*). White wines produced from white grapes quickly turned yellow and lost their delicacy after a year, but those produced using the pinot noir stayed white and maintained their quality for three or four years. As Paul notes, nature therefore "compensated for the crudity of the technology of vinification, which were then incapable of preventing the oxidation of the wine."[10] Special care had to be taken, however, so the skins did not color the wine, and yields from the pinot noir were significantly smaller than those of white varieties.

Dom Pérignon was also extremely successful at blending the grapes that were received as tithes or produced on the abbey's estate. Blending was essential, as the quality and characteristics of the grapes produced on each plot varied significantly, as well as from one harvest to the next. Another consideration was the need to add the correct amount of sugar: sufficient to produce the secondary fermentation and create the sparkle, but not too much, which would lead to excessive numbers of broken bottles.[11] Finally, it appears that Pérignon took advantage of the improved bottle-making techniques that were being introduced in England during the 1660s. Stronger glass bottles, hermetically sealed by using corks, allowed more of the carbonic acid gas to be retained without breaking the bottles. According to the writer Michel Dovaz, champagne production can be divided into a "before" and "after" Pérignon: "B.P.: few bottles, no or hardly any bubbles, no corks, none of the excellent *blanc de noirs*, no wine that kept, etc. A.P.: thanks to him and also to others, especially brother Oudart, the sparkling wine of the Champagne came into existence, sold well and dear, and was put on the road to a real commercial career."[12]

This early momentum associated with Dom Pérignon was not maintained. The high production and transportation costs and the fickle nature of demand made most growers and négociants reluctant to switch from producing red table wines to sparkling ones. Yet by the end of the eighteenth century this was beginning to change. The historian Thomas Brennan has argued that Labrousse's thesis on the flourishing state of French viticulture before the 1770s does not hold for the Champagne region, and the decline in trade between Épernay and Paris forced growers to develop new products and markets and thereby challenge the

---

[9] Paul (1996:204–5).

[10] Ibid., 201.

[11] By contrast the blending of wines with different qualities to produce a standardized product was regarded as fraud in the eighteenth century. Ibid., 205.

[12] Dovaz (1983:18), cited and translated in Paul (1996:206).

TABLE 6.1
Sales of Champagne, 1850–59 to 1900–1909

	Annual sales (thousands of bottles)			Total exports 1844/1850 = 100	French market (% of total sales)
	France	Export	Total		
1785		300			
1844–50	2,074	5,002	7,076	100	29.3
1851–60	2,560	7,440	10,000	141	25.6
1861–70	2,839	9,958	12,797	181	22.2
1871–80	2,632	16,730	19,362	274	13.6
1881–90	2,916	17,980	20,896	295	14.0
1891–1900	5,927	19,569	25,496	360	23.2
1900–1913	10,537	21,676	32,213	455	32.7

Sources: Forbes (1967:150); Faith (1988:209).

brokers who dominated the old trade.[13] As a result, exports grew from about 300,000 bottles in 1785 to five million by the 1840s. Yet as late as 1832 the Marne produced only 50,000 hectoliters of sparkling wines, compared with 310,000 hectoliters of ordinary red wine for local use, and 120,000 hectoliters of quality red for export.[14] On the eve of the First World War this had changed, and champagne sales were over 32 million bottles, or roughly 240,000 hectoliters (table 6.1), while the market for table wines had almost disappeared in the face of competition from the Midi.[15]

Behind this growth in champagne sales there were significant changes in the methods of wine making and the nature of the wine itself. One major advantage was the introduction of disgorging to remove the sediment and create a clean, transparent wine. The secondary fermentation produces sediment composed of a variety of substances, including dead yeasts, gum arabic, tartrate, calcium, and bicarbonate of potassium, which had to be removed with the minimum amount of loss of pressure inside the bottle.[16] Traditionally the wines were decanted and rebottled, but this process was beginning to change in the mid-1810s, as John MacCulloch noted:

[13] Brennan (1997:240–46).
[14] Loubère (1978:109, 281).
[15] Maizière predicted this massive growth in trade in 1846: "sparkling wines have made fortunes for twenty merchants, ensures an honest living for a hundred more, and provide a prompt and profitable outlet for the product of every class of grower; yet the present state of the trade, already ten times as lucrative as the old, is only in its infancy and can multiply tenfold within a few generations." Maizière (1848), cited in Faith (1988:55).
[16] Forbes (1967:322–23).

To remove this [sediment] and to render the wines marketable, those of the best quality are decanted clear into fresh bottles in about fifteen or eighteen months, when the wine is perfected. A certain loss, amounting to one or two bottles in a dozen, is sustained by their explosion previous to this last stage. Another process is sometimes adopted for getting rid of the sediment without the trouble of decanting in this mode; the bottles are reserved in a proper frame for the purpose during a certain number of days so as to permit the foulness to fall into the neck; while in this position the cork is dexterously withdrawn and that portion of the wine which is foul, allowed to escape, after which the bottle is filled with clear wine, permanently corked and secured with wire and wax.[17]

The new system was a two-stage process, known as *remuage* and *dégorgement*. The bottles were first placed in the riddling frame, which comprised two planks of wood placed in position on the floor in the form of an inverted V. Each plank was pieced, and up to sixty bottles of champagne were placed neck-down into the slots. Skilled workers rotated, oscillated, and tilted the bottles slightly each day for several weeks until the sediment was all concentrated in the bottle's neck next to the cork. The temporary cork was then carefully removed from the bottles to allow the carbonic acid gas to escape and with it the sediment. The quick movements of skilled workers resulted in little wine being lost. In 1889 an easier method was developed, the *dégorgement à la glace*, which involved freezing the head of the bottle in calcium chloride and allowing the sediment to be expelled as an ice pellet. With both methods, a small amount of wine and sugar was then added to create the desired type of champagne (*brut, sec, doux*, etc.).[18]

Wine quality was also improved when André François showed in 1837 how to determine the exact amount of sugar needed to obtain a satisfactory secondary fermentation. According to André Jullien in his treatise of 1816, bottle losses varied between 15 and 20 percent, and sometimes even 30 or 40 percent; in 1833 Moët & Chandon lost 35 percent and in 1834, 25 percent.[19] The greater control of the secondary fermentation, together with the continuing advances in bottle-making techniques, reduced the losses from breakage from about 25 percent in the late 1850s to 10 percent by the 1870s.[20]

---

[17] MacCulloch (1816:117), quoted in Simon (1934a:40–41).

[18] Forbes (1967, chap. 28) gives a good description of the various processes, on which this is based. Producers also learned to exploit more skilfully the differences in temperature found in their cellars: the warmer areas were used to start the secondary fermentation in the spring, and the cooler parts to slow it down if the bottles started breaking. By the late nineteenth century selected yeasts and sugar were added to help the secondary fermentation.

[19] Jullien (1824:18); Moreau-Bérillon (1922:123).

[20] Guy (1996:18). Vizetelly (1879:54), reported in the late 1870s that while some houses lost 7 or 8 percent, the figure for some others was less than 3 percent.

Economies of Scale, Brands, and Marketing

Property was highly fragmented in the Marne, with 14,430 growers each owning less than a hectare of vines, 3,202 having between 1 and 5 hectares, 89 between 5 and 10, and only 18 with more than 20 hectares.[21] The vast majority of the growers sold their grapes to a very small number of firms that possessed the necessary capital and skills to produce and market champagne. In the 1880s 83.4 percent of production was exported, and the figure for the best champagnes was probably even higher. The Syndicat du commerce des vins de Champagne, established in 1882 by the largest houses to promote their interests, never had more than a hundred members, and it was reported in 1908 that nine-tenths of the exports of fine champagne were sold by just thirty-four houses.[22] The growing market power of these houses, especially concerning the setting of grape prices, worried growers even before the appearance of phylloxera.[23]

One of the reasons why champagne was more expensive than other wines was that it required greater labor inputs. The topsoil was shallow, and every four years or so a mixture of sand, clay and the *cendres noires* (a form of impure lignite, containing iron, sulfur, and oligo-elements) was quarried and spread on the vines. The result was that the thickness of the topsoil was maintained over the centuries and has become "an artificial creation, a creation . . . of such complexity that it is never likely to be successfully reproduced or imitated elsewhere."[24] On the prephylloxera vineyards, up to fifty thousand wooden stakes were driven into the ground each spring to support the grapes and removed after the harvest to avoid them rotting in the ground. Pruning restricted the yield of the vine and ensured that the grapes were grown close to the ground, allowing them to capture more of the heat, but obliging the harvesters to bend double to collect them. Grape growing in Champagne was also especially risky, with grapes in danger of damage from late spring frosts or summer hail, as well as downy mildew and *pourriture grise* produced by excessive summer moisture. Most growers reduced their exposure by working the vines only part-time. Finally, and unlike other French wine-producing regions, only a limited number of growers pressed their own grapes.[25] The champagne houses selected their grapes from a wide variety of vineyards to obtain grapes with different characteristics. According to one contemporary:

> The aim is to combine and develop the special qualities of the respective crûs, body and vinosity being secured by the red vintages of Bouzy and Verzenay, softness and

---

[21] *Le Vigneron Champenois*, March 29, 1911.

[22] Ibid., July 5, 1908.

[23] Opposition centered on René Lamarre, the author of the pamphlet *La Révolution champenoise*, published in 1890.

[24] Forbes (1967:24–25).

[25] Vizetelly (1879:25).

roundness by those of Ay and Dizy, and lightness, delicacy, and effervescence by the white growths of Avize and Cramant. The proportions are never absolute, but vary according to the manufacturer's style of wine and the taste of the countries which form his principal markets.[26]

To achieve this, the leading firms established crushing and pressing facilities in the main villages. Moët & Chandon, which was unusual in that it owned 365 hectares of vines itself, had facilities in seventeen different villages. These vines produced the equivalent of about 1.3 million bottles, but because this was "utterly inadequate" to cover the firm's demand, large quantities of grapes were also bought from local growers. As the wine was stored in bottles rather than casks, the champagne cellars also had to be considerably larger than those found elsewhere, and the need to control carefully the temperature of the wines during the secondary fermentation required the cellars to be much deeper. The chalk soils of the region were simple and inexpensive to dig out, but sufficiently strong for a cellar. Moët & Chandon's cellar in Épernay had an aggregate length of over 10 kilometers and stored between eleven and twelve million bottles and 25,000 casks of wine, with the firm employing 20 clerks and 350 cellarmen, although the total payroll came to 1,500.[27]

The champagne houses, just like those in Porto and Jerez, blended wines to meet the different demands of the consumer. However, unlike port and sherry, champagne had to be bottled before being shipped, and this allowed the *maisons* to control the wine from production to consumption and their brands were considered an important guarantee of purity. The champagne houses developed their house wine by careful blending and kept extensive stocks in order to supply wines of a "similar type and of uniform quality, year after year, irrespective of the vagaries of the weather."[28] Labels and branded corks informed the consumer that they were drinking not just champagne, but Bollinger or Roederer. Selling champagne was therefore relatively easy for retailers. The successful branding of champagne by the leading houses led to an increase, not a reduction, in prices for consumers, and even Britain's giant retailer Gilbeys started selling producer-branded champagnes rather than its own from 1882.[29]

Product development was another major area of success. Champagne was originally sweetened significantly, in order to hide "the sharpness of youth which time alone could have toned down."[30] British retail merchants made a number of attempts to sell dry champagne in the mid-nineteenth century, but it only became important in the late 1860s, and it was not until the 1874 vintage that *brut*

---

[26] Ibid., 50.
[27] Ibid., 109, 113. The 900 acres (about 550 hectares) of vines provided employment for 800 laborers and vinedressers, a doctor and chemist, which were provided free of charge by the firm, and sick pay and retirement pensions for workers.
[28] Simon (1934a:66).
[29] Faith (1988:73).
[30] Simon (1934a:45).

champagne was generally available and sold by most of the leading houses.[31] According to André Simon, the introduction of dry champagnes made it fashionable in some circles to drink it at dinner and helped to increase consumption in England from three million to over nine million bottles between 1861 and 1890. Sweet dessert champagnes remained the rule throughout France, together with some export markets, such as Russia.[32]

A taste for dry champagnes allowed the creation of "vintage" champagne, which consisted of a select blend of wines produced in a specific year that was noted for its excellence and could be marketed at considerably higher prices than the ordinary house wines. Vintage champagnes became fashionable and prices rapidly increased, greatly outstripping the costs of storing the wines. Thus the Clicquot 1884 vintage sold for 94 shillings a dozen when it was released on the market in 1891 but reached 165 shillings by 1898; Moët & Chandon's vintage of the same year increased from 88 to 260 shillings between the same dates.[33] The shippers allocated wines to retailers before they were actually released, and these often changed hands several times before the wine was delivered. Wine merchants encouraged consumers to purchase vintage wines, and inevitably some shippers were tempted to increase the supply, eventually leading to a decline in their novelty.[34] André Simon, writing in 1905, noted that "the 1889 and 1892 vintages being excellent wines, and sold on the market at a highly favourable time, mark the apogee of the vintage Champagne boom from the point of view of the public."[35] Consumption in Britain reportedly then declined in the "clubs and the less fashionable Hotels and Restaurants" and was also negatively influenced by the growth in entertaining in restaurants rather than private houses.[36] Once more an increasingly critical press argued that poor-quality grapes were being used in champagne production, which affected demand, especially for the cheaper wines.[37]

Although exports to the major Britain market started falling in the late nineteenth century (fig. 4.4), total sales of champagne continued growing until the First World War, driven by the growth in French demand (table 6.1) and heavy advertising. Mericer, an important producer of cheap wines, received considerable publicity when it constructed the largest barrel in the world, containing the equivalent of 200,000 bottles, for the 1889 Universal Exhibition and had it towed to Paris by twenty-four white oxen in a journey that took three weeks. The same house commissioned one of the first advertising films for the 1900 Exhibi-

---

[31] Simon (1905:99, 137). Simon notes that the taste for sweet champagne lasted longer outside London. For example, the shipper George Goulet exported sweet champagne with 16 percent liquor to a merchant in Birmingham, but with only 2 percent to London.

[32] Simon (1934a:43, 50).

[33] Ibid., 74.

[34] Simon (1905:147).

[35] Ibid., 146.

[36] *Ridley's*, June 1909, p. 558.

[37] Simpson (2004:99).

tion and received widespread publicity when a giant balloon broke loose and transported nine unsuspecting drinkers and their waiter from Paris and deposited them unhurt sixteen hours later in the Austrian Tyrol, where the company was fined twenty crowns for importing illegally six bottles of champagne.[38] Other marketing strategies were less dramatic. As large quantities of champagne were drunk in hotels and restaurants, some shippers paid corkage fees to waiters, amounting to a fee for every bottle of their wine that was sold.[39] Hoteliers, however, learned that the system could also work to their advantage, and a certain Messrs. Gordon demanded a payment of £500 to stock a particular brand of champagne in their five hotels.[40] Brands became everything, and one writer noted in 1890 that "within ten years we will no longer recognise the name of champagne, but only those of Roederer, Planckaert, Bollinger, without any idea what the wines will be made out of."[41]

Champagne sales increased almost five times from the late 1840s to the 1900s, partly as a result of an increase in the area of vines and better wine-making skills, but also because of the use of outside wines, leading to a decline in quality.[42] Kolleen Guy notes that fraud changed from "a peripheral concern in the 1890s" to "the central issue for producers of both ordinary and fine wines at the turn of the century."[43] The best champagnes continued to be sold under the manufacturer's brand name and vintage and were not adulterated, but the reputation of other houses was questioned. The situation was made considerably worse between 1908 and 1910 when the local grape harvests failed, encouraging the widespread use of cheap white wines from other regions. The irony was not lost on local growers that some houses, which had gone to court to protect the "Champagne" name, were now themselves using outside wines. Unlike Bordeaux, the debate on establishing a regional appellation took place against the background of the destruction caused by phylloxera, and not overproduction.

THE RESPONSE TO PHYLLOXERA

Phylloxera was first discovered in the Champagne region in 1890, making it the last major wine-producing area in France to be infected. Despite this delay, the question of how to respond to the disease created fierce conflicts in the local

[38] Faith (1988:68).

[39] Ridley's, August 1893, p. 436. In this case the payment was 4d. a cork. Simon (1934:112–13) argues that four major houses refused to participate in this type of scheme.

[40] Ridley's, December 1890, p. 655.

[41] Lamarre, cited in Faith (1988:78).

[42] The decline in red wine production released some land, while better control of the secondary fermentation limited wastage.

[43] Guy (1996:211). However, Tovey noted in 1870 that "even at Epernay and other places in Champagne it is well known that there are houses which use but little of the wine of the district in the manufacture of their wines" (19).

community. Like the leading growers in Bordeaux, the big champagne houses were concerned about the quality of wine produced from the new grafted vines and therefore wanted to do all that was possible to preserve the traditional French rootstock. Unlike the Bordeaux growers, however, most champagne houses were not major producers of quality grapes themselves but depended on hundreds of small growers. How the small winegrowers responded to phylloxera would have major implications for the big champagne houses.

A *comité d'étude et de vigilance contre le phylloxera* was created in 1879 to develop a plan of action, consisting mostly of "agricultural professors, large landowners, and négociants." Several visits were organized to the Midi, and this led some to argue that the spread of phylloxera was caused by poor cultivation, and therefore it would never be a threat to the growers of quality grapes for champagne.[44] The appearance of diseased vines on the edge of the Marne in July 1889 challenged this particular belief, and Gaston Chandon of Moët & Chandon purchased and destroyed the infected vineyard, while the Syndicat du commerce proceeded to do the same when a further 2 hectares of diseased vines subsequently appeared.[45]

Even if immunity to phylloxera was now recognized as impossible, the experience of fine wine producers in Bordeaux suggested that the destruction could be delayed, perhaps indefinitely, by the use of chemicals. In Bordeaux the large growers had used their own resources to protect their own vines and were relatively indifferent to the plight of small growers, whose wine they did not buy. The dependence of the large champagne houses on the thousands of small growers required them to keep the whole region disease free for as long as possible. This in turn required a collective response, as all diseased vines had to be uprooted and destroyed, and the land treated with chemicals to be sure that no aphids remained. To be successful, all growers had to participate, as phylloxera would become quickly established if some remained outside the syndicate. The institutional arrangement to oblige all growers to participate, and thereby remove the potential problem of free-riders, was the law of December 1888. This allowed a *départemental syndicate* to be created if two-thirds of growers possessing three-fourths of all vines, or three-fourths of all growers with two-thirds of all vines, supported the project. In June 1891 it was claimed that 17,370 growers (67.5 percent of the total) and 9,772 hectares of vines (76.2 percent) had agreed, and a local Association syndicale autorisée por la défense des vignes contre le phylloxera was authorized the following month. All growers in the Marne were obliged to pay dues, and the syndicate's officials had the legal right to enter any property and uproot and destroy any diseased vines. Growers were to be compensated.

[44] Guy (2003:92).
[45] Lheureux (1905:47).

Disputes began immediately. A number of growers claimed they had shown interest only in the possibility of establishing a syndicate, rather than supporting a specific project.[46] Others claimed that some of the signatures were not those of vine growers, or that they belonged to deceased growers. Conflicts also arose over the annual fee, with smaller growers demanding that the better, and hence more valuable, vineyards should pay more than the less valuable ones. However, the major area of dissent was over the methods used to tackle phylloxera. The small growers wanted to encourage the use of experiments with new American vines so that they could replant as quickly as possible, while the large champagne houses wished to keep American vines (and hence possible infection) away from the Marne for as long as possible. This difference in strategy quickly became apparent, and the list of candidates presented by the large landowners (and champagne firms) for the steering committee was challenged successfully by René Lamarre, whose own list predominantly comprised small growers.[47] Yet success for the growers was short-lived: when the committee met in May 1892, the original twenty-five members found that their numbers had grown to eighty-three. Additional financial contributions had been obtained from the Reims Chamber of Commerce, the Syndicat du commerce, and other institutions, and this legally allowed them to official representation, thereby snatching control away once more from the small growers.[48] Officials were assaulted when they tried to remove diseased vines, and the police estimated that in 1894 nearly 75 percent of vignerons were actively opposed to the association.[49] Small growers now drew up a new petition calling for its dissolution.

Why did the association fail? Lucien Lheureux, a writer generally favorable to the négociants' position, noted in 1905 that if the syndicate had operated as planned, it might have delayed the spread of phylloxera by a decade or so.[50] In fact, although phylloxera had been found on the edge of the Marne in 1889, only 659 hectares were infected, or 4 percent of the total area, as late as 1900. Yet growers had several worries concerning the Association syndicale. First, the big champagne houses were relatively immune to phylloxera. If their own vines, or those of growers from whom they habitually bought, were destroyed, they could look elsewhere for their grapes. More important, they also carried large stocks of wines in their cellars, and they continued to do business successfully during the worst of the epidemic. By contrast, those small growers whose vines were destroyed immediately lost their livelihood. However, these fears clearly belonged to the future, as the area infected in the whole of the Marne was still minimal

---

[46] Guy (2003:100–101).

[47] Rather than unite growers, therefore, the Association syndicale divided them. Lamarre attacked its work from his newspaper, the *Révolution Champenoise*.

[48] Lheureux (1905:55–56); Piard (1937:57).

[49] Cited in Guy (2003:110–11).

[50] Lheureux (1905:67).

when the Association syndicale disappeared in 1895. A more immediate preoc-
cupation was the high profile of the Syndicat du commerce within the Association syndicale, which created considerable unease among many growers. Many
believed that the Syndicat du commerce had displaced the market by acting as a
monopoly and unilaterally establishing grape prices and controlling sales. The
difficulties faced by the villagers of Damery in establishing a producers' cooperative at this time, for example, were attributed to the supposed boycott of its wine
by the champagne houses.[51]

Finally, by the 1890s American vines were being successfully planted in large
areas of France. Nationally the area replanted jumped from 75,000 hectares in
1885 to 436,000 in 1890, and by this date the area of replanted American vines
outnumber by more than four to one the total area that was being treated chemically or flooded (table 2.3). Small growers in the Champagne region demanded
public institutions to experiment and find which vines were most suitable to the
chalk soils of the Marne, and to determine which French scions produced the
best wines. A policy limited to just the elimination of vines was unpopular with
small growers because, and as one petition to the mayor of Épernay put it, "Phylloxera is at our doors and a large part of the vines are anaemic and sterile. We
need to regenerate our vines ... and to profit from the experience of others before we lose everything."[52] In the face of widespread opposition, the Association
syndicale was quietly disbanded in 1895.

The disappearance of the association did not imply the end of collective action against phylloxera. It was quite the opposite. As in Bordeaux, small growers
formed local village syndicates to help fight the disease. In Bouzy in 1900, for
example, growers who owned 186 of the village's 215 hectares of vines belonged
to the local syndicate. A further 14 hectares belonged to Chandon, who not only
treated these vines privately but contributed 800 francs to the local syndicate.[53]
In March 1898 the Association viticole champenoise (AVC) was established as
an umbrella organization on the initiative of twenty-four of the leading champagne houses to provide funds and practical help to individual local syndicates.
As article 3 of its statutes noted, the AVC was established not only to fight phylloxera and other vine diseases, but also to help replant diseased vines.[54] A research station was established in the village of Ay, and the AVC, together with
Moët & Chandon, was at the forefront in providing workers with the skills to
grow the new vines.[55] For example, in 1902 seven syndicates participated in a
competition for grafting new vines organized by the AVC, a figure that doubled
the following year and included 112 participants. Prizes were awarded according

---

[51] Guy (2003:108).
[52] Letter of 1903 from growers to the mayor of Épernay, cited in ibid., 110.
[53] Le Vigneron Champenois, January 31, 1900.
[54] Lheureux (1905:75); Piard (1937:59).
[55] Lheureux (1905:92).

to the quality of work, rather than the speed.[56] By 1900 almost four thousand growers grouped together in 39 syndicates were receiving help from the AVC, and they were treating chemically about a tenth of their vines. These figures increased to almost 11,000 members and 117 syndicates in 1911 (table 6.2).

Small growers were happy to accept help from the AVC because, unlike with the Association syndicale, they retained full control over local decision making, and from an early date the AVC helped with replanting rather than just destroying diseased vines. It remained active until 1938. The area of vines in the Marne drifted lower, from 15,640 hectares in 1898 to 13,870 in 1908, before falling to 12,790 hectares in 1913. The shortage of chemicals and the disruptions caused by the First World War finally allowed phylloxera to spread rapidly, and the area slumped to 6,900 hectares in 1919.[57] The Syndicat du commerce initially played no role in the AVC, although it contributed 400,000 francs in 1908 to help growers whose harvest had been devastated by mildew, and who lacked funds to purchase the necessary chemicals.[58]

Yet if the local syndicates and the AVC helped absorb the shocks caused by phylloxera, growers soon faced another, even greater problem. While growers' production costs were increasing because of the need to purchase chemicals to fight disease, their incomes declined steeply as a result of a combination of low grape prices and poor harvests. The continual growth in champagne sales was met by outside wines being sold as champagne, and local growers turned to the 1905 legislation on adulteration and pure food to recapture some of their lost market power.

ORGANIZATION OF A REGIONAL APPELLATION

The wine crisis in the Marne was brought about by the destruction of local harvests. Augé-Laribé estimated 350,000– 400,000 hectoliters as a "normal harvest," a figure that was reached seven times during the decade 1897–1906 but not once over the next five years (table 6.3).[59] It is true that during the five years between 1888 and 1892 harvests fluctuated between only 127,000 and 278,000 hectoliters, but the grapes in these years were of a high quality, and growers were compensated by prices two or three times higher than normal years (fig. 6.1 and table 6.3). By contrast, late rains rotted part of the 1907 harvest, and that of 1908 was devastated by mildew. The 1909 harvest was small but of good quality, but the 1910 harvest was almost totally lost because of

---

[56] *Le Vigneron Champenois*, February 10, 1904. Moët & Chandon had also established its own workshop for grafting. Ibid., July 10, 1904.

[57] Piard (1937:66).

[58] *Le Vigneron Champenois*, March 10, 1909.

[59] Augé-Laribé, in ibid., March 29, 1911.

TABLE 6.2
Phylloxera and Local Syndicates Receiving Help from the AVC in the Marne

	Local syndicates and phylloxera							All departments	
				Total area in hectares					
	Syndicates	Members	Vines	Infected	Treated	Uprooted	Replanted	Area infected	Area replanted
1898	36	3,300	3,000					38	20
1899	39	3,716	3,184	276	368	33	24	90	42
1900	40	3,871	3,246	364	315	35	24	659	178
1901	67	5,710	4,513	451	611	44	40	1,335	272
1902	69	6,217	4,624	816	746	76	53	2,041	367
1903	71	6,461	4,826	935	1,415	81	63	2,036	435
1904	76	7,134	5,069	2,256	2,764	169	140	2,421	698
1905	86	7,992	5,463	3,922	3,698	259	169	3,317	810
1906	97	8,392	5,762	4,695	4,508	306	211	4,339	1,032
1907	99	8,850	6,189	4,610	4,900	229	251	4,987	1,200
1908	105	10,081	6,711	5,261			248	5,561	1,424
1909	111	10,330	6,836	6,103			324	5,985	1,808
1910	117	10,941	7,409					6,515	2,000
1911	124	11,322	8,039					6,500	2,000
1912	128	15,594	7,987					Na	Na
1913								Na	2,518

Sources: Lheureux (1905:79); *Le Vigneron Champenois,* March 10, 1909, and March 8, 1910; and Piard (1937:61, 66).

TABLE 6.3
Production, Prices, and Wine Sales in the Marne, 1889–1911

	Annual harvest (thousands of hl)	Grape price—Verzenay (60 kilos)	Grape price—Rilly (60 kilos)	Grape price—Ville-Dommange (60 kilos)	Merchant sales of champagne (millions of bottles)	Merchant purchases minus Marne harvest (millions of bottles)
1889–92	205	185	160	120	23.6	+3.3
1907–11	169				34.6	+17.8*
1892–96	463	79	66	54	23.3	−8.5
1897–1901	447	68	52	36	28.3	−10.0
1902–6	484	61	45	28	32.0	−8.6
1907	300	75	60	45	33.7	+21.5
1908	127	75	60	45	32.0	+17.5
1909	268	85	65	45	39.3	+3.4
1910	10		100		38.6	+29.2
1911	139				29.4	

Sources: Production and merchants' sales—Ridleys, various years; prices—Le Vigneron Champenois, April 7 and August 23, 1911; final column—see text.
*1907–10 only.

mildew and cochylis. Finally, drought caused the 1911 harvest to be very small, although of excellent quality.[60]

As table 6.3 suggests, the low prices and small harvests were only part of the story. Merchants were actually selling more wine, even though the harvests for the five years between 1907 and 1911 were only a third of those between 1902 and 1906. This was achieved in part by the champagne houses drawing on the stocks of fine wines that they had in their cellars. Gilbey, for example, was often at pains to argue in the British press that neither phylloxera nor mildew threatened the supply of fine champagnes for consumers. In 1904 it was noted that the Marne produced 348,000 hectoliters of wine, equivalent to 84,260,000 bottles. Even if this figure is reduced to 60 million because of production losses, it still dwarfs the sales of 25 million bottles that year. Some therefore argued that there was no need to worry about phylloxera.[61] Yet this assumption was mistaken. In the first instance, figures for the total harvest, or merchants' sales, provide little information concerning quality (and hence prices). Henri Sempé, for example, notes that only half the 1897 harvest could be classified as vins supérieurs, which sold for 125 francs per hectoliter, while vins ordinaires

[60] Bara (1998:172–73).
[61] Le Vigneron Champenois, August 13, 1904.

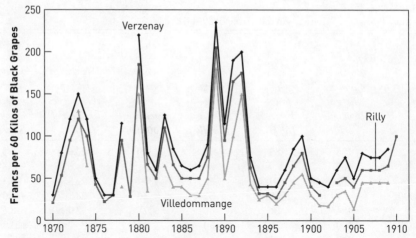

Figure 6.1. Grape prices in the Champagne region. Source: *Le Vigneron Champenois,* April 7 and August 23, 1911

sold at about a third this figure.[62] André Simon, writing in 1905, noted that the production on average of *grands crus* was 9 million bottles (produced on 3,100 hectares), while that of *premier crus* was 11.5 million (3,867 hectares), giving a total of just over 20 million bottles.[63] Finally, there was about 8,000 hectares more "which yield too rough a wine to be used for sparkling *champagne*; much of this wine is, and the whole of it should be, drunk on the spot as a *vin ordinaire.*"[64] When the quality of the local harvest was poor or insufficient, the temptation for producers of cheap champagne to seek alternative supplies from outside the region was strong. Indeed, as in Bordeaux, merchants argued that without the flexibility to choose the most suitable grapes regardless of their origin, neither they nor the local growers would be able to sell their wines in bad years.[65]

The increased demand for champagne also hid an important change in the market because, if total sales increased by 51 percent between 1889–90 and 1907–11, exports grew by a relatively modest 18 percent. By contrast, consumption in the French domestic market tripled, from 4.4 to 13.2 million bottles. French consumers preferred the cheaper *vins de deuxièm choix*, and smaller firms, catering to this market, became increasingly important.[66] Between 1905 and 1909, 48 percent of the sales in Épernay (6.9 million bottles) and 23 percent (4.3 million) from Reims were sold in France.[67]

[62] Sempé (1898:44).
[63] Quoted in Faith (1988:64).
[64] Simon (1905:119).
[65] Vitu (1912:64).
[66] Guy (2003:79).
[67] Calculated from *Le Vigneron Champenois*, June 1, 1910.

The worsening economic situation of many growers became apparent after 1907. An estimate of "outside wines" bought by merchants is shown in the last column of table 6.3. The annual wine production in the Marne has been converted into an equivalent to that used by merchants, by using the same coefficients as given in *Le Vigneron Champenois*.[68] This can be considered as the potential annual supply of local wine for merchants. Actual purchases by merchants have been calculated by adding to domestic and foreign sales the changes in annual stock levels (the year runs from April to March). The figures for the potential supply have been subtracted from the actual purchases, with a positive figure in the table implying that merchants were buying from outside the Marne, and a negative figure meaning that some local wines were being left unsold. This estimate is not without its problems, not least because many local wines were not used for champagne, as André Simon noted, and the fact that the agricultural year was different from the commercial one being used. However, table 6.3 suggests that massive purchases were made by merchants from outside the region after 1907, and that this was a relatively new situation, both with respect to years of relatively abundant local harvests (1892–1906) and to previous years of local harvest shortages (1889–1892).

The prosperity of the industry encouraged the champagne houses to be energetic in protecting both their private brands and the general reputation of champagne. The Syndicat du commerce was established in 1882 precisely to help monitor the industry, and the champagne houses prosecuted individuals who infringed their own brands under the 1824 law. Another area of concern was to convince the courts that the word "champagne" was not a generic name for any sparkling wine. The decision in 1889 by the Cour de Cassation, France's highest tribunal, made it illegal for producers from outside the region to use the words "champagne" or "vins de champagne" to describe their wines, but the *maisons* were still legally allowed to buy wine from outside the region for bottling at Reims or Épernay, to be sold as champagne. Another activity that remained uncontrolled was the purchase of must in the Marne for manufacture in Germany, where import duties on grape juice were lower than on champagne.

The Syndicat du commerce noted in 1904 that its statutes required that all wines using the name champagne be produced from grapes grown and fermented in the region.[69] Yet growers argued that the use of outside wines was common, and that the statutes of the Syndicat du commerce provided only a voluntary code of conduct. Therefore, while the firm Mercier was expelled from the Syndicat du commerce when it refused to stop sales of wines to Germany and Luxembourg where they underwent their second fermentation, this decision was not made public.[70] Another problem was that if the Syndicat du commerce

---

[68] Allowing for losses in production, this implies sales of 87 bottles per 100 liters of grape juice produced.

[69] Guy (2003:122).

[70] Ibid.,79. There was logic in Mercier's action, as the increased import duties on champagne made it uncompetitive in Germany.

represented the large houses, there were a significant number of smaller ones supplying the growing market for cheap champagnes who were not members, precisely because this was not in their interests. The rapidly growing French market for cheap champagne also attracted the attention of some of the leading houses, although they took care not to sell cheap wines under their own brand. The persistence of poor harvests between 1907 and 1911 forced many houses, including those of the Syndicat du commerce, to look elsewhere for supplies for this new market.

The growers' response to phylloxera, namely, the creation of local syndicates and the AVC, provided a useful rehearsal to confront the new situation of persistent low prices, devastated harvests, and wine purchases by champagne houses from outside the Marne. In August 1904 a new growers' organization was established, the Fédération des syndicats viticoles de la Champagne, with the specific aim of controlling fraud. Its first leader was the moderate Edmond Bin, and perhaps in normal circumstances a compromise over wine purchases would have been reached with the merchants. The Syndicat du commerce was worried itself about the declining reputation of the name "champagne" and agreed with the Fédération that the boundaries of the "true" Champagne were the three arrondissements of Épernay, Reims, and Châlons-sur-Marne. Representatives of the Marne were instrumental in attaching an amendment to the 1905 legislation on consumer protection a clause which included reference to establishing regional appellations.[71] Despite this agreement between the two major organizations, the following years saw bitter conflicts, exploding in violence in 1911. This was caused by two very different factors. First, the machinery for establishing the boundaries of the appellation was left to the Ministry of Agriculture, and political opposition was immediate from those growers excluded and the merchants who stood to lose from its creation. Second, the economic conditions in the Marne and the Aube made growers desperate, especially after the poor harvest of 1908 and the nonexistent one of 1910.

Unlike Bordeaux, there were major disputes over what constituted the natural region of Champagne. The modern department of the Marne contains Épernay and Reims, the two major centers of production, but the old province of Champagne was much more extensive and included almost exactly the present-day departments of Aube, Haut-Marne, and Ardennes. Growers in the Aube were particularly incensed at being excluded from the first boundary proposal, as they claimed they had replanted low-yielding varieties (half pineau and half gamay) after phylloxera to guarantee grape quality to make champagne, and as a result they believed they would be uncompetitive against the high-yielding producers of sparkling white wines in the Loire and elsewhere.[72] The Fédération replied that the Aube growers were themselves low-cost producers of inferior wines, as

[71] Ibid., 135, 140.
[72] Vitu (1912:58). Pineau is a synonym for the pinot family of grape varieties and better-quality wines, while the gamay is a high-yielding variety.

TABLE 6.4
Production Costs of Grapes in the Marne and Aube

		Yields (per hectare)	Production costs (per hectare)	Production costs (per hectoliter)*	Farm gate price (per hectoliter)
Marne	Grands crus	20–25	3,000–3,500	144	
Marne	Moyens and petits crus	40–45	2,000–2,500	53	
Aube		80–100			20–25

Source: Cited in Vitu (1912:62).
* Yields and production have been averaged.

production costs were much higher in the Marne because the greater density of the vines made the use of plows impossible (table 6.4). Given the wide range of wines produced in most French departments it was easy to cite unrepresentative cases, and this seems to have been the case with the Fédération, as average yields in the Marne increased from 24 hectoliters per hectare in the 1890s to 30 hectoliters in the 1900s, while those in the Aube increased from 16 to only 24 hectoliters.[73] The real threat from the Aube perhaps lay more in the future, as phylloxera had advanced faster there than in the Marne, so there was the possibility of significantly higher yields after replanting. Yet table 6.5 suggests that only very little wine could have entered the Marne from Aube between 1907 and 1911 because the collapse in its harvest was even greater than in the Marne, leaving growers who had invested heavily in replanting as desperate as their neighbors. The harvest in the Aube in 1910 was just 0.8 percent of what it had been in 1892–96, so it was little wonder that many champagne houses were able to deny that they were buying supplies there. The presence of large number of growers in both departments encouraged a political compromise, and the final decree of June 1911 created two zones, the Marne and L'Aisne, areas which had been initially included in the 1908 proposal, and another including Aube, Haut-Marne, and Seine-et-Marne, whose growers could still sell their grapes to the champagne houses, although this information had to be shown on the bottle.[74]

Although the new regional appellation shifted some power back to the growers, it was less significant than in the Midi or Bordeaux. Producers of quality grapes were always going to find a market because the leading champagne houses needed to create an exclusive product, although many of these took the opportunity provided by phylloxera to integrate backward by buying diseased vineyards cheaply. The voluntary regulations followed by the Syndicat du commerce

[73] Calculated from Lachiver (1988:598–99, 608–9).

[74] Vitu, *Délimitations régionales*, 36. For the conflicts, see especially Bonal (1994); Faith (1988); Forbes (1967); and Guy (2003).

TABLE 6.5
Harvest Size in the Aube and Marne, 1889–1911

	Aube	Marne
1892–96	371,600 hectoliters	462,600 hectoliters
	Index 1892–96 = 100	
1882–86	112	83
1887–91	84	59
1892–96	100	100
1897–1901	86	97
1902–6	59	105
1907–11	19	36

Sources: Table 8.3; Gabriel (1913:182–83).

before 1911 were now extended to all producers, including those selling cheap champagne in the domestic French market. The widespread economic difficulties of growers and the legislation of 1905 provided a political opportunity to restrict the market power of those houses specializing in cheap champagne, but only the creation of cooperatives would allow them to establish some independence from the major, independent champagne houses.

Today's major champagne houses were already established by 1840, and their early success was closely linked to marketing. As first movers, they were able to consolidate their success by product development in the form of "vintage" champagne, and by the second half of the nineteenth century the well-known brands commanded a significant premium. They were more successful than producers in Bordeaux, Jerez, and Porto in controlling the quality of their product. Champagne had to be bottled at source, and, although fraud was a major concern in the late nineteenth century, significant advances had been made by individual firms and the leading champagne producers' pressure group, the Syndicat du commerce des vins de Champagne, in controlling the illegal use of brand names and the appropriation of the word "champagne" for other sparkling wines in France. The 1889 High Court decision declared that only wines made in the small geographical area of the Marne could be described as champagne, thereby banning the use of champagne as a generic term for sparkling wines in the French market. Legislation also protected the champagne houses from the appropriation of their brands by others. Secure in the knowledge that the legal system would protect their brand, firms such as Moët & Chandon and Perrier-Jouët invested heavily in the making and advertising of their fine wines and developed a sufficient reputation to allow them to sell at higher prices than their competitors did. The wine shortages caused by phylloxera and other diseases after 1900 did not seriously threaten this market. The leading firms not only enjoyed extensive stocks of old wines and could afford not to buy in poor years, but they could charge consumers higher prices due to the scarcity of quality champagne.

A major new market developed among the emerging middle classes, leading to a 50 percent increase in world sales during the 1888s and 1890s, which encouraged new entrants into the industry to supply cheap champagnes.[75] With the collapse of local grape production, they were forced to look outside the traditional area of supply, and the bitterness of the events of 1911 can be explained by the desperation of growers in the Marne and the Aube as they faced both disastrous harvests and low prices, at a time when sales were booming.

[75] As Henry Vizetelly noted (1879:61): "now-a-days the exhilarating wine graces not merely princely but middle-class dinner-tables, and is the needful adjunct at every petit souper in all the gayer capitals of the world."

CHAPTER 7

Port

> No wine in the world can produce such a combination of colour,
> aroma and tastes as a fine vintage port.
>
> —Geoffrey Murat Tait, 1936:46

WINE ACCOUNTED FOR about half of all Portugal's exports during the nine-teenth century, with port assuming by far the greatest share. The best and most expensive was made by British merchants to meet consumer demand in their home market, where it enjoyed a privileged position for at least a century and a half after the signing of the Methuen Treaty in 1703. Port is a fortified wine, produced from grapes grown along the steep valley of the River Douro and blended and matured in the cellars (lodges) of the shippers in Vila Nova da Gaia, opposite Porto in northern Portugal. Increasing the alcoholic content of the wine to 17 or 18 percent makes it microbiologically stable, and brandy was traditionally added before shipping which considerably reduced the risks of damage caused by transportation or as a consequence of poor storage.[1] By the mid-eighteenth century, brandy was also added to stop fermentation prema-turely, thereby creating a naturally sweet wine. As fortified wines kept longer on opening, port and sherry decanters became a common feature on the side-boards of many middle- and upper-class households during the nineteenth century.

The addition of alcohol in the production of dessert wines influenced the na-ture of the commodity chain, as not only did it solve the problems associated with making wine in a hot climate, it also offered economies of scale in the prep-aration of wines for export and marketing.[2] Fortified wines were relatively easy to blend to create large batches of house wines with similar characteristics from one year to the next, facilitating the creation of brands, and allowing British re-tailers to obtain repeat orders that they could then sell under their proprietary name, or communicate shifts in the nature of consumer demand. Shippers con-solidated their positions by gaining greater control of the chain and its profits through product development in the form of vintage port, which was a high-

---

[1] Amerine and Singleton (1977:150); Tait (1936:33–36). Wines could still spoil, and Tait noted that "it is necessary to emphasise the importance of keeping casks and vats quite full. This is too little understood in Great Britain, with the result that even if the wines do not actually 'go wrong,' they become 'flat,' woody, and out of condition."

[2] Economies of scale were also found in distilling to make the spirits and brandy required in the production and strengthening of wines.

quality, expensive wine produced only after an exceptional harvest. These factors helped create entry barriers, as potential new competitors in Porto needed both access to large stocks of wine to meet retailers' demands and their own distribution networks in the British market.

This chapter begins by looking at the development of port wine for the British market, and the geographical separation between grape production in the Douro valley and the maturing and exporting houses in Porto. This is followed by discussion of the development of different types of port wine and how the sector responded to the challenges of the second half of the nineteenth century, namely, the problems of maintaining supplies and product quality during the phylloxera epidemic, and the opportunities and difficulties faced by producers in creating a mass market for cheap ports. Finally, the chapter considers the conflicts between the British exporters and Portuguese growers over regulation and regional appellations from the eighteenth century. Winegrowers in the Douro valley, unlike those who produced sherry in Jerez, enjoyed government support for a regional appellation, while shippers were able to influence negotiations so that the new Anglo-Portuguese Commercial Treaty in 1916 recognized that port and madeira came only from Porto and Madeira.

PORT AND THE BRITISH MARKET

When war broke out with France in the 1690s, the British merchants looked to Portugal for wines, and this trade received a major boost with the Methuen Treaty of 1703. This left duties on Portuguese wines at a maximum of two-thirds of those paid on French wines, in exchange for removal of restrictions on the entry of certain types of woolen cloth. The British market remained by far the most important both in terms of volume and especially value, and port wine accounted for over a third of all Portugal's exports as late as the 1870s, and still almost a fifth at the turn of the twentieth century (fig. 7.1, table 7.1).

The early exports came from Portugal's coastal regions, but this wine proved to be too thin and astringent for the British market. By contrast, those from inland along the banks of the River Douro were dark red, the result of a fast and furious fermentation at high temperatures.[3] The earliest reference to alcohol being added to port before shipping to avoid it deteriorating was 1678, but by the early 1700s it was also added during fermentation to change taste, and according to Sarah Bradford, the eighteenth century is the story of how the "despised red portugal" was developed into a wine "cherished and revered" by its British consumers.[4] The degree of sweetness depends on the amount unfermented saccharine, which in turn is the result of the initial sugar content of the

[3] Robinson (2006:540).
[4] Tait (1936:33); Bradford (1969:43).

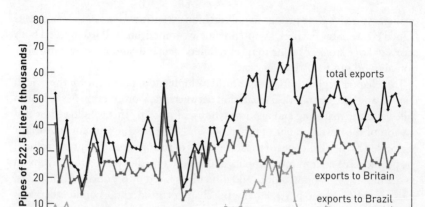

Figure 7.1. Portuguese exports of port wine. Source: Andrade Martins (1990:*Quatro* 66)

grapes and the moment when the brandy is added. Alcohol was added again to the wine before it was sent down river to mature in Porto, and again prior to export, and even in Britain itself. The amount of brandy varied significantly but increased over time as the nature of the wine changed. The *Agricultura das Vinhas*, a handbook on viticulture published in 1720, recommended that 13.5 liters of brandy be added to a pipe of about 525 liters to strengthen and improve the wine's quality. This was insufficient to significantly change the wine's flavor, and most growers appear to have continued to use less than 25 liters before 1820.[5] In the late 1860s A. Girão noted that the better growers used 25– 50 liters of brandy per pipe and another 50 to 76 liters after fermentation.[6] The historian Norman Bennett writes that before 1914 the quantity of brandy ranged from as little as 5 percent to occasionally over 50 percent, depending on the nature of the original wine must, the price and availability of alcohol, and the type of port being produced.[7] The general increase over time reflects the greater skills in wine making and changes in consumer taste. Geoffrey Tait, for example, argues that there was a switch in fashion around the mid-nineteenth century from making dry ports with a complete fermentation to making them stronger and sweeter by arresting fermentation with larger quantities of brandy.[8]

The addition of brandy allowed wines to last considerably longer without deterioration, but it made them hard and crude so that they had to be matured much longer before they could be drunk. The smaller quantities of alcohol added

---

[5] Martins Pereira and Almeida (1999:37), cited in Bennett (2005:40).

[6] Bennett (2005:41).

[7] Ibid., 27, 41. The availability and price of brandy was at times crucial, and on occasion the wine had to be fermented without any being added at all.

[8] Tait (1936:81–82).

TABLE 7.1
Wine as a Share of Total Portuguese Exports, 1840s–1900s

	Port wine	Madeira	Other wine	Total
1840–49	37.7	7.1	6.6	51.4
1850–59	37.7	3.7	10.5	51.9
1860–69	37.1	1.3	6.5	44.9
1870–79	35.5	2.1	7.9	45.5
1880–89	26.6	2.6	22.8	52.0
1890–99	24.4	2.8	16.4	43.6
1900–1909	17.7	2.3	13.5	33.5
1905–14	17.8	1.9	14.7	34.4

Source: Lains (1992), table 3.4.

during most of the eighteenth century resulted in the wine being ready to drink within the year of making, but by the end of the eighteenth and early nineteenth centuries the average period for maturing port according to the historian A. D. Francis "was creeping up to three years and often more."[9] Certainly John Croft in the late 1780s was drinking wine that had been kept four years in the casks followed by a further two years in bottles, while Henderson in 1824 noted that wines were being kept ten or fifteen years.[10]

Grape production and the manufacture of port wine were separated geographically as no other major wine region prior to the railways. The grapes were grown inland, some 100 kilometers up the Douro valley, and the wine was then shipped downriver to the wine lodges of the *exportadores* in Vila Nova, opposite Porto. The British consul described the wine region in 1884:

> The district lies on either side of the River Douro, commences about 60 miles from its mouth at Oporto, and reaches with a varying breadth of from 3 to 15 miles, to the Spanish frontier; that is, for about 27 miles. The soil is a brown, slaty schist, the mountains are lofty and precipitous, and the land is ill-watered; most of the lateral valleys having no more possibility of irrigation than is afforded by a tiny brook, a torrent in winter, and in the height of summer a dry watercourse. Roads, in the true sense of the word, do not exist; and it would be safe to say that no equal portion of the earth's surface, so rich is agricultural wealth, possesses means of communication so extraordinarily bad: chiefly because no plant but the vine would grow in places so rugged and so inaccessible.[11]

The steep valleys and the need for terracing made grape production labor intensive, while the vines produced low yields. At the outset of our period there existed a strict separation of functions between wine production, and the subse-

[9] Francis (1972:260).
[10] Bradford (1969:56).
[11] United Kingdom Consular Reports (1884), no. 79, p. 112. Report by Consul Crawford.

quent maturing, blending, and exporting of the wines in Porto. The shippers benefited from knowledge and connections in the British market, but they often lacked sufficient capital and local connections in Portugal and were inexperienced at appreciating the potential of young wines.[12] In the mid-eighteenth century growers were independent and sold their grapes or wines to local *comissários* (brokers). Around 1800 these brokers acted for a single shipper and reported on the state of the vintage, prepared and monitored contracts with the growers, and arranged for the transfer of the wines downriver to Porto.[13] In this way, exporters were able to overcome their lack of knowledge of viticulture "by exchanging patronage, power, and market access for the judgment and access to social networks possessed by their rural agents."[14] As the century advanced, Porto's merchants were themselves leasing property and controlling their suppliers more closely in the vineyard and winery.[15] Merchants established ties with a few trusted growers, who received loans and supplies of brandy and, in cooperation with the merchant's representative, followed their advice in wine making.[16] Forrester noted in 1852 that the merchants preferred to buy their grapes and make the wine themselves, while by the late nineteenth century some of the best estates, such as Quinta do Zimbro, Quinta da Boa Vista, and Quinta da Roeda, were owned by British merchants.[17] Yet fine wine was only one segment of the trade. John Peter Gassiot, a partner of the export house Martinez, Gassiot & Co., noted in 1862 that by buying from these "Portuguese speculators," "our operations are very simple, and we could carry on a large business with a comparatively moderate amount of capital, because the large trade in this country is not a trade in fine wine, but for good, useful, honest, sound wine—wine that will not turn out bad, but which has no pretensions to be called superlative wine."[18] The role of these houses, as with sherry in Jerez, was absorbed by the exporters over the century, as they integrated backward into wine making and maturing.

Port shippers initially often traded in a wide variety of goods, and between 1727 and 1810 they belonged to the Factory, an institution that acted to protect the collective interests of the British merchants in Portugal. The historian Paul Duguid has shown that the successful export houses in the nineteenth century, in contrast to the previous century, were those that specialized exclusively in port wine.[19] Many of the export houses had partners in London, and these collected orders that were forwarded to Porto, thereby providing a supply of infor-

---

[12] Duguid (2005b:515). For sherry, see chapter 8.

[13] Bennett (1990:226).

[14] Duguid (2005a:463).

[15] Vizetelly (1880); Duguid (2005b:525).

[16] Bennett (1994:254).

[17] United Kingdom. Parliamentary Papers (1852), Forrester, p. 25; *The Times*, October 4, 1899. Offley & Co. took a lease on the Quinta da Boa Vista as early as the 1830s and bought it in 1866 and the Cachucha Quinta in 1889. Croft bought the Quinta da Roeda in 1889. Bennett (1994); Duguid (2005b:525).

[18] Shaw (1864:177).

[19] Duguid (2005b:510–11). Warre & Co., for example, one of the leading eighteenth-century

mation on the shifts in demand for particular types of wine, and a major part of the long-term success of these companies can be explained by their control of the British market. Indeed, leading Porto houses such as Sandeman & Co. and Cockburn & Co. were set up by established wine firms in Britain.[20] Without access to a network of retailers, exporters often found that the only outlets for their wines were auctions, where prices were often unpredictable.[21]

Unlike in Bordeaux, Jerez, or Reims, much of the foreign merchant community did not change their nationality, and intermarriage with the Portuguese was relatively rare. This makes David Ricardo's choice of the trade in wool and wine between Portugal and England, which has since been repeated ad infinitum in students' economic textbooks, a surprising example to use in his explanation of comparative advantage.[22] First, England could hardly be regarded as a wine producer at this time. Second, writers such as David Hume believed that without the preferential duties, relatively little port would have been drunk in Britain, suggesting that the merchants competed on an unfair playing field.[23] Finally, not only did British manufacturers produce the woolen goods to sell in Portugal, but a large number of them also matured the wine from Porto and shipped it to Britain. The gains from trade clearly benefited one side considerably more than the other.

This is certainly what the Portuguese government believed. While it was aware that the British merchants were necessary for exporting the wine (not to mention the political and military support that Britain provided for Portugal in its dealings with Spain and France), it provided a pretext for earlier government intervention in the wine market and to a much greater extent than in either France or Spain. In particular, it created the world's first regional appellation that attempted to bypass British merchants by exporting wine directly to foreign markets.[24]

PRODUCT DEVELOPMENT AND THE DEMANDS OF A MASS MARKET

Wine making and blending were highly skilled tasks, and an expert taster allowed a firm to produce ordinary ports with the same characteristics indefi-

---

houses that traded in a variety of commodities, including wine, saw its annual exports fall from 2,423 pipes in 1792–1801 to 370 pipes in 1832–40.

[20] Ibid., 517.

[21] *Ridley's*, although not an impartial observer, noted that "we have frequently alluded to what we consider the futility on the part of Portuguese Growers at attempting to create a Market for their Wines by consigning them to Public sale on this side." It continued that "shippers knew the tastes of consumers, growers did not" (December 1909, p. 1066).

[22] The theory of comparative advantage appears to have been first proposed by Robert Torres in 1815 (O'Brien 2004:210).

[23] Hume (1752:88).

[24] There were earlier ones in Tuscany in 1716 and Tokay in 1737, but these lacked the regulatory bodies that existed in the Douro (Moreira 1998:31).

nitely.[25] Port wine falls into two main categories: wood ports, which are those matured in casks, and vintage ports, which are matured in bottles. Port matures better in bottles than in wood, as the wine's aroma and color are not lost and a highly desirable crust is deposited. Maturing wines in wooden barrels also involves a financial loss because of evaporation and leakage. Vintage ports consist of a selection of wines from distinct vineyards vatted together after a particularly good year and then kept separate from other wines. Each individual shipper produces distinct wines, has its own opinion as to what constitutes quality, and decides whether to declare a harvest a "vintage," which would affect the reputation of the house.

Port was traditionally sold under the retailer's name, and it was the retailer's reputation and skills that were rewarded with greater sales and profits. The creation of vintage port changed this as it allowed the shippers in Porto, rather than the retailer, to brand the wine. Prior to phylloxera the wine was kept for three years in wood before it was bottled, but it was not drunk for at least another three years, and often as much as fifteen or twenty years or more.[26] Wines were shipped in casks to Britain to be bottled and then left in the retailer's or the customer's cellar—"well-to-do Farmers" purchased a pipe or two whenever there was a good vintage—thereby pushing the significant costs of storage and ageing "down the supply chain."[27] The local wine merchant was often expected to advance some credit for the purchase and take responsibility for bottling the wine in the consumer's cellar. However, by the late nineteenth century a combination of agrarian recession and the lack of town cellars in the growing urban sector increased the demand for bottled vintage wines ready for immediate consumption, rather than wines to lay down for several decades.[28] Berry Brothers in 1909, for example, advertised not only a number of old house wines, but bottled vintage ports under the shippers' brands for 1881, 1884, 1887, 1890, 1896, and 1900. The most recent vintage wine that was available was that of 1904, having been bottled in 1906 and produced by five different Porto shippers (Croft, Taylor, Martinez, Warre, and Fonseca).[29] Retail merchants required not only considerable capital when dealing with vintage ports, but also the necessary skill to bottle it, although a carefully selected bottled port reputably afforded "their holders far less anxiety as to their sale than do many other Wines which their cellars

[25] Tait (1936:61–63). Tait argued that a "good taster can easily make a firm, just as a poor one can mar it."

[26] Ridley's, September 1894, p. 530, suggested three years at least. The time depended on the quality of the vintage. Simon (1934b:87) noted, for example, that wines from 1896 were "now at the top of their form."

[27] Ridley's, November 1905, p. 130. Duguid (2003:437). Exports to Britain in bottles were described as 'infinitesimal." United Kingdom Consular Reports (1893), no. 304, p. 7. An agreement by the Port Wine Shippers in 1907 required that all vintage port be shipped within seven years of the date of the quoted vintage (Ridley's, February 1907, p.100).

[28] Ibid., September 1904, p. 639.

[29] See appendix 2.A.5.

contain."[30] Few merchants had the capital to carry large stocks of bottled vintage port, and *Ridley's* believed in 1894 that there were "no more" than eight Houses with more than "20,000 dozens."[31]

The greatest volume of port in Britain, however, comprised wood ports, which arrived for immediate consumption, being sold straight from the barrel in public houses, or bottled. Tawny port differed from vintage port in that it was a blend of wines from a number of years and was matured in the wood for several years. Over time it lost some of its color, hence its name, and was bottled immediately prior to sale. Because of the time it took to make, tawny port was also expensive, but a cheaper version could be achieved by simply mixing red wines with a portion of white.[32]

Other types included crusted ports which, like vintage ones, were of a particular year but of an inferior quality, and ruby ports, a blended wine that was sold before it had matured into a tawny.[33] Classification was complicated and changed over time, as noted by the English writer Henry Vizetelly when he visited Porto in about 1880:

> One great disadvantage under which shippers of Port wine labour is the frequent change of fashion with regard to the style of wine demanded in England. So constant are the changes and so endless the varieties now-a-days that it has been said there are almost as many styles of Port wine as shades of ribbon in a haberdasher's shop. At one time a deep-coloured, heavy wine will be in vogue; at another a wine paler in colour and lighter in body, but rich in favour. Sometimes dry wines are in request, and latterly the fashion has set in for thin wines of a light tawny tint, the result of their resting for many years in the wood—the kind of wines, in fact, which the Oporto shippers invariably drink themselves.[34]

As *Ridley's* noted, "a Wine that takes twenty years in bottle before it is fit for consumption, is scarcely a commercial article," and attempts were made to shorten the maturing process.[35] Supply-side changes helped, as the postphylloxera wines matured quicker, with two instead of three years in the wood becoming the norm and the time required in the bottle shortening, with consequent

---

[30] *Ridley's*, September 1894, p. 530.

[31] Ibid. In total *Ridleys* suggest that there were "upwards of 750,000 dozens of Wine of this description in London alone," but that the "industry is divided amongst a great many hands, most of whom, however, are what are known in the Trade as 'old-fashioned' Merchants."

[32] Simon (1934b:9–10); *Ridley's*, September 1894, pp. 531–32; January 1900, p. 53.

[33] Simon (1934b:10) writes that "Ruby Ports are to Tawny Ports what Crusted Ports are to Vintage Ports: first cousins, but not poor relations."

[34] Vizetelly (1880:145).

[35] *Ridley's*, September 12, 1894, p. 531. The report continues:

> a Wine which, with three years or so in bottle, will show age and bottle character, and at the same time leave a profit if sold about the limit of 30s, is within the reach of all. During the last few years, many attempts have been made to displace bottled Wines with colour, by tawny Wines from the wood, many of which, with age have especial merit.

savings for merchants. As a result, wines from the 1880s and 1890s were reported as to have "come to hand so rapidly" compared with those of the 1860s and 1870s, and "in many cases vintage Wines of the last fifteen years have been able to do duty for what previously required blending for early use."[36]

The arrival of phylloxera created shortages of cheap wines in the Douro district, a problem solved in the short term by using those from the southern Portuguese provinces. These wines were traditionally used to make the brandy that was added to port, but they were now exported as port wine itself, which *Ridley's* feared risked undermining consumer confidence for fine wines, such as happened with sherry in a classic "lemons" scenario.[37]

By creating a reputation for fine vintage wines, the leading port shippers were less willing to try to compete in the cheap section of the market where in any case they had no comparative advantage.[38] Instead it was the British importers and retailers who looked elsewhere for cheap generic ports, such as Spanish red from Tarragona in southeastern Spain, that could be sold in London for about half the price of young ports of the Douro (figure 4.6). Growers and merchants in this coastal region of Catalonia had for centuries adapted their products to changing demand, whether distilling spirits for the Dutch market, fortified wines for export to the colonial market, or cheap wines for Barcelona.[39] British demand, a cheap peseta, and the arrival of phylloxera in Portugal offered growers not only the possibility of exporting red wines to be blended with those from Porto, but also a market for "Tarragona Red," sold in bottles that became popularly known as the "Big Bob Bottle" (figs. 7.2, 7.3). Prior to the mid-1880s, exports were minimal, but they then grew rapidly, so that in 1896 more Spanish red was imported into Britain than sherry and Spanish white.

The supply of cheap "port wines" from Tarragona could only be met after a number of technical problems in wine production were resolved. In particular,

---

[36] Ibid., 530. The report continues, noting that "without in any way decrying the value of old tawny Wines, we believe that their merit is for Lodge purposes, and not for direct consumption, and we should be sorry indeed to see the day when lightness took the place of colour in popular demand" (532).

[37] However, although the leading shippers might be *sans reproche*,

> once let the idea get abroad that the quality of Port Wine is going down, and past experience tells us what will be the result. Those whose vinous education is sufficient to give power of discrimination will of course not be effected—knowing that the Shipping Houses in whom they have been accustomed to trust may still be relied upon. But with that class of consumer who knows Port only as Port, and to whom such names as Sandeman, Martinez, Cockburn, Silva and the rest, convey no meaning, the result in the end is inevitable.

Ibid., March 12, 1888, pp. 133–34.

[38] During the phylloxera epidemic, however, *Ridley's* (January 1882, pp. 6–7) noted that "too much" cheap wine from southern Portugal was exported as port, and "one of the best known and oldest established firms at Oporto" sold "as a specimen of 'Port Wine' a lot that was described 'as bad as anything ever so misdescribed.'"

[39] Vilar (1962).

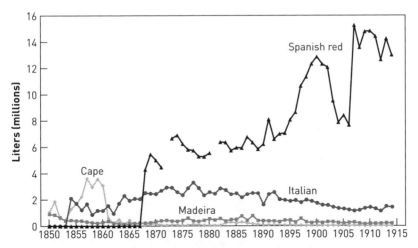

Figure 7.2. UK imports of generic port wines. Source: Wilson (1940:362–63)

better fermentation methods were required to rid the wine of the earthy flavor and coarseness that so often characterized it.[40] One local development was the artificial maturing of bottled wines, called insolation. This was a modification of Pasteur's system, with direct sunlight being used rather than artificial heat, in a similar fashion to what had been used in Madeira a century earlier. Insolation not only helped prolong the life of the wine but also accelerated the ageing process cheaply. According to the type required, the wine was left in clear, white, glass bottles and exposed for between ten days and a month. The color changed from a "purple tinge" to pass "successively through every shade of ruby, from ruby to *rancio*, and from *rancio* to light tawny if left long enough."[41]

Yet the British demand for Tarragona wines was brief, as exporters failed to maintain the quality of their product in the face of rising prices caused by the devastation of local harvests by phylloxera and the rapid appreciation of the peseta after the conclusion of the Cuban war. Merchants in Tarragona looked for cheaper, and inevitably inferior, Spanish wines to remain competitive. The end result was not unexpected, as "when genuine Tarragona was offered to the Public, it was quick to appreciate it, and the consumption rapidly increased. But when flagons of rubbish, bearing the name Tarragona, were substituted, the discriminating reverted to some more trustworthy beverage."[42] Exports of Spanish

[40] *Ridley's*, January 1897, p. 9; Roos (1900:208–9).
[41] *Ridley's*, January 1900, pp. 114–15.
[42] Ibid., August 1901, p. 556). Five years later *Ridley's* noted that

the consumption of Tarragona Wine in the British markets have considerably declined since Importers decided to neglect the genuine product for a blend which was cheaper. Not only have the distribution of the "Big Bob Bottle" brought an honest Wine into disrepute and crippled a lucrative brand of their trade, but their example has engendered the production of a still baser

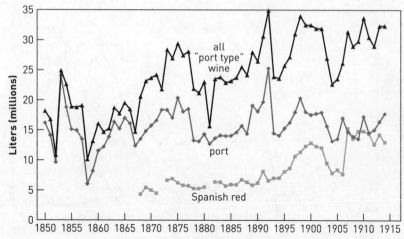

Figure 7.3. UK imports of port and Spanish red. Sources: Wilson (1940:362–63) and *Ridley's* (various years). All "port type" wine includes port, Spanish red, cape, madeira, and Italian.

red remained strong, especially in the years up to the First World War, but a significant proportion was described as "Basura de Batalla," namely, rubbish wines produced in Valencia, Alicante, and La Mancha, rather than the "choicest products of the Priorato."[43]

RENT SEEKING, FRAUD, AND REGIONAL APPELLATIONS

The creation of the Companhia Geral da Agricultura das Vinhas do Alto Douro in 1756 to control the port wine trade was just one of a number of wide-ranging initiatives by the future Marquês de Pombal to regulate the Portuguese economy. The background to the Companhia was a steep drop in exports and low prices that followed an exchange of correspondence between the British Factory in Porto and the Douro brokers, who mutually blamed each other for adulterating the wine and thereby causing a decline in its reputation.[44] This crisis not only affected the Douro growers and the Porto merchant community but also led to a fall in tax revenue for the government and the worsening of the country's balance of payments. The Companhia was given the power to regulate the production; sales, prices, transportation, storage, and exports of wines and brandy from the Douro region; the distilling and sale of brandy in northern Portugal; and a

---

type. The cheapeners filled their bottles with Wine which was certainly not legitimate Tarragona, and their pupils and followers are now selling a composition which is not even legitimate wine (November 1906, p. 928).

[43] Ibid., June 1900, p. 395.
[44] Tait (1936:74); Francis (1972:195, 202–5); and Shaw (1998:142–43).

monopoly on selling wines in public houses in Porto and the surrounding districts, as well as the important export trade to Portugal's Brazilian colony.[45] The area that supplied grapes for port was strictly defined, and wine was divided into four categories: the *legal* wine, the only kind that could be officially exported to Britain; *separado*, which was set aside for export to the rest of Europe and Brazil; *ramo*, a wine that was reserved for local consumption; and *refugado*, which was used for distillation. According to its statutes, its objectives were "to uphold with the reputation of the wines the culture of the vineyards, and to foster at the same time the trade in the former, establishing a regular price for the advantage alike of those who produce and who trade in them, avoiding on the one hand those high prices which, rendering sales impossible, ruin the stocks, and on the other such low prices as prevent the growers from expending the necessary sum on the cultivation of their vineyards."[46] It was the Companhia tasters who determined what wine could be exported, and the British merchants claimed that the regulations increased costs but not the quality of the wine.

The Companhia was more than just an attempt by a mercantilist state to regulate the country's leading economic sector, as it was also a joint-stock company whose merchants had important trading privileges associated with the buying and selling of wines in the Douro valley. As such, the original Companhia reflected the interest of a predatory state, as well as acting as a private rent-seeking monopoly and a regional appellation. The Companhia was heavily criticized from the first moment in the British press, which repeated the complaints of its merchants in arguing that it was an institution to create rents and enrich the Portuguese growers and local merchants at their expense.[47] However, while the Companhia was not a disinterested regulator, the British merchant community, despite their considerable and consistent protests, perhaps did not suffer unduly.[48] The Companhia controlled some of the worst abuses in wine production, and the period between 1781 and 1807 has been described as a "golden age" for port.[49] The Companhia helped reinforce a market that was already segmented, so that by around 1840 British companies accounted for about two-thirds of exports to their own domestic market, but only one among the leading ten firms that traded with Brazil, and vineyard ownership by British firms was still negligible.[50] The appearance of new exporting houses after the Napoleonic Wars (Sandeman, Martinez, Cockburn, and Grahams) might suggest a changing of the guard took place, as some old, established houses declined, but the port trade in

---

[45] Moreira (1998, chap. 2); Duguid and Lopes (2001:4); and Tait (1936:77).

[46] Cited in Vizetelly (1880:105).

[47] The protests by local tavern owners and growers excluded from the appellation were brutally suppressed in 1757 (Barros Cardoso 1996).

[48] Duguid and Lopes (1999:88).

[49] Croft (1788); Bennett (1990).

[50] Duguid and Lopes (2001:14, 15, 19).

general enjoyed relative stability.[51] Likewise, it is probably more correct to see the Companhia not as an attempt by the Portuguese state to protect its citizens against the rapacity of foreign exporters, but rather as a monopoly created by a weak fiscal state that created sinecures for a small domestic elite in exchange for revenues, making it hated as much by other Portuguese citizens as it was by the British merchants.[52] Finally, although more of the profits remained in Portuguese hands than would otherwise have been the case, abuses were limited by the fact that port wine was traded internationally and had to compete with other alcoholic beverages in the British market, and by the fact that the Portuguese government depended on the sector for foreign exchange and taxes.

The Companhia's privileges were abolished in 1834 following the Liberal victory in the 1828–34 civil war but were reestablished once more with a new charter for twenty years in 1838. Its powers were now reduced and wine could be freely sold. Although the Companhia still determined what could be exported, it was also expected to promote wines from outside the region and provide credit to growers. The Companhia continued to act as a trading company after its other privileges were abolished. Between 1852 and 1865 the Comissão Reguladora da Agricultura e Comércio das Vinhas do Alto Douro was created, composed of elected members from the growers and Portuguese merchants. Its powers were limited and proved short-lived, but it represented more closely the interests of the sector, rather than just providing sinecures.[53]

A new regulatory institution was proposed in the late 1880s, the Real Companhia Vinícola do Norte de Portugal, which one critical trade journal described as being "a body of eight Portuguese gentlemen," whose "interest in the Port Wine Trade is by no means extensive," and the Portuguese government.[54] The Companhia Vinícola was to receive an annual subsidy equivalent to £3,385 over five years and give a guaranteed 6 percent interest per annum for thirty years to the shareholders, who would provide capital of £450,000. The company was to sell its wines with a certificate of origin and promote them in foreign markets. Once again the British shippers in Porto felt threatened, claiming that as the new company had no stock of its own, it would sell poor-quality young wines under the "Government Arms" and generally discredit the trade. To widen their negotiating base, the British shippers collected the signatures of 2,500 Douro farmers and attracted the backing of the influential periodical *O Commerico do Porto*.[55] Between January 21 and February 3, all but two of the wine lodges closed their doors in protest, and when the unemployed workers and military came into "collision," as *Ridley's* reported, the shippers wrote to the foreign secretary,

[51] Duguid (2005b).
[52] Bennett (2005:30–31).
[53] This paragraph is based on Moreira (1998, chap. 2).
[54] *Ridley's*, February 1889, p. 91.
[55] Ibid., March 1889, pp. 148–49. See Cockburn (1945:25).

TABLE 7.2
Demand and Supply for Douro Wine under "Free-Market" Conditions, 1865–1907

Sources of demand	Sources of supply
Export markets:	Douro valley:
Vintage port—United Kingdom	Demarcation
Ordinary port and other wines	Outside demarcation
United Kingdom	Other sources
Brazil	Central and southern Portugal
France	Foreign imports of brandy and alcohol
Local markets	Industrial alcohol
"Consumo" wine	
Brandy	

Lord Salisbury, to demand protection. The Portuguese government for its part reportedly censured all telegraphs leaving Porto to stop news of the events from reaching British newspapers. All the lodges locked out their workers again for eight weeks in May and June with the exception of three, whose premises had to be protected by the municipal guard. The standoff ended only when the government guaranteed that the new company's mark would carry no implication of origin or quality.[56]

While *Ridley's* accounts are clearly biased in favor of the British shippers and against this group of "eight Portuguese gentlemen," the reality was that the divisions ran much deeper than simply Anglo-Portuguese rivalry, as the wine industry was suffering from major structural problems. The ending of domestic restrictions in 1865 created what was effectively a free market for wine and brandy in the Douro region, and at the same time Portugal's new rail network linked the region to wine and brandy producers in other regions.[57] Table 7.2 shows in a schematic form the supply and demand for wines under "free trade" conditions that were operating between 1865 and 1907. In addition to British demand for port wine, France and Brazil imported large quantities of poorer quality ones, and cheap wines (*consumos*) were drunk locally and others used for making brandy to fortify port. Figure 7.1 shows port exports increasing fourfold from the low point in the late 1850s to the mid-1880s. Not only did the local supply of wine in the Douro fail to keep up with this growth, but the destruction caused by phylloxera led to it falling by at least a third between the 1860s and mid-1880s.[58]

[56] *Ridley's*, May 1889, pp. 255–56; June 1889, p. 309; July 1889, p. 352. The lodges of Martinez Gassiot and the two Sandeman firms remained open.

[57] Bennett (2005:90).

[58] Andrade Martins (1991:656).

Phylloxera first appeared in 1867 in Sabrosa (Douro), and Sandeman's agent noted in 1886 that "good genuine port" was very scarce as the greater part of the firm's best Douro vineyards had been destroyed.[59] The high price paid for ordinary wines for local consumption encouraged proprietors in southern Portugal, whose estates had been devastated earlier, to replant with American vines, which, "with a prolific soil and a ready market for the produce," led to yields "being sensibly increased." Industrial alcohol was used to fortify the cheaper wines.[60] However, the supply of wine in the Upper Douro was also, "somewhat elastic," as traditionally only a small area of the delimited district was planted with vines, and only a proportion of the wine—less than half—made into port, the rest being used for making brandy or for local consumption.[61] High prices now led to these "other wines," to be "brought into a state of perfection," helping growers to make up the deficit.[62] Fine wine remained in short supply, but as early as 1887 there was a recovery in production of the cheaper ones in the Upper Douro, and the following year the harvest was 25 percent greater.

It was against this background of better harvests that the growers in 1887 formed a new pressure group, the Liga dos lavredores do Douro, and lobbied their government to demand that the "Port" brand be limited exclusively to wines produced in the Douro, and exclude the use of outside wines to make port wine and the brandy used in their production. The shippers feared that their ability to export cheap wines that had been produced outside the Douro, which allowed them to sell competitively outside the British market, would now be ended. They argued that only they were responsible for quality, and that creating "artificial" boundaries to restrict supply would not only exclude them from using some fine wines from outside the delimited area, but encourage growers to plant high-yielding grapes on unsuitable land within it. In addition, they demanded inexpensive brandy to compete with cheap generic ports in the British market, which could not be obtained in the Douro itself.

On February 5, 1906, a committee appointed by the government, representing growers and shippers, met to consider the problem. A proposed new law created a new delimitation, but this increased the Douro district from 40,000 to 600,000 hectares. It was almost immediately reduced to about 200,000 hectares, "to something like that given by Baron Forrester's Map."[63] To placate southern growers, brandy for fortifying wines could not be produced in the Douro district.[64] Only wines from this Douro appellation could now be sold as port in Portugal, but this did not protect the collective brand in foreign markets. As one

[59] Cited in Bennett (1994:273).
[60] Ridley's, December 1894, p. 698.
[61] Tait (1936:16).
[62] Ridley's, March 1889, p. 149.
[63] Ibid., January 1909, p. 17; Tait (1936:84–85).
[64] Ibid., June 1907, p. 434.

"experienced" shipper noted, if southern wines were not allowed to come north, they would be shipped directly from Lisbon, as they had been in the eighteenth century.[65]

The Anglo-Portuguese Commercial Treaty of 1914, which allowed Britain to trade on most-favored-nation terms in exchange for restricting the use of the names "port" and "madeira" to all Portuguese wines, threatened to open the floodgates of large quantities of Lisbon wines being exported as port. The Douro growers rose in rebellion and troops appeared on the streets once more, leaving fourteen laborers killed and another twenty injured in Lamego.[66] When the treaty was finally signed in 1916, the controversial clause 6 was amended to make it illegal to sell in Britain any port or madeira that had not come from Portugal and Madeira respectively, and these wines had also to be accompanied by a certificate issued by the "Competent Portuguese Authorities." Consequently only wine produced in the designated area of the Douro could use the name 'port' in both the Portuguese and British markets. This gave the Douro growers a greater measure of protection than was achieved in any wine-producing nation and led André Simon to claim that "this Anglo-Portuguese Treaty is quite as important, if not more, as the famous Methuen Treaty of 1703."[67]

Both port and sherry were traditionally produced for the British market, and despite attempts at diversification, the demands of this market were the key to regional prosperity until late in the twentieth century. Fine wines were expensive to produce, and shippers naturally wanted to participate also in the growing British market for cheap wines after 1860. In theory the demand for cheap wines could be met domestically, with the wines exported by the traditional houses in Porto and Jerez, or they could be imported from other wine-growing regions by merchants based in Britain. The port shippers found it harder than those in Jerez to outsource cheap substitutes themselves, both because there were relatively few regions in Portugal that were able to provide a sufficient supply of wines at a price that might have been acceptable to the British consumer and because of local grower opposition. As a result, it was the British retailers who made "port" cheaper by blending it with generic wines from places such as Tarragona, while the sherry houses themselves competed strongly in the British market with cheap substitutes. Product innovation in Porto involved the creation of vintage port, which gave shippers a major incentive to invest in private brands. While Cockburn, Martinez, or Sandeman might not have been household names, they had a reputation as suppliers of fine vintage ports and consequently were much less interested in selling cheap wines that might damage their brand.

[65] Tait (1936:85).
[66] *Ridley's*, August 1915, p. 530.
[67] Simon (1920:250).

A major peculiarity of the port commodity chain was the nationality of the shippers, which gave the local growers greater influence over the industry than might otherwise have been expected. This resulted in a number of attempts to regulate trade by limiting the freedom of foreign shippers through intervention and regional appellations. Initially, this was simply the concession of a monopoly by the government to a small number of favored individuals in exchange for tax revenue. However, by the late nineteenth century regional appellations were seen as a means of creating a wider distribution of wealth and benefiting the numerous Portuguese growers. By contrast, the British shippers found it much easier to engage with their own government to obtain privileges under the guise of the Anglo-Portuguese alliance or turn to the Foreign Office for help in their relations with the Portuguese government, such as in 1889.

CHAPTER 8

From Sherry to Spanish White

> Sherry is a foreign wine, made and drunk by foreigners; nor do the
> generality of Spaniards like its strong flavour, and still less its high
> price, although some now affect its use, because its great vogue in
> England, it argues civilisation to adopt it.
>
> —Richard Ford, 1846/1970:177

SHERRY IS A NAME GIVEN to a wide variety of wine types produced in the
geographical region around the Bay of Cadiz, including Jerez de la Frontera,
Sanlúcar de Barrameda, and El Puerto de Santa María, although for brevity the
region is referred to here as Jerez.[1] Exports to the British market grew rapidly
from the late 1820s to a peak in 1873 at about 30 million liters and accounted
for 37 percent of the wine drunk in that market. Sales then declined even more
rapidly and were no greater in the early 1890s than they had been seventy years
earlier. This chapter shows the nature and limits of organizational change in
the production and sale of sherry over the century. Despite an apparent flexibility in responding to increased demand in international markets, the rapid
drop in sales was caused by the decline in the reputation of sherry, as merchants
in Jerez and especially Britain sold adulterated and cheap imitation wines. Although there was much talk about protecting the name of sherry in Jerez, this
proved difficult because of the diversity of interests within the producing region itself. The big export houses responded to weaker demand for their fine
sherries by moving down-market to achieve volume, and they were better able
to weather the economic difficulties and appearance of phylloxera than either
the growers or the traditional winemakers. While the political influence of
small growers in France allowed them to capture market power from the merchants by establishing regional appellations and cooperatives, this did not happen in Jerez. Instead the large shippers consolidated their economic power and
maintained the right to export whatever wines they wished as "sherry," arguing
that as the British government was unwilling to recognize sherry as being exclusive to Jerez, they had to compete with producers of "sherries" from countries such as South Africa.

---

[1] Other local wines from Chiclana, Chipiona, Puerto Real, Rota, and Trebujera were also
sometimes sold as sherry, as were those from el Condado (Huelva), Lebrija and the Aljarafe
(Sevilla), and Montilla (Córdoba).

TABLE 8.1
Trade in Sherry and Spanish White Wine

	Exports of sherry from Jerez and Puerto Santa María* (millions of liters)	Imports of Spanish white into the United Kingdom	
		Total imports (millions of liters)	Spanish white (as % of total imports)
1821–25	5.75 (1821–82)	6.67	20.5
1826–30	10.15	10.18	28.2
1831–35	13.20	11.15	31.1
1836–40	15.92	13.36	32.5
1841–45	15.04	11.83	33.6
1846–50	17.82	12.54	34.1
1851–55	22.54	14.72	34.7
1856–60	23.26	15.34	36.5
1861–65	28.62	23.21	38.2
1866–70	33.71	27.66	30.9
1871–75	39.53	29.69	33.9
1876–80	27.35	21.62	26.9
1881–85	25.09	15.19	21.6
1886–90	24.17	12.15	17.4
1891–95	17.27	9.88	13.8

Sources: Exports—Maldonado Rosso (1999:312), and Montañés (2000:253–54). Imports—Wilson (1940) and *Ridley's*, various years. Prior to 1868 it has been assumed that white wines were equivalent to 84 percent of total Spanish imports.

*Exports to all countries, not just Britain

The Organization of Wine Production in Jerez

Richard Ford perhaps exaggerated the contribution of foreigners to sherry production but not the importance of the British market, which accounted for 87 percent of all exports at the height of the region's prosperity in the early 1860s.[2] A study in the 1880s estimated that more than half the region's wines were exported, and this would include virtually all the fine sherry.[3] There are few figures for production for the nineteenth century, but movements in exports (table 8.1) and prices (fig. 8.1) show the general trends. Exports to all markets rose from an annual average of less than 10 million liters in the 1820s to almost 40 million in the early 1870s, and the price of must in Jerez tripled between 1850–53 and

[2] For the role of foreigners, see especially Maldonado Rosso (1999:264–69) and Fernández-Pérez (1999:77–79; also *Ridley's*, various years.

[3] In the mid-1880s the province of Cadiz produced 450,000 hectoliters, of which Jerez contributed 175,000, El Puerto de Santa María 120,000, Sanlúcar 65,000, and Chiclana 62,000. Local consumption was between 180,000 and 200,000 hectoliters (Archivo Ministerio de Agricultura, Madrid, *legajo* 82.2).

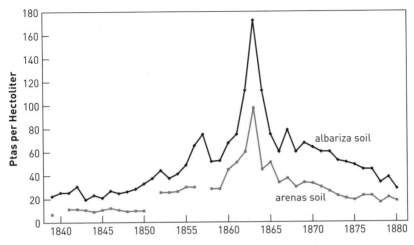

Figure 8.1. Price of must in Jerez, 1840–1880. Source: Gonzalez y Álvarez (1878:39–55)

1860–63. In 1863 Thomas G. Shaw calculated the value of sherry imports to Britain at ten million pounds,[4] and Spanish trade statistics suggest that sherry accounted for an eighth of all exports in the late 1860s.[5] Prices from the late 1860s then declined rapidly, and sherry began to lose market share in Britain after 1873, especially to French wines.

Although both port and sherry producers depended heavily on the British market, there were important differences in the organizational structure and distribution of market power along the two commodity chains. The Jerez wine trade was much older, with sack being popular in Britain from in the sixteenth century if not before, although by the second half of the eighteenth century wine imports were small.[6] At this time the sherry trade was limited to two types of young wines: those shipped while still on their lees in October or November, and those shipped after they had been racked in March and April. In both cases the wines blended with older and stronger wines or strengthened with wine spirit.[7] Institutional arrangements reflect this trade in young wines. The Gremio

---

[4] Quantities shipped were 66,321 butts (of 500 hectoliters each), and the price in Jerez was 40–250 pounds per butt (Shaw 1864:235).

[5] Sherry represented a fifth of the country's total exports in terms of value in 1850–54 (almost three times more than ordinary table wines), while sherry and other fortified wines (*vinos generosos*), accounted for half of all wine exports in terms of volume (Prados de la Escosura 1982, table 7; Dirección General de Aduanas, various years).

[6] The early history of sherry "sack" is covered in Jeffs (2004, chap. 3).

[7] Referred to in Jerez as *mosto* and *vino en claro*, respectively (Maldonado Rosso 1999:48–54). This was done by one of three different ways: *arropado* (or *abrigo*), the addition of concentrated grape syrup for making sweet wine; *cabeceado*, using older wines of higher strength; and *encabezado*, adding spirits to strengthen white wines.

de Vinatería de Jerez fixed annual prices for grapes and must, allowed growers with over 10 *aranzadas* (4.5 hectares) of vines the possibility to establish retail premises for the local sale of their wines, and restricted the sale of wines produced outside the region.[8] Local growers could not be merchants, and the storing of wines by merchants was prohibited. The growth in British demand (and perhaps the development of the solera, a new wine-producing technique), encouraged John Haurie and a group of growers and exporters to successfully challenge several of the guild's privileges and led to important organizational changes in the industry from the last third of the eighteenth century.[9] Large city bodegas (cellars) were built to store and mature wines. By the mid-nineteenth century there were three very different groups involved in the sherry trade: the *cosecheros* (responsible for growing and crushing the grapes); the *almacenistas* (for the fermenting and maturing of wines); and the *extractores* (for exporting). Vertical integration of the different activities into a single enterprise remained limited.[10] While brokers (*corredores* or *encomenderos*) in the eighteenth century linked sellers in Jerez with foreign buyers, it was the sherry firms themselves who established exclusive contracts with agents in the British market in the nineteenth century. These agents acted as wholesale merchants, selling to local retailers who often sold sherry under their own names (brands). As the distance between agent and exporter was considerable—a letter between London and Jerez took sixteen to twenty days in the 1830s[11]—agency agreements were carefully crafted and sometimes led to formal partnerships or even intermarriage. Robert Blake Byass, the London agent of the company Gonzalez, Dubosc & Co., became a partner in 1855, and the company was changed to Gonzalez Byass some years later. A granddaughter of the firm's founder, Manolo González, married one of Alfred Gilbey's sons in the 1870s, and two of Gilbey's granddaughters would marry González boys in the 1920s.[12]

Grapes were grown on three major soil types. The best was the *albarizas*, a brilliant white, chalky soil, which had an excellent capacity for storing the winter rains. According to one nineteenth-century expert, Parada y Barreto, the wines produced on this soil were "fine, clean and strong, but of scarce production."[13]

---

[8] Ibid., 60. Other guilds were established in Sanlúcar and El Puerto de Santa María.

[9] Haurie's legal victory was accompanied by the creation of a more liberal wine market by the *real orden* of 1776. Ibid., 133–39.

[10] One major exception was Domecq, who in 1840 owned 460 *aranzadas* of vines in the Marcharnudo region. Visitors such as Busby, Ford, and Vizetelly have left accounts of Domecq's business. In his study of thirty probate inventories of bodega owners for the period 1793–1850, Maldonado Rosso (1999:222–27) finds that twenty-one owned some vines, but only seven of these had more than 10 percent of their assets invested in vineyards.

[11] Ibid., 333.

[12] Faith (1983:45); Fernández-Pérez (1999:72–87); and Maldonado Rosso (1999:288–96).

[13] Parada y Barreto (1868:71) estimates between one and half and two botas per *aranzada*. 5,000 hectares and a yield of 22 hectoliters per hectare, implies a maximum of 110,000 hectoliters, a figure that needs to be reduced by a sixth because of losses occurring during the wine-making

On the *barros* or clay soils, the yields were twice as much, but quality was less and labor inputs were greater because weeds grew in profusion. Finally, the *arenas* or sandy soils had good yields and were easy to cultivate but produced the poorest wines.[14] A detailed survey at the beginning of the nineteenth century lists forty-three different varieties of grapes to be found in the Sanlúcar–Jerez region, but even at this early date a handful of varieties predominated. The *palomino* or *listan*, a white grape, was found on about 50 percent of the albarizas soil, and the *perruno* on another 20 percent, and sixty years later Parada y Barreto noted an even greater concentration of the palomino.[15] By the 1840s there were perhaps 6,000 hectares of vines in Jerez, and this would increase by at least a third over the next quarter century before stabilizing at about 8,500 hectares.[16] This growth in the area of vines was much less than the increase in wine exports. Part of the difference was made up by drawing on stocks of old wines or employing more labor in the vineyard to increase yields, but in particular there was a significant increase in the use of wines produced outside the Jerez region, either for blending with local wines or for exporting directly as "sherry."[17]

As in other wine regions, there were many small plots of vines, and on the arenas and barros soils six hundred growers had an average of 3 hectares (6.8 *aranzadas*) each in the mid-1860s. Larger holdings that required wage labor were more common on the better albarizas soil, and as early as 1840 some 85 percent of vines were found on holdings of between 4.5 and 45 hectares, with four holdings even larger. At the other extreme 40 percent of the growers (171) owned less than 4.5 hectares, each worked predominantly by family labor, although these accounted for only 11 percent of the albarizas vines (table 8.2).

In the first half of the nineteenth century the grapes were crushed and fermented on the farm, but by the second half burned sulfur was used to impregnate the grapes so they could be brought to Jerez for crushing.[18] The wine was usually racked off the lees between January and March and fortified with 12–36 liters of alcohol per butt (equivalent to 2–6 percent of the volume).[19] Tales of travelers drinking perfectly good unfortified wines from the region were no doubt sometimes true, but, as with all wines, production conditions could vary significantly from one year to the next. Sherry requires a minimum strength of

---

process, maturing, and transportation. Even if all this wine was exported, it would have accounted for only a quarter of total sherry exports in the early 1870s.

[14] Parada y Barreto (1868:59, 79).

[15] Ibid., 114, 116.

[16] López Estudillo (1992:50–53) provides a detailed discussion on the accuracy of contemporary estimates of the area of vines.

[17] Simpson (1985b:174–84) considers the changes in grape varieties, increased labor inputs, and improved vineyard techniques on yields.

[18] Jeffs (2004:178).

[19] The bodega barrels were about 600 liters, compared with the 500 liters for those used in shipping. One-sixth of the wine was lost as ullage. Old sherries especially can reach naturally high levels of alcohol, if only because about 5 percent of the wine is lost each year through evaporation.

TABLE 8.2
Distribution of Holdings in Jerez de la Frontera

	Hectares	Number of owners	Average size of vineyard (hectares)
Albariza soils			
1840	4,488	442	10.2
1864	5,207	421	12.4
1877	5,106	311	16.4
Barro and arenas soils			
1840	1,660		
1864	1,800	594	3.0
1877	2,113	639	3.3

Sources: 1840—United Kingdom. Parliamentary Papers (1852), Gorman, pp. 710–38; 1864—Revista Vinícola Jerezana, no. 12, June 25, 1866; 1877—Archivo Municipal. Jerez de la Frontera, Archivo Memoranda, no. 6, folleto 9.

Note: The 1864 and 1877 figures probably underestimate the total area of vines.

15.5 percent alcohol for its proper development, and this strength was not achieved if weather was unfavorable at the time of the vintage or if there was an incomplete fermentation.[20] After racking, the wine was kept for several years in contact with the air. The high alcohol content and the special conditions found in the sherry bodegas produces a film made up of microorganisms living on the surface of the wine (the *flor*), which protects it from acetification and allows the wine to develop its distinctive flavor.[21] Adding alcohol, often in greater quantities than that noted above, also allowed shippers to export younger wines, which helped undermine the product's reputation.

The exact nature of the trade in old sherries is difficult to establish. Price lists indicate that sherry was sold on occasion by the vintage and estate in the first half of the nineteenth century, but this became increasingly rare as the century progressed.[22] In fact, even the so-called vintage sherries probably contained wines produced from different years. The inability of winemakers to control sufficiently the chemical changes that took place during fermentation using traditional, unscientific methods led to an "enormous variety" of sherries being produced.[23] Chance could result in two butts producing very different types of wine

[20] González Gordon (1972:143).

[21] Ibid., chap. 5; Jeffs (2004, chap. 10).

[22] Jeffs (2004:214–15); Montañés (2000:263). Appendix 2.A.5 shows that "old" sherry was still common, but the "Old Golden, Very Fine 1847" (Gonzalez Byass?) probably refers to a solera started in that year, rather than a vintage wine.

[23] González Gordon (1972:105–6). This writer notes that "this was undoubtedly caused not only by the yeasts in the must itself but also by microorganisms existing in the vessels used."

despite having come from the same vineyard, been pressed together, and subsequently been stored side by side. It also implied significant annual variations: a butt of must laid down each year on one of William and Humbert's vineyards between 1933 and 1936 produced the following wines: an *amontillado*, a *fino viejo*, a *palo cortado*, and an *oloroso*.[24] Consequently, unblended sherry could not be finally characterized until it was three or four years old, when it was divided into two main groups: *finos* (including *finos, manzanilla*, and *amontillados*) and *olorosos* (including *palos cortados, olorosos*, and *rayas*). Sweet wines, *pedro ximenez* and *moscatel*, were also produced locally. While this allowed producers the possibility to develop wines for different drinking occasions, it created considerable confusion for foreign consumers: "The public knows so much now of vintages and growths of Champagnes, Ports, Clarets, and to a lesser degree of Burgundies and Hocks. . . . With Sherry the case is different, and the consumer knows nothing but vague names such as Vino de Pasto, Amontillado, Oloroso, etc; the result is that he has less means of judging what price he ought to pay."[25]

The fact that sherry is a mixture of wines of different vintages makes it difficult to date accurately the shift from exporting young wines to older ones. Thus in 1833 James Busby noted that "the higher qualities of sherry are made up of wine the bulk of which is from three to five years old, and this is also mixed in various proportions with older wines," although he also noted that cheaper sherries were being exported after only two years.[26] The *solera* system was gradually developed from the very late eighteenth century through to the mid-nineteenth century and consisted of wines of a similar type, but at different stages of development, stored in rows of large casks. As Busby noted, "what is withdrawn from the oldest and finest casks is made up from the cask which approach them nearest in age and quality, and these are again replenished from the next in age and quality to them. Thus, a cask of wine, said to be fifty years old, may contain a portion of the vintages of thirty or forty seasons."[27] If only small quantities of wine were removed, the new wine assumed the characteristics of the older one in a few months. Initially the solera did not significantly reduce the skills or capital requirements, as fine sherries entered the solera only after four years.[28] However, soleras were also developed for producing cheap, young wines. The solera system was crucial as it reduced storage costs associated with maturing the wine and allowed exporters to sell wines with the same characteristics year after year regardless of the nature of the harvest, which was essential if brand names were to be

---

[24] Jeffs (2004:209). Sherries change their character with age, and this classification refers to 1959.

[25] *Ridley's*, March 1892, p. 165.

[26] Busby (1833:15). For the early development of the solera, see Maldonado Rosso (1999, chap. 9). The workings of the solera are described in González Gordon (1972:118–23). and Jeffs (2004:212–21).

[27] Busby (1833:15).

[28] Parada y Barreto (1868:129) and Vizetelly (1876:105) both give four years.

created. It was also particularly useful for production of the lighter sherries, the *finos*.

In the first half of the nineteenth century there had been a strong preference in England for wines that were heavy and sweet to be drunk after meals, such as port, madeira, malaga, and sherry. Around the middle of the century there was a drift toward lighter, drier wines, often consumed as an aperitif or with meals.[29] This change in fashion is impossible to quantify, but it did not go unnoticed by contemporaries. Denman, writing in 1876, noted that the

> general public taste has so manifestly altered that the wine trade is being revolution-
> ised. The strong old Sherries and Ports of the past are gradually being supplemented by
> lighter qualities, which our fathers would scarcely have recognised as wines. Instead of
> strong draughts derived from added alcohol, and cloying sweetness from added sac-
> charum, persons are looking for wine flavour, and bouquet and cleanness upon the
> palate.[30]

Although storing the wine for a long time could produce excellent finos, it required a considerable outlay of capital. The solera system, on the other hand, produced an equally acceptable drink after only a few years, and the slow decline in the importance of the *almacenista* can be attributed to both the growth of the solera system and the switch in demand for younger, lighter wines.

A few export houses dominated the trade. Between 1852 and 1865, for example, the houses of Garvey, Domecq, and González (Byass) accounted for a third of all exports from Jerez, and the leading five houses rarely exported less than half of all the sherry. Although at times there were suspicions of collusion between the largest firms,[31] the fragmentation of business in England and the low entry costs to exporting implied that shippers were required to compete on price and provide a wide range of different-quality sherries to their foreign agents. However, while the dozen or so leading houses could not stop other individuals from shipping wines, these often lacked adequate trade connections in the British market to distribute their wines.

SHERRY AND THE BRITISH MARKET

As sherry and port accounted for approximately three quarters of British imports in the 1850s, wine merchants naturally looked first to these regions for sources of cheap wines after the reduction in import duties in the early 1860s.

---

[29] Drummond and Wilbraham (1958:337) note the custom of taking an aperitif before dinner was probably introduced in England in the early nineteenth century, although they do not specifically mention sherry.

[30] Denman (1876:3). For a general survey of the problems, see Tovey (1877:158–61).

[31] This is hinted at in *Ridley's*, February 1856, p. 3.

Ridley's thought that "with regard to white wines suitable for British requirements, we have at present seen nothing capable of competing successfully against
Sherry: French productions, except at very high rates, are for the most part of a
very indifferent quality, and, unless fortified up to 30 per cent, will not be safe as
regards keeping properties."[32]

Sherry sales increased, and the price of must in Jerez tripled between 1850–53
and 1860–63 (fig. 8.1). This was not caused so much by structural problems associated with normal supplies being inadequate for the increased British demand, as Sir James Tennent and others had feared, but rather by the short-term
effects of vine disease (powdery mildew) and drought. Between 1853 and 1856
harvests fell from about 60,000 or 70,000 butts, valued at £7 each, to 18,000 or
20,000 butts, which were sold for £16–£20 each, even though quality was poor.[33]
This temporary sharp drop in the supply of fine sherry led to high wages and
considerable prosperity in Jerez for those with wine stocks. It did not last, and
prices fell quickly after 1863, when the postoidium wines became sufficiently mature to export.

Jerez was traditionally a high-cost wine-producing region because of low
yields, high labor requirements, and taxation.[34] After 1860 the shippers therefore looked for supplies of cheaper wines elsewhere, a process helped by the
opening of new railways, and especially the direct link between Madrid and
Jerez, which was completed in 1866.[35] The transport costs between the white
wine–producing region of Montilla in Córdoba and Jerez, for example, were cut
from £8 to £2 per barrel, encouraging the exporter Gonzalez Byass to established
wine-making facilities in that city.[36] Established shippers increased volume and
looked to reap economies of scale by supplying cheap wine because, as one newspaper noted, it was irrelevant to the exporter if "he buys at four and sells at six,
than if he makes the purchase at eight and realises it at ten."[37] Some of these inferior wines were mixed with sherry, but others were exported after only a few
months through Jerez's commercial networks. *Ridley's* responded to this change
by adding prices for a sixth wine to its list—"Sound Cadiz White Wine"—which

---

[32] Ibid., May 1860, p. 5.

[33] In 1860 it was noted that exports since 1855 had been "supplied from the large stocks on hand,
and not by late vintages" (ibid., February 1860, p. 5).

[34] United Kingdom. Parliamentary Papers (1878/79), p. 176; and *Ridley's*, April 1887, p. 166.

[35] *Revista Vinícola Jerezana*, 1/9 10.5.1866. This newspaper continued,

the exporters . . . not being able to pay . . . the high prices then being asked, had to look to other
. . . areas; as a result, Montilla wines began to compete with our better wines, and those from the
Condado (Huelva) and Sevilla, with the poorer ones. This novelty led to fierce competition in
our region . . . and . . . was the origin of the adulteration of Jerez's wines.

[36] United Kingdom. Parliamentary Papers (1878/79), p. 169; also p. 119. See also Espejo (1879–
80, 4:651–52).

[37] *Revista Vinícola Jerezana*, 1/6 25.3.1866, p. 44.

TABLE 8.3
Production Costs of "Elbe" Sherry

	Quantity and unit cost	Total cost
Proof potato spirit	40 gallons @ 1s. 4d.	£2 13s. 6d.
Water	56 gallons	
Capillaire	4 gallons @ 5s.	£1
Wine	10 gallons @ 4s.	£2
Cask		12s.
Labor and shipping		10s.
Commission		2s. 6d.
Discount for cash		4s.
Total		£7 2s.
Wholesale price		£8

Source: Ridley's, January 7, 1864, p. 5, and text.

sold at £14–£16, compared with the five traditional qualities of sherry, which ranged from £22 to £250 a butt.[38]

Jerez producers were not the only ones who supplied imitation sherries, and in 1863 a group of twenty-nine Jerez shippers gave *Ridley's* a gift of £100 for "discovering, exposing, and frustrating traffic in spurious Wines"—wine that had been shipped from London to Cadiz and returned as "Sherry."[39] The British retailer Gilbey's was one of the first to import "sherry" from South Africa shortly after the Crimean War, which until 1860 benefited from colonial preference.[40] Australia, likewise, began producing an imitation. However, the product that did the most harm was Hamburg "sherry," which was manufactured from cheap industrial alcohol produced from sugar beet or potatoes and, according to the *Medical Times and Gazette* was "in its original state, . . . a light German wine of poor quality, not possessing in that condition sufficient preserving powers to render it suitable for shipment, or, indeed, for consumption as a natural wine in its own country."[41] An attempt was made to overcome these defects by adding spirit and saccharine. One recipe for "Elbe" sherry is given in table 8.3, and, as the author noted, the return of capital of around 13 percent could be considerably increased if more water, and less wine, was added.

Although from 1865 this type of wine was prohibited from entering Britain

[38] *Ridley's*, February 1867, p. 2.

[39] Ibid., September 1863, p. 16. George Ridley was subsequently given the Cross of the Order of Carlos III in 1870. Ibid., March 1870, p. 4.

[40] Waugh (1957:6).

[41] Quoted in Tovey (1883:7). Emphasis in original.

for reasons of public health, it is clear that similar wines were being sold.[42] Indeed, Gilbey's, which in 1875 was responsible for 5 percent of all wine sold in Great Britain, dropped "Castle Hambro Sherries" from its lists only in 1877.[43] As one senior partner noted, "for some years past we felt that we could not give that assurance of their genuiness as wine as we wished to do, and we decided some years ago not to ship them. In fact, we thought that they rather interfered with the status of our business generally."[44] The reputation of sherry had been damaged much earlier, as Thomas G. Shaw observed in his important book in 1863: "Sherry has long been the favourite wine, but the quantity of bad quality now shipped and sold under its name has already injured its reputation; while the high prices of any that is good and old offer an opening for the introduction of another white kind."[45]

One of the reasons why adulteration was relatively easy was the high alcoholic content of sherry. From the late 1870s wine shortages caused by phylloxera in France encouraged sherry exporters to use the cheaper industrial alcohol to fortify their less expensive wines to remain competitive. The extent to which these were prejudicial to health is difficult to assess, but the public concern hastened the decline in popularity of the drink.[46] By contrast, the better sherries, when properly matured, required less alcohol than those that were exported after only a few months.

In Jerez the shippers explained the decline in British imports after 1873 to the higher duty that their wines had to pay because of their greater strength compared with French wines, and they launched a major publicity campaign to reduce them.[47] Not all were convinced in Britain, and it was argued that a reduction in duty from half a crown to a shilling a gallon would have only a marginal

---

[42] United Kingdom Consular Reports (1865), no. 53, p. 657. Jerez produced its own version of Hamburg sherry according to the British consul's report of 1865:

> During the past year large quantities of wines have been introduced into the district from Malaga and Alicante; but these wines have not proved serviceable or usable, their peculiar, earthy and tarry character being impossible to overcome; as, although mixed with other wines but in small quantities, the unpleasant flavour and "smell" is always distinguishable to a judge of wine.

This did not stop the wines being used, however. "The low spirituous compounds are made up with molasses, German potato-spirit, and water; to which some colouring matter, and a small quantity of wine are added; much in the same manner that the 'Hamburg-sherries' have been manufactured to which of late the London Custom-House has, very properly, refused admission."

[43] Faith (1983:12); *Ridley's*, January 1877, p. 3. "Castle" was the brand name of Gilbey's. Given Gilbey's insistence on quality, it seems unlikely that their version was prejudicial to health. However, it created confusion for consumers.

[44] United Kingdom. Parliamentary Papers (1878/79), H. P. Gilbey, p. 149.

[45] Shaw (1864:217).

[46] United Kingdom. Parliamentary Papers (1878/79), pp. 121, 122, 170, 266.

[47] See especially Pan-Montojo (1994:103–10).

TABLE 8.4
Alcoholic Strength of UK Wine Imports, 1856–1893 (percent)

	Wine imports of 14.8 degree or less, 1875	Wine imports of 14.8 degrees or less, 1882	Wine imports of 17 degrees or less, 1893
Spain	3.0	10.7	62.7
Portugal	1.6	1.7	3.9
France	91.68	99.0	99.8
Other countries	16.86	25.1	81.1
Total	24.22	35.3	61.6

Source: United Kingdom. Parliamentary Papers (1878/79), pp. 315–17; *Ridley's*, August 1883, p. 241, and June 1894, p. 361.

impact on retail prices and might not even be passed on to the consumer.[48] Duty on imported wines was finally cut in 1886, when the shilling duty was extended to wines of between 14.8 and 17 degrees. As table 8.4 shows, by the early 1890s almost two-thirds of Spanish wines paid the lower duty, although red wines, mainly from Tarragona, now accounted for 70 percent of total exports. The fact that sherry exports did not recover with the lower duties after 1886 can be explained once more in part by adulteration, with producers illegally using German spirits and salicylic acid as preservatives in their attempts to reduce the alcoholic content of the cheaper sherries so as to enter the lower tax bracket.[49] Although perhaps the quantities adulterated were small, it reinforced once more in the public mind the notion that cheap sherries were dangerous.

The widespread concerns about adulteration also threatened the reputation of fine sherry. *The Times* in 1873 carried a letter from a Dr. Thudichum that drew attention to the supposed health hazards caused by the addition of gypsum in the crushing of the grapes and sulfur in the fumigation of the casks, activities that were common everywhere in Jerez. Thudichum's claims were

> disseminated throughout the length and breadth of the land by a local Press ever hungry for copy and not too careful either as to its accuracy or the mischief that would naturally accrue from its publication. Thus it has happened that, outside the Trade

[48] The British authorities were concerned primarily with the implications on revenue, as a too liberal duty risked encouraging an increase in consumption of strong wines at the expense of spirits. In terms of alcohol content, spirits were already taxed more than wine (although tax on beer was less). There was also concern that lower duties would result in strong wines being imported to be illegally distilled, even though cheaper alternatives, such as alcohol produced from maize, oats, sugar, and other products, existed. United Kingdom. Parliamentary Papers (1878/79), pp. 26, 55, 162, 182.

[49] *Ridley's*, February 1888, p. 58, 62, 70; October 1888, p. 474.

Papers themselves, little or nothing has been published to disabuse the public mind of the wrong impressions under which it labours as to the supposed unwholesome character of Sherry.[50]

Six years later Robert Houldsworth, a partner of Gonzalez Byass, complained that doctors, "a very powerful section of the community," "have been running down sherry lately."[51] The asymmetries of information, where the buyer has insufficient knowledge of the quality of a product prior to purchase, were now accompanied by supposedly informed advisors arguing against drinking even fine sherry bought from reputable merchants, and this situation was ended only when the *Lancet* vindicated the drink in a report published twenty years later.[52] This failure to adapt to the more impersonal markets of the late nineteenth century and the inability of the legislation in Britain to reduce the threat of adulteration suggest the need for a closer look at the shippers' interests in Jerez, and the difficulties in developing self-enforcement mechanisms to control quality.

PRODUCT INNOVATION AND COST CONTROL

The independence of the almacenistas, who bought young wines from growers to mature in their cellars for future use in blending, was eroded even before the fall in prices after 1863, as the large shippers integrated backward and purchased their businesses to guarantee stocks of fine wines.[53] Those almacenistas that remained independent found a much more competitive market after 1863, in part because of their overvalued stocks bought at the height of the boom, but also because of the spread of the solera system, which shortened the maturing time required for wine production. Exporters, by contrast, were able to limit the effects of lower prices by selling large volumes of cheap, young wines, often without their brands. These were brought directly from growers and the shippers prepared them in their cellars, without the need for the almacenistas.[54] The widespread publicity given to the annual shipping lists, which ranked shippers according to the volume of exports regardless of the quality of the wine, provided an additional incentive for them to move down-market. Under

---

[50] *Ridley's*, November 1898, p. 762.

[51] United Kingdom. Parliamentary Papers (1878/9:176.

[52] See Jeffs (2004), pp.93–6 and 174–7.

[53] John Haurie (nephew) told his buyers in 1857 that the firm had secured the stocks of *the three largest Almacenistas* in Jerez, which added to *our own*, enable us to offer sound wines as will maintain the reputation of our former shipments" (*Ridley's*, October 1857). Emphasis in the original. In 1862 Gonzalez Byass bought 92 butts of old sherry for £10,000. Ibid., July 1862, pp. 2, 3.

[54] Domecq, for example, shipped large quantities of "light low" wines in 1864 and 1865 that did not carry the firm's brand. Ibid., January 1866, p. 7. Likewise, Gilbey's imported unbranded sherry and white wine from Gonzalez Byass. For the impact of this cheap wine trade on a shipper's profits, see especially Montañés (2000).

pressure from the London wine trade, the publication of these lists was ended in 1878.[55]

The solera system allowed wine quality to remain constant from one year to the next, which facilitated the development of brands. Gonzalez Byass, for example, had a number of old soleras, such as Matusalén, Apóstoles, or Tío Pepe, expensive wines that sold in the late 1870s for five shillings or more a bottle, about five times the price of "cheap" sherries.[56] Harvey's famous Bristol Cream dates from about the same time,[57] while Berry Brothers listed William and Humbert's Dry Sack at four shillings a bottle in 1909, equivalent to the price for a bottle of Scotch whisky.[58] The relatively high price of these sherries saw them being sold through traditional distributional channels of specialist wine merchants rather than by grocers who owned an off-license. They were for immediate drinking as, unlike port, sherry does not improve once in the bottle. Genuine sherry was simply too expensive to brand for the mass market, and in any case many British retailers preferred to sell their own brands. Trade was "hand-to-mouth," with retailers keeping low stocks and then ordering small lots of highly specific wines directly from the Jerez shippers.[59]

The decline of sales in the British market after 1873 was partly offset by rising French demand for cheap wines because of phylloxera, and between 1886 and 1892 France imported about 110,000 hectoliters of sherry and white wines a year compared with Britain's 140,000 hectoliters. Total exports therefore remained strong until the late 1880s, but the growth in demand of cheap, young sherries was accompanied by a major decline in the quantity of mature wine to be found in Jerez, and by 1895 it was estimated that there were no more than 8,000 butts of wine older than thirty years in "the whole district." This switch to young wines also affected relative grape prices, and in 1882 it was noted that "prices for new wines from the best soils have declined, whereas wines from inferior soils have fetched fair prices."

The large sherry houses not only benefited from the growth in French demand, but also looked to develop new markets. Around 1872 the firm Santarelli

---

[55] *Ridley's*, February 1878; also March 1871; January 1874.

[56] Gonzalez Byass's 1878 price list for Cadiz, reproduced in Montañés (2000:264). Tío Pepe was introduced at the lower duty of 1s a gallon. Retail prices calculated from Simpson (2005b, table 1).

[57] Harrison (1955:106) gives 1882 as the date for Bristol Cream, and it is not shown in Harvey's 1867 price list reproduced opposite Harrison's p. 115. However, *Ridley's*, July 12 1880, p. 209, cites an auction of one of Harvey's customers where twenty-two dozen cases of Dark Gold Sherry, Bristol Milk, bottled December in 1862, and twenty-three cases of Old Pale Sherry, Bristol Cream, also bottled in 1862, selling at around a pound sterling per bottle.

[58] Appendix 2.A.5.

[59] "Instead of selling to market Houses parcels of several hundred butts at a time, and leaving the latter to distribute them, the majority of comparatively small Wine Merchants go to the Shipper direct and order their butt, hogshead or even quarter-cask from head-quarters. Each has his own fancy as to the proportion of "dulce," colour or style as the case may be, and each parcel, however small, has to be made up separately on these lines." *Ridley's*, March 1892, p. 164.

Hermanos began to sell bottled sherry in Spain,[60] and markets were found especially in Madrid and new export markets created in Cuba, Puerto Rico, the Philippines, and South America.[61] Another market for cheap wines was the commercial production of brandies from the mid-1870s, using French stills and experts brought over to provide technical assistance.[62] In 1887 the British consul noted that production was still in its "infancy"[63] but two years later reported that "portable stills are being distributed amongst the vineyards of his district by an influential company recently formed, and an increased amount of capital is yearly being invested in this new brand of Jerez commerce," while in Jerez itself, "several Houses" established large scale distilleries using the "Charente system."[64] Brandy for the first time gave the large sherry houses a major product to sell in the domestic market, but they found much cheaper wines for its production in La Mancha.

There were a number of even more ambitious attempts to diversify output, including an attempt in the early 1880s at producing sparkling sherry, which was "practically the typical sherry in an aerated form—a sort of concentrated sherry and seltzer." It failed, as did the attempt a decade later to produce champagne using the *champenois* method with imported skilled labor from Épernay.[65] Attempts to produce and promote nonalcoholic sherry were no more successful.[66]

The boom in Jerez in the mid-1860s had created business fortunes and widespread prosperity, with local wages doubling.[67] Labor requirements were high, at an estimated annual 182 days work per hectare on the albarizas soil, 107 days on the barros, and 93 days on the arenas soil.[68] Following the drop first in prices and then in exports, there followed a succession of business closures, widespread unemployment, and major social unrest. Employers cut costs by paying lower wages

---

[60] *El Guadalete*, April 22, 1908. González Gordon (1972:36) suggests another company, J. de Fuentes Parrilla, as being the first, between 1871 and 1873. The first bottle factory in Jerez was not established until 1896 (ibid., 172).

[61] *Revista Vitícola y Vinícola*, November 16, 1884, p. 7; and Fernández de la Rosa (1909, 2:261).

[62] González Gordon (1972:183). Spirits (*aguardiente*) had traditionally been produced in the region for fortifying wines, but the low import duties led to large quantities of German industrial alcohol being imported. In Jerez itself there were eight modern alcohol factories by 1886, with a capacity of 30–40,000 hectoliters, although production was limited at the time to 8,000 hectoliters. Nevertheless, these 8,000 hectoliters required about 70,000 hectoliters of poor quality wine. Archivo Ministerio de Agricultura, Madrid, *legajo* 82–2.

[63] United Kingdom Consular Reports, Cadiz (1888), no. 103, p. 135.

[64] Ibid. (1890), no. 714 (Cadiz for 1889), p. 4; and *Ridley's*, January 1890, p. 34.

[65] *Ridley's* (June). Three Jerez firms marketed the drink at the high price of 52s. per dozen f.o.b. *Ridley's*, April 1881, pp. 106–7; June 1891, p. 409.

[66] Ibid., July 1881, p. 207.

[67] Ponsot (1986) gives an increase from slightly under 10 *reales* (2.5 pesetas) to 19 *reales* from the early 1850s to the early 1860s. In Catalonia, by contrast, wages increased from 2 pesetas to 2.2 pesetas (Garrabou and Tello 2002:629). See also Simpson (1985b:180–82); López Estudillo (1992:48).

[68] As these refer to tax estimates for 1857, they are probably a minimum. Montañés (1997:69).

as well as using less labor. A group of workers petitioned the mayor of Jerez in the summer of 1866, noting that while the wages of day laborers had declined by half from the early 1860s, food prices had not fallen.[69] In August 1882 *El Guadelete*, a local newspaper, wrote that "the majority of the vineyards in Jerez are found badly cultivated or virtually abandoned, covered in weeds or handed over to tenants who, with few resources, cultivate themselves obtaining a very small salary for their work with the selling of the grapes."[70] Some owners experimented with sharecropping, with "day labourers who, for the consideration of receiving from one-third to one-half of their produce, agree to cultivate and to properly keep up their possessions."[71] Although sharecropping had a long and successful history in some wine regions, such as Beaujolais and Tuscany, this would not be the case in Jerez, where the contract was primarily used by landowners to avoid having to negotiate with a militant labor force.[72] The short-term nature of these contracts encouraged the sharecropper to maximize output with no concern for the long-term future of the vines.

There were also a number of possibilities for introducing labor-saving technologies. The substitution of the secateurs for the pruning knife in the 1870s and 1880s reduced the skills required for pruning and hence wages and was bitterly opposed by the workers, many of whom owned small plots of vines themselves but depended on seasonal employment on the large estates.[73] *Ridley's* noted in 1883 that a foreman had been murdered, "and the reason is supposed to be that he had been using scissors for pruning."[74] Elsewhere, labor-intensive methods continued to be used, with tillage being carried out using the *azada*, a type of hoe that was considerably more labor-intensive than plows. There were few changes in wine making, as the grapes were crushed by feet rather than machines, and, unlike in most other wine-producing regions, women were rarely employed in the vineyards before the turn of the century.[75]

Some in Jerez even regarded the approach of phylloxera as a blessing and

---

[69] *Revista Vinícola Jerezana*, 1/9, 10.5.1866, p. 70. In December the same newspaper (1/23, 10.12.1866, p. 179) noted that "the cultivation of our vines, the basis of our wealth, is decaying visibly."

[70] *El Guadelete*, August 12, 1882.

[71] United Kingdom Consular Reports (1885), part 6 (Cadiz for the year 1884), p. 928. See also Crisis Agrícola y Pecuaria (1887–89, 3:168; 4:34), both cited in Zoido Naranjo (1978:69).

[72] Landowners preferred the less radical workers as tenants. See López Estudillo (1992:67–68). For sharecropping and viticulture, see Carmona and Simpson (forthcoming).

[73] Fernández de la Rosa (1909:466); *Revista Vitícola y Vinícola* (1884), no. 3, p. 6; Crisis Agrícola y Pecuaria (1887–89), vol. 4, no. 236, p. 36.

[74] *Ridley's*, December 12, 1883, p. 358.

[75] Vizetelly (1876:16) noted that "advocates of woman's rights will regret to hear that the labours of the softer sex are altogether dispensed with in the vineyards of the South of Spain," but the socialist newspaper *El Viticultor* (October 6, 1900, no. 56) noted that women were being employed at the harvest at between three quarters and one peseta a day, compared with the three pesetas earned by men.

hoped that wine shortages would produce "abnormally high prices."[76] When it finally appeared in 1894, it spread quickly, destroying virtually all the vines by 1909, by which date only a third of the area (2,640 hectares) had been replanted and the rest of the land was being used for cereals.[77] The failure of prices to improve and financial hardship was one explanation for the slowness of replanting, but there were also the difficulties associated with finding suitable American roots because of the high limestone content on the albarizas soil, resulting in the early vines performing badly and leading to the vine disease chlorosis. The area of vines in the province of Cadiz fell from 21,000 hectares in 1882 to 6,000 hectares in 1904, leading to a loss of between three and four million days employment.[78]

WINE QUALITY AND THE DEMAND FOR A REGIONAL APPELLATION

Jerez was the only region of the four discussed in this section that had failed to establish a regional appellation by the early twentieth century, even though as early as 1844 Jacobo Walsh complained about the "deplorable state" of trade with England and the need for growers to establish in Jerez a "Gran Compania" that would export only local wines.[79] A second, much wider debate began in the mid-1860s, when a newly created newspaper representing the interests of the growers and almacenistas claimed that exporters had "abstained from buying old wines from the almacenistas," and later that "22 or 24,000 botas" of wine had been bought from outside Jerez and El Puerto de Santa María by train.[80] Death threats were made to shippers for using wines produced outside these regions as sherry, and in 1871 it was noted that "within the last few days much damage has been *wilfully* caused in the vineyards by the discontents; in some places by cutting off all the branches and shoots, and in others by sweeping with a hard broom all the fruit."[81] The response of the exporters was not to deny claims of using outside wines, but rather to blame the poor local harvests and the need to be able to export very cheap wines to compete in the British market.[82] This market for cheap wines led the agronomist Fernández de la Rosa in 1886 to propose two distinct "town brands": one for local wines, and one for wines that had been

---

[76] United Kingdom Consular Reports (1896), no. 1839 (Cadiz), p. 3.

[77] Archivo Municipal, Jerez de la Frontera, *legajo* 523.

[78] Quevedo y García Lomas (1904:57–58). The number of workers in Jerez's vineyards fell from an estimated 5,000 in the 1880s to 2,400 in 1921 according to Montañés (1997:139).

[79] Walsh (1844:3–5). Walsh envisaged a public company that would require no privileges and remain independent of the state.

[80] *Revista Vinícola Jerezana*, January 1866, p. 1; June 1867, p. 283. This was equivalent to 110,000–120,000 hectoliters.

[81] *Ridley's*, May 1871, p. 7.

[82] Cabral Chamorro (1987a:178).

produced outside the city.[83] Growers and small winemakers continued to demand an exclusive sherry brand for locally produced wines, while the exporters argued that businesses located in Jerez or El Puerto should also be included, regardless of where wines came from. In 1914 the shipper Domecq suggested that local wines should be called "Jerez" rather than "Sherry," as the "latter expression has become, to some extent, a generic term used to denote all manner of imitations of the original and genuine product."[84]

This failure to create a regional appellation despite widespread local support can be explained by the lack of political voice of growers, and the fact that phylloxera helped strengthened the exporters' control of the commodity chain as they integrated backward. By 1912–13 some forty houses owned 1,000 hectares of the 2,500 hectares that had been replanted in Jerez and were responsible for 30,000 botas of the 34,000 exported.[85] When in 1935 a controlled appellation was finally established—the *Consejo Regulador de la denominación de origen Jerez-Xérès-Sherry*—the shippers were still able to buy outside wines when harvests were poor or prices high.[86]

The region of Jerez saw some of Europe's worst rural violence in the half century prior to the First World War, with the anarchist "black hand" creating panic among property owners in the 1880s, and the city itself being stormed by unemployed rural workers in 1892. In France the Waldeck-Rousseau law of 1884 legalized trade unions in an attempt to encourage moderate workers to participate and reduce the influence of extremists. No such law existed in Spain, and the short periods when workers' organizations were legalized often coincided with bitter strikes, followed by the inevitable repression. In rural Andalucía there was an increasing spiral of social conflicts occurring from the 1880s, followed by problems in 1902–3, 1918–20, and 1931–33, while during the Second Republic (1931–36) the agrarian problem and attempts at land reform, especially in areas such as Jerez, have been identified by historians as the single most important cause of the Civil War.[87]

It is tempting to argue that the increasing inequality and the lack of a legitimate political voice among field workers, small growers, and bodega workers forced them into radical nonparliamentary politics and violence, while the shippers used their political influence to block any attempts to establish a local appellation and thereby restrict their profits. Indeed, Temma Kaplan has claimed

---

[83] Ibid., 181.

[84] *Ridley's*, February 1907, p. 103.

[85] Cabral Chamorro (1987a:191).

[86] Ibid., 193–94. The 1935 law provided a minimum length for maturing wine and a minimum alcoholic strength, controlled grape yields, fixed minimum export prices, and restricted exports to 60 percent of a shipper's stock in a single year. As elsewhere, there have been frequent modifications of the clauses. Fernández García (2008:195–96).

[87] Preston (1984:159–60); Malefakis (1970). The level of conflict varied significantly within Andalucía.

that "the evolution of anarchism in Northern Cádiz Province was inextricably tied to the declining prosperity of independent winegrowing peasants, pruners, and coopers after 1863, and their collective response to their condition."[88] Yet neither the winegrowers nor the bodega workers were attracted to radical politics. Laborers in the vineyards remained skilled and even in the 1900s earned at least double what day laborers were paid in cereal farming.[89] It was workers in this second group—laborers who drifted from farm to farm for seasonal employment over the wide expanse of the Jerez plain and experienced a major decline in their employment opportunities—who were attracted to anarchism.[90] By contrast, those who worked in the vineyards and bodegas belonged to the labor aristocracy and were more likely to follow moderate republicanism (and, when legalized, the moderate Union general de trabajadores) rather than preparing for revolution.[91]

The decline in demand for old, fine sherries was to be permanent. In 1895, even as harvests suffered because of phylloxera, the shipper Gonzalez Byass auctioned a considerable quantity of old sherry, which they considered in excess of requirements.[92] A few years later Edward VII disposed of some sixty thousand bottles of sherry from the royal cellars.[93] The shift toward cheaper wines was underlined by the Asociación Gremial de Exportadores, the shippers' pressure group founded in 1910, which argued that for consumers sherry was not linked to a geographic area but was a generic name used for certain wines throughout the world. It therefore opposed any move to establish a regional appellation, claiming it would increase costs and reduce their competitiveness.[94]

By the late nineteenth century shippers were increasingly competing in the British and French cheap wine markets. Brand names for cheap sherries were weak, and the sale of large quantities of foreign wines as "sherry" implied that Jerez's merchants had little option but to buy from low-cost producers such as La Mancha if they wished to compete in this segment of the market. The several

[88] Kaplan (1977:12). This claim has been convincingly challenged by Cabral Chamorro (1987a). See also Montañés (1997).

[89] In 1905–6 vine growers earned 2–3.5 pesetas a day, compared with the 0.5–1.25 earned by agricultural workers. This second group, however, was given food equivalent to 0.75 peseta. Archivo Municipal, Jerez de la Frontera, Protocolo Municipal 404 (1905–6).

[90] Cabral Chamorro (1987b) and Montañés (1997). For the segmented local labor markets and their influence on conflicts in Andalucía, see Carmona and Simpson (2003, chap. 3).

[91] Cabral Chamorro (1987b).

[92] In total, some 2,500 butts (*Ridley's*, March 1895, p. 159). *Ridley's* noted in May 1895 that "the very fine Wines fetched good prices, and the medium Wines fair ones, although, on the other hand, the cheaper descriptions, which showed signs of a hurried preparation for shipment, cannot be said to have sold particularly well" (304). The sale made £66,000, an average of £26 per butt.

[93] Jeffs (2004:103).

[94] Cabral Chamorro (1987a:192).

thousand small growers in Jerez lacked the political support to create a regional appellation, as did the shippers to achieve legislation in London that limited the use of the word "sherry" to wines produced from Jerez, along the lines incorporated in the Anglo-Portuguese Commercial Treaties of 1914 and 1916.[95] Trade did eventually recover in the twentieth century, but for cheap, rather than old mature sherries.[96]

---

[95] Shippers recognized the importance of the collective brand, resulting in the creation of a lobby in London in 1910, a fact particularly welcomed by *Ridley's*, which hoped it would "stop the rot" (January 8, 1913, p. 9). The Asociación de Exportadores de Vinos was founded in 1885, and the Asociación Gremial de Criadores Exportadores de Vinos in 1889 (González Gordon 1972:41).

[96] Fernández (2010); Jeffs (2004).

PART IV

The Great Divergence: The Growth of Industrial Wine Production in the New World

WINE PRODUCTION was of little importance in the New World until the late nineteenth century. This was despite the considerable interest shown by colonial governments as wine was considered a valuable addition to local diet; was a necessity for the Catholic Church for celebrating Mass; and offered a potential source of taxation. In addition, viticulture, in contrast to ranching, required permanent, densely settled communities of small farmers. Wine exports also promised to reduce the dependence of Britain and the Netherlands on French and Iberian imports. Hernán Cortés introduced vines into Nueva España (present-day Mexico) in the 1520s, and they were reportedly being grown as far south as Chile by the second half of the century, and in Alta California (present-day California) when the Spanish missions arrived in the late eighteenth century.[1] There were also unsuccessful attempts in the sixteenth century by French Huguenots in Florida, and slightly later by British settlers in Virginia, to make wines from native American grapes.[2] Elsewhere, the Dutch produced small quantities of wine and brandy in the Cape Colony, and vine cuttings were on the ships that arrived to establish Australia's first colony in January 1788.[3]

The limited success of these early initiatives can be attributed to a variety of factors, including vine disease, poor wine-making systems, the very low population densities in those areas most suitable for specialized viticulture, and the huge distances that separated them from their potential markets. These restrictions were relaxed with the rapid development of the global economy in the late nineteenth century, which brought massive immigration, improved transportation, and a better understanding of the needs of viticulture and wine making. Production conditions were very different from those found in Europe. Factor prices were reversed, as land was relatively cheap and abundant, and labor expensive. Wages for unskilled workers in Argentina in 1910, for example, were 34 percent more than in France, in Australia 90 percent more, and in the United

---

[1] Unwin (1991:218, 301).
[2] Pinney (1989, chap. 1).
[3] Unwin (1991:247).

States 139 percent more.[4] Although capital markets were weak for wine producers everywhere, profits from gold mining in both California and Australia benefited the local industries, and the greater scale of production in the New World by the late nineteenth century allowed producers better access to banks than their European competitors enjoyed.

Another difference was terroir, and in particular climate. Vines were easy to grow and the risks from disease limited in the New World, but the hotter climates produced grapes with a high sugar content that caused fermentation to end prematurely, leading to the development of bacterial diseases and ruining the wine. A number of colorful pioneers, such as Agoston Haraszthy in California and James Busby in New South Wales, traveled to Europe to collect vines and learn at first hand production techniques, although it was only in the 1890s that European technology started to resolve the problems associated with wine making in hot climates. This revolutionized the sector and allowed the possibility of producing large quantities of homogenous cheap wines that could be sold under brand names.

High import tariffs and the arrival of millions of immigrants, together with the New World's high fertility, gave the new industry a rapidly growing potential market. Population increased elevenfold in Australia between 1850 and 1910 (from 0.4 to 4.4 million); sixfold in Argentina (from 1.1 to 6.8 million); fourfold in the United States (from 23.3 to 92.8 million), and almost threefold in Chile (from 1.4 to 3.4 million).[5] Immigrants arriving in the New World brought with them their own drinking customs, and wine was the favorite alcoholic beverage for many of the Italians and Spaniards settling in Argentina and Chile. By contrast, most consumers in Australia and the United States drank beer and spirits, increasing marketing costs for merchants as they had to educate drinkers about wine and compete with other sectors of the alcohol industry, which underwent even greater changes in this period, especially in standardizing quality and reducing adulteration.

Producers responded to these differences in factor endowments, terroir, technology, and market demand by creating an industry that was very different from the artisan nature of European production.[6] In particular, the new, large-scale wineries producing often in excess of 10,000 hectoliters were dependent in a way they were not in Europe on specialist grape producers, and wine producers became much more involved in marketing. In Europe, grape production and wine making were usually integrated in a single business, and producers therefore had to expand capacity in both sectors to respond to market upturns.[7] In the New

---

[4] Williamson (1995:178–81). Figures are for unskilled urban building workers.

[5] Maddison (1995:104, 106, 112).

[6] Simon (1919:105).

[7] In the Midi a separation was becoming apparent in some areas by 1900, resulting in a similar problem of a lack of cellar space. The problem was resolved by the creation of producer cooperatives (chapter 3).

World this was not the case, and growers and wine owners faced different investment incentives, which in the case of Argentina especially led to periodical major downturns. Important differences also emerged among New World producers concerning who put the brand and controlled the value chain. In California a handful of San Francisco's merchants created a hierarchical organization, integrating horizontally and vertically and investing heavily in advertising and brands to sell to distant consumers, controlling at one time about 80 percent of the state's wine sales and making it one of the world's largest wine businesses. By contrast, it was a British importer who created a buyer's organization that distributed two-thirds of Australian exports and controlled the value chain. Finally, the Argentine industry was dominated by a dozen or so winemakers, but these had only limited success at creating brands and integrating forward into marketing. Instead they sold huge quantities of wine in a market where annual per capita consumption reached 60 liters by 1914, compared with just 2 in the United States and 5 in Australia. By 1913 Argentina was the world's seventh largest producer, Chile the ninth, the United States the tenth, and Australia the eighteenth. This section looks at the development of the industry in California in the United States, South Australia and Victoria in Australia, and Mendoza in Argentina.

CHAPTER 9

Big Business and American Wine:
The California Wine Association

> The California Wine Association would "cultivate more vineyard acreage, crush more grapes annually, operate more wineries, make more wine, and have a greater wine storage capacity than any other wine concern in the world."
> —Ernest Peninou and Gail Unzelman, 2000:33

WINE HAS TWO very different histories in the United States. On the East Coast the attempts to plant European vines (*Vitis vinifera*) failed repeatedly because of the excessively harsh winters for the cold-sensitive European vine or endemic cryptogammic diseases such as downy and powdery mildew, anthracnose, and black rot, which thrived in areas of high humidity, as well as phylloxera.[1] From the late eighteenth century a number of new vines appeared east of the Rocky Mountains that were either domestications of native American species (*V. labrusca* and *V. rotundifolia*) or naturally occurring hybrids between American and European varieties, such as concord or catawba. These varieties were often resistant to the fungus diseases and could withstand the harsh winters, but they produced a strong, disagreeable, "foxy" flavor in the wine. Some of the *labrusca* hybrids were also not very resistant to phylloxera. As the grapes were naturally low in sugar and high in acidity, sugar and water were added, and in the second half of the nineteenth century merchants blended these wines with those produced in California.

By contrast, by 1901 California accounted for nine-tenths of the nation's wine production, but had just 2.4 million of its 91.2 million inhabitants. Viticulture was introduced into Baja California by the missions, the first of which was founded in San Diego in 1769, although the earliest reference to the planting of grapes dates from 1779, at the San Juan Capistrano mission.[2] After the missions were secularized in 1833, the vineyards were neglected and eventually abandoned. Privately owned vineyards began to appear as early as 1818, but the size of the industry was small as California's population was just fourteen thousand inhabitants on the eve of the gold rush. By 1852 it had jumped to a quarter of a million. As powdery mildew was devastating Europe's vineyards, the *California*

---

[1] Amerine (1981:1–3), on which this paragraph is based.
[2] Pinney (1989:238) believes that viticulture probably was not practiced before this date.

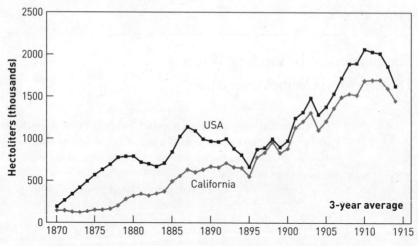

Figure 9.1. U.S. and California wine production, 1871–1912. Source: California State Board of Agriculture (1912:191) and Shear and Pearce (1934: tables 6 and 10)

Farmer declared in 1855 that "California may become the vineyard of the world," and the California Horticulture Society noted that the state possessed 10 million acres (approximately 4 million hectares) of potential vineyards, and that the European industry employed directly and indirectly five million people.[3] Viticulture grew rapidly, so although the United States imported thirty-eight times more wine than it produced in 1840, the consumption of domestic wines was nearly four times greater than consumption of imports by 1886.[4] California wine output quadrupled between 1866–69 and 1883–86, doubling again by 1898–1901, and once more in the following decade (fig. 9.1). On the eve of the First World War, output stood at 2.16 million hectoliters, of which 44 percent was dessert wine (wines that had been fortified with brandy) and 56 percent dry table wines. By 1913 California had 134,000 hectares of grapes, 69,000 of which were for wine, 45,000 for raisins, and 20,000 thousand for table grapes. An estimated 15,000 heads of families were directly engaged in viticulture, with many more being employed in wine making and ancillary industries.[5] The United States, or more correctly California, had gone from producing virtually no wine to producing as much as Germany. Technological change was also important, but while California farmers developed new technologies in many branches of agriculture that made them more productive than their European competitors,[6] the wine industry remained dependent on French scientific knowledge and

[3] Leggett (1941:68); Pinney (1989:262).
[4] Peninou, Unzelman, and Anderson (1998:255–56); Crampton (1888:184).
[5] California Board of State Viticultural Commissioners (1914:2, 4).
[6] For the rapid development of technology and market organization in California, see especially Morilla Critz, Olmstead, and Rhode (1999).

equipment before 1914. European contemporaries were astonished not by the rapid growth of the sector or the nature of the technology used, but by the distinct organizational structure of the industry. The creation of a wine trust that controlled the industry was unique and even inspired imitations in the Midi, although these came to nothing.[7]

This chapter looks at how the industry developed. The first part shows how producers adapted grape growing and wine making to local conditions; the next considers the relationship between growers, winemakers, and San Francisco's merchants that led to the creation of the California Wine Association (CWA); and the third part examines the difficulties in selling to consumers accustomed to drinking beer and whisky rather than wine. Despite the success of the CWA, consumption of dry wines was strictly limited outside the small group of immigrants from southern Europe, and it was dessert wines that proved to be the most dynamic sector in the decade or two prior to Prohibition.

CREATING VINEYARDS AND WINERIES IN A LABOR-SCARCE ECONOMY

The initial growth of viticulture was slow, and in 1850 the main center of production was on fertile land around Los Angeles, using the mission grape variety, which was well suited for the production of fortified sweet wines in the hot southern climate, especially when wine-making skills were poor, but not for table wines. The population of Los Angeles as late as 1880 was just 11,183 inhabitants—very small compared with the major market of San Francisco, which saw its population increase from 34,776 to 233,959 between 1850 and 1880.[8] Local merchants tried to increase their wine supplies to compete with the expensive European imports. The firm Kohler & Frohling, which dominated the Los Angeles trade at this time, together with two other promoters, established a cooperative at Anaheim, some 32 kilometers south of Los Angeles.[9] This allocated fifty plots of 8 hectares to each settler and set aside more land for housing, schools, and other public use. Between 1857 and 1859 the cooperative paid $20,000 for 22,789 days of field work in the preparation of the vineyards, fencing, and digging irrigation ditches. The vine growers were from Germany and were a diverse group, with little or no practical knowledge of grape growing. The cooperative was ended once the vineyards were established, but the colony continued to maintain a strong communal spirit. A further 600 hectares were added in the early 1870s to meet the needs of the growing population, with a large part of the land being planted with vines. Kohler & Frohling purchased most of the wine, and output jumped from 11,000 hectoliters in 1864 to 47,000 in 1884. This

---

[7] See chapter 3.

[8] Carter et al. (2006, 1:110).

[9] Kohler & Frohling already bought the produce of some 142 hectares of vineyards each year. Pinney (1989:254–56); Carosso (1951).

was the last good harvest before a mysterious new vine disease, subsequently named Pierce's disease, destroyed the vines and forced growers to switch to walnuts and oranges instead.[10]

The rapid growth of San Francisco following the Gold Rush encouraged local merchants to search for wine supplies nearer to the market, with one of the most important ventures being the Italian-Swiss Colony. This was founded by Andrea Sbarboro, a native of Genoa, and originally conceived as a grape-growing cooperative, ostensibly to solve unemployment problems among Italians and Italian-Swiss in the city.[11] Sbarboro proposed to pay the workers part of their wages in company shares, thereby making them partners in the project. In 1881 some 600 hectares of land were bought north of Sonoma in a village named after the Piedmont wine town of Asti. The scheme failed as a cooperative because the workers rejected their shares and insisted on receiving full wages, so the business was run instead as a joint stock company, but with a strong paternalistic tendency.[12] The Italian-Swiss Colony prospered at wine making and integrated forward into distribution, selling its chianti-type wine under its "Tipo" label in raffia-covered bottles imported from Italy. Pietro Rossi, a Piedmontese chemist, was the driving force behind the colony, and he established one of the first laboratories in California to control wine quality.

There were also a number of very large private estates. Agoston Haraszthy's Buena Vista vineyard by the early 1860s included over 160 hectares of vines, which Haraszthy believed to be the "largest in the United States" at the time.[13] Buena Vista was relatively modest, however, compared to Leland Stanford's Vina Ranch between Red Bluff and Chico, which had 1,450 hectares of vines by 1887, although all 64,000 hectoliters of wine produced in 1890 had to be distilled because of its poor quality.[14] Stanford's original purpose was to produce a cheap, light, dry wine, but the land was too fertile and climate too hot so instead the ranch specialized in the production of brandy and sweet wines, with the former winning a high reputation.[15] Other large vineyards included the Italian Vineyard Company, which had 800 hectares of vines in a single field near Cucamonga in the early 1890s, and the Riverside Vineyard Company, with 1,000 hectares of vines.[16]

California and other parts of the New World suffered a shortage of both skilled and unskilled labor, and the amount of physical labor required for planting new vineyards, constructing wineries and cellars, and building roads was im-

[10] Carosso (1951:60–67); Pinney (1989), pp.285–94.

[11] Palmer (1965:251–75).

[12] Pinney (1989:328).

[13] Ibid., 277. It failed financially after it was turned into a joint-stock company.

[14] Ibid., 321–25.

[15] Husmann (1899:561). The Vina Ranch became part of the endowment of Stanford University and was eventually sold and the vines uprooted.

[16] Husmann (1903:417).

mense. New investment was poured into the industry at moments of high wine prices, leading to short periods of rapid expansion. For example, between 1877 and 1887 California's production jumped from 150,000 hectoliters to 579,000, and in Napa alone the number of wineries increased in a decade, from 49 (in 1881) to 166 (in 1891).[17] In 1889 Edward Roberts estimated that the five thousand vineyards in California employed some thirty thousand to forty thousand men in "cultivating, picking, storing, pressing, bottling and in otherwise caring for the crop and preparing the wine for the consumers."[18] Until the last decade or so of the nineteenth century, a significant part of this labor force was provided by Chinese.

The Chinese had a reputation as being cheap, hard working, and quick to learn. By the late 1880s they had come to dominate the state's wine labor market, and both Haraszthy and Stanford used them extensively on their estates. Controversy had surrounded their use from the beginning, but it was widely recognized that some form of cheap labor was required if California was going to develop its huge agricultural potential. However, the passing of the Chinese Exclusion Act of 1882, which both restricted the entry of new Chinese and prohibited the naturalization of those that had arrived, together with the growing hostility toward them, effectively ended the flow of new immigrants, and the Chinese population in California dropped from seventy-five thousand in 1880 to forty-six thousand by 1900.[19] The problem was especially acute on the large estates: one contemporary suggested that growers were in the same predicament as the cotton planters had been after Civil War and believed that many vineyards would end up being subdivided and worked as small farms.[20] By the late 1880s Chinese labor had become scarce and was no longer considered cheap, so growers looked to a new group of immigrants—Italians, who sometimes also brought vine-growing skills with them. Between 1890 and 1910 California's Italian population quadrupled, from fifteen thousand to over sixty thousand, which also led to an increase in demand for grapes to produce wine.

From the start, California's vineyards and wineries were designed to save on labor. To facilitate cultivation, vines were often located on valley bottoms and planted in long, straight rows. The rich soils had the advantage of producing higher yields than those on the hillsides, a fact that growers exploited as the price differential between the shy and heavy bearers was not sufficient in most areas to encourage the production of better wines. Vines were extensively planted, with about 2,500 to the hectare, and the two rows of 2.1–2.4 meters each allowed a horse to plow with two or three furrows in both directions. By the end of the century this had changed, so that on most modern vineyards the vines were being planted with one row about 1.8 meters wide and the other diagonal row 3

---

[17] Heintz (1977:16, 18).
[18] Roberts (1889), cited in Heintz (1977:36).
[19] Heintz (1977:50).
[20] Hilgard (1884:5).

meters, permitting the use of double plow teams on the wider row.[21] The rapidly growing farm-implements business of California led to new specialized equipment being produced for the vine, including gang plows with up to four shares to stir the middle of the row. A variety of methods for training the vines were adopted, but not those that obstructed cultivation, such as were popular in Germany, France, and other countries.[22] According to the historian Vincent Carosso, viticulture in California actually used less labor per hectare than many other crops.[23] By the turn of the century a single man was able to work a good, well-cultivated vineyard in full production of 8 hectares, with hired labor needed only for the harvest. The winegrower George Husmann wrote in 1883 that wine production was simple: "The very ease of the pursuit, which allowed anyone, even with the simplest culture and the most common treatment, to raise a fair crop and make a drinkable wine, has led many, in fact a large majority, to embark in grape growing who knew but little about it, and did not try to learn more." Yet he continued by noting that "easy as are grape culture and wine making here, there is a vast field for improvement; and nowhere else perhaps are rational knowledge and proper skills more needed."[24]

The first task was to find a more suitable grape variety than the mission, and growers such as Antoine Delmas, Charles Lefranc, and Emile Dresel were instrumental in bringing large quantities of cuttings into the state. The Hungarian immigrant Agoston Haraszthy made a highly publicized visit to Europe in 1861 and claimed to have collected 1,400 different grape varieties—perhaps 300 was more likely—and he also had 100,000 cuttings and rooted vines sent to his Buena Vista vineyard and winery at Sonoma.[25] Any growers' resistance to change in the old vineyards was ended with the spread of phylloxera, which destroyed large numbers of mission vines that otherwise might have remained in production. According to Haraszthy's son Arpad, the mission variety declined from about 80 percent of California's vine stock in 1880 to 10 percent a decade later, when an estimated 90 percent of the state's wine grapes were of "the best foreign varieties."[26] However, the transformation had only just begun. Professor (later Dean) Eugene W. Hilgard of the University of California's College of Agricul-

---

[21] Husmann (1896:194). See also Charles Krug in California Board of State Viticultural Commissioners (1888:45).

[22] Husmann (1896:219).

[23] Carosso (1951:19).

[24] Husmann (1883:iv).

[25] Pinney (1989:279). The vineyard failed because the vines were too close, the attempts at producing sparkling wines were unsuccessful, and phylloxera destroyed many vines (Amerine 1981:9). Agoston Haraszthy was proclaimed by the state the "father" of California viticulture in 1946, but Thomas Pinney, among others, has challenged some of his supposed successes, showing that he was not the first to bring superior grape varieties to California—Jean Louis Vignes and Kohler & Frohling had already played an important role in this—or to have introduced the zinfandel grape (Pinney 1989:263, 184; Sullivan 2003, esp. chap. 6).

[26] Haraszthy, cited in Carosso (1951:132).

ture noted in 1884 that "among the most important and at the same time most difficult questions still to be settled for California viticulture, is the special adaptation of grape-varieties to local climates and soils, and to desirable blends; and before these points are settled, many heavy losses and disappointments will be sustained."[27] Economic incentives, as elsewhere, generally encouraged growers to plant heavy bearers rather than those appropriate for fine wines.

California may have had one or two of the world's largest vineyards, but these were exceptions. Just as in Europe, the vast majority belonged to small, family farmers, and originally many dedicated part of their land and labor to other activities.[28] In 1888 George Husmann noted:

> We have thousands, perhaps the large majority of our wine growers . . . who are comparatively poor men, many of whom have to plant their vineyards, nay, even clear the land for them with their own hands, make their first wine in a wooden shanty with a rough lever press, and work their way up by slow degrees to that competence which they hope to gain by the sweat of their brow. Of these, many bring but a scanty knowledge to their task.[29]

In 1889 there were an estimated 65,000 hectares of vines and 5,000 growers producing grapes for both wine and the table, giving an average holding of 13 hectares.[30] When those used for wine production alone are considered, the 1891 survey shows that there were approximately 2,750 growers with more than 2 hectares (5 acres) of vines covering an area of 35,000 hectares.[31] For Napa and Sonoma, which at this time accounted for 46 percent of the area of California's vines destined for wine production, the average holding was 12 hectares, slightly less than the state average. In these two counties there were 531 holdings larger than 8 hectares, but these accounted for three-quarters of all vines, and the largest 220, with more than 16 hectares, occupied half the vineyard area (table 9.1). In total there were 41 vineyards with more than 50 hectares (125 acres), representing a fifth of the total area, and 85 percent of them had their own wineries.[32] At the other extreme, 61 percent of holdings accounted for 26 percent of the vines, but only 1 in 10 had their own wineries. If growers with less than 2 hectares were included, the extremes in property

[27] Hilgard (1884:3).

[28] Pinney (1989:337).

[29] Husmann (1888:iii).

[30] Roberts (1889:199). The figure includes vines for table grapes and raisin. Sullivan (1998) gives a figure of 36,500 hectares for vines at this time.

[31] California Board of State Viticultural Commissioners (1891). My calculations. The figures are clearly approximate, as the size of many vineyards is rounded to 10, 15, 20 acres, etc. In addition, there are few vineyards that specifically state that they had young vines not yet in production (these are excluded from table 9.1). Some growers also switched between producing grapes for the table and grapes for wine, according to market prices.

[32] Tulveay Vineyard is excluded as no information is provided concerning whether it had winemaking facilities.

TABLE 9.1
Vineyards and Wineries in Napa and Sonoma Counties, 1891

Size of holdings (hectares)	Number of holdings	Total area of vines (hectares)	% of all vines in this category	Number of wineries	% of holdings with winery	Hectares per winery
0–4	398	1,252.6	7.7	26	6.5	48
4–8	445	2,947.0	18.2	56	12.6	53
8–16	311	3,843.7	23.7	88	28.3	44
16–25	100	2,134.4	13.2	32	32.0	67
25+	120	6,033.6	37.2	82	68.3	74
Total	1,374	16,211.3	100.0	284	20.7	57

Source: California, Board of State Viticultural Commissioners (1891); my calculations.

ownership would be even greater. In Sonoma, for example, there were 9,068 hectares of vineyards cultivated in 815 holdings of over 2 hectares, but these statistics exclude "at least five hundred acres of small vineyards of less than five acres (two hectares), planted for family use."[33]

The advantages of the family grape-producing farm, namely, low monitoring and supervision costs associated with labor and the absence of incentives for land abuse, were offset by their lack of scale in wine production and their weakness in marketing the wine. When production was just for local consumption, these problems were unimportant for family growers. However, the rapid growth in California's wine production after 1880 was not based on the expansion of rural, local markets but was linked to urban demand, both in San Francisco and in distant markets such as New York, Chicago, and New Orleans. The distinct economic advantages offered by small, family grape producers in association with large-scale wine making facilities were appreciated by Percy T. Morgan of the California Wine Association, who, while noting in 1902 that merchants were organizing "great vineyard companies with ample capital for the laying out and planting of vast tracts of wine grapes," added that "it is not from these great tracts that the larger portion of the tonnage for wine purposes will be derived. It is from the small vineyardist, cultivating from ten to fifty acres (four to twenty hectares); cultivating and looking after his lands individually, and thereby obtaining from 30% to 50% more tonnage to the acre than is possible from the great vineyard tracts, that the very remunerative results will accrue."[34]

[33] De Turk (1890:45).

[34] Morgan (1902:95), which reproduces an earlier article in the Sacramento Bee. Morgan probably exaggerates the competiveness of the small farm, not just because his organization would benefit most from an increase in the supply of grapes, but also because the political support of grape producers, numerically by far the largest sector of the industry, helped justify the CWA's demands for high tariffs and federal legislation to control adulteration and fraud.

If grape growing in a hot climate was easier than in cooler regions such as Champagne and Burgundy, wine making presented more problems. The first obstacle in California was the lack of practical knowledge shown, for example, by the fact that because most Europeans and "all Germans," picked their grapes as late as possible, many Californian winemakers initially did the same. Late picking maximizes the grape's sugar content, an important consideration in northern Europe, but hardly a problem in Southern California, or indeed anywhere else in the state. Instead growers in hot climates needed to pick early to conserve the grapes' acidity, which helped preserve the wine, as well as learn not to mix ripe with unripe grapes, to exclude moldy or rotten grapes, to choose a vessel with adequate size for fermentation, to clean and disinfect all the utensils, and so on.

The early 1880s saw a rapid increase in the construction of commercial wineries using labor-saving designs. They were cut deep into the hillsides, and grapes were taken by horses and wagons to the top floor for crushing. Where this was not possible, grapes were moved to the top floor by an elevator. Once the grapes were crushed, the juice fell by gravity to the fermenting vats on the floor below and finally to the bottom of the winery to mature. Hand pumps were labor-intensive, and winemakers started investing large sums at the turn of the century to introduce gasoline and then electricity generators. Pumping was crucial, not just to move wine from one barrel to another, but also to provide the cool water needed to reduce the temperature of the fermenting must.[35]

The development of viticulture in the cooler coastal regions around San Francisco made it much easier, at least in theory, to produce dry table wines. Indeed, as the hot and dry summers allowed the grapes to mature sufficiently, winemaking was easier than in Missouri and the eastern states, or in central and northern Europe. George Husmann's wines produced on the Talcoa Vineyard in Napa were fully fermented after six days and cleared enough to be sold to merchants six weeks later for maturing; they were then shipped "as far East as Connecticut, when not more than a year old, and arrive in perfect condition."[36] Yet many winegrowers were not so successful. Eugene Hilgard, in a famous article in the *San Francisco Examiner* of August 8, 1889, blamed the serious overproduction at that time on the poor quality of wines being produced. According to Hilgard, the reasons for the poor wines included (1) the wrong choice of grape varieties; (2) growers cultivating too large an area so that grapes were not harvested at the correct moment; (3) grapes not properly sorted prior to crushing; (4) fermentation tanks that were kept too full so that there was no room for a protective cover of carbon dioxide; (5) the use of excessively large fermentation tanks and hot grapes; and (6) improper handling of the fermentation and subsequent activities.[37] The solution for Hilgard was the development of scientific wine making,

[35] The Inglenook winery is an example of the first type, and Eshcol winery of the second. Both buildings still stand today. Heintz (1999:102, 211).

[36] Husmann (1883:284).

[37] Cited in Amerine (1981:11–12). See also Husmann (1883:288) for a similar list of complaints.

and this could only be achieved through the construction of large industrial wineries.

Charles Wetmore, a leading member of the Board of State Viticultural Commissioners, strongly disagreed with Hilgard on the need for larger wineries and in particular the separation of grape-growing activities from wine production. He believed that small wineries could produce fine wines, and that if growers stopped making their own wines, their economic incentives would be to use grape varieties that maximized output rather than quality.[38] Other writers, such as Husmann and Bioletti, also insisted that fine wines were best made in smaller cellars, "under the constant, watchful supervision of the proprietor, who must himself be an expert connoisseur in wines and must know just how to handle the product at every stage during, as well as after, its manufacture."[39] However, these fine wines accounted for perhaps only 5 percent of the state's production, and there is no doubt that by the turn of the twentieth century, when skilled enologists and laboratories in wineries were very scarce and market demand was for cheap dry and dessert wines, these were most economically made in large, modern wineries.[40]

Change happened fast. The 1891 survey reports over six hundred wineries, but Edward Adams wrote a few years later that there were only between two hundred and three hundred winemakers that were "equipped with the plant necessary to produce good wine on any commercial scale."[41] The Census of Manufactures gives a lower-bound estimate of the factory share of the value of wine output as just 13 percent in 1880, but this rises rapidly to 38 percent in 1890 and 78 percent by 1900.[42] The *Pacific Wine and Spirits Review* also noted that while previously nearly every vineyard had had its own fermenting house and storage cellar, by 1900 most growers sold their grapes to winemakers, "except to a limited extent in some of the older districts."[43]

PRODUCTION INSTABILITY AND THE CREATION OF THE CALIFORNIA WINE ASSOCIATION

There were five major players in the California wine trade in the late nineteenth century: grape producers; winemakers; San Francisco shippers; the East Coast bottlers and distributors; and consumers, who were themselves deeply divided

---

[38] Wetmore (1885:27). The Board of State Viticultural Commissioners and the University of California clashed over resources and research priorities frequently during the 1880s (Carosso 1951:141–43; Pinney 1989, chap. 13).

[39] Husmann (1899:559).

[40] Bioletti (1909:384).

[41] Adams (1899:520).

[42] Olmstead and Rhode (2010:280).

[43] *Pacific Wine & Spirits Review*, December 1906, p. 43.

between those who drank alcohol and those who were temperate. The development of new markets after 1880 created incentives to expand production, but new coordination arrangements were required between the different links in the commodity chain if an adequate return on the high levels of capital investment now required was to be achieved. In particular, commercial growers needed a market for their grapes; winemakers required both a supply of grapes, to benefit from the potential economies of scale associated with the new wine-making equipment, and a ready market for the wine; and local merchants found it necessary to maintain their competitive position and block out East Coast merchants dealing directly with local producers. While some of these problems were not dissimilar to those found in Europe at this time, they were solved in very different ways. The weak bargaining power of small growers in the California wine industry made it impossible to obtain concessions from government to create new institutions such as those found in France. By contrast, the favorable business climate for large enterprises led to the creation of a trust that integrated both horizontally and vertically, from the production of grapes through to the distribution of wines.

The California wine industry switched from boom to bust on a number of occasions. After a period of rapid growth, which saw the area of vines increasing by 50 percent between 1873 and 1876, the industry plunged into depression because of the difficulties associated with marketing outside California.[44] In Napa, savings banks refused to make loans on vineyard property as they considered that vines did not increase the land's value, and elsewhere farm animals were allowed to wander in the vineyards before the vines were pulled up and sold for firewood.[45] The recession was ended by phylloxera in France, which led to a sharp increase in wine prices, and prosperity was reinforced by higher tariffs in 1874 and then again in 1883. Imports fell from 372,000 hectoliters in 1873 to 129,000 in 1885.[46] Between 1877 and 1887 the area of vines in Sonoma Country increased from 2,800 to 8,780 hectares, while in Napa the increase was from 1,360 to 5,840 hectares. For Eugene Hilgard, writing in 1884, only the appearance of phylloxera in California itself and the impact of the Exclusion Act were obstacles to future growth.[47] The boom lasted until 1886, when a large, inferior crop caused prices to collapse, but harvests continued to grow because of the excessive plantings in the early 1880s.[48] The wine crisis resulting from the panic of 1886 continued well into the 1890s as the economy suffered the severe downturn of 1893.

The wine crisis of the decade from 1886 to 1895 was typical of those affecting the New World in this period, and not dissimilar to the *mévente* of the Midi in

---

[44] Carosso (1951:74, 95).
[45] H. W. Crabb, quoted in Husmann (1883:171); and Hilgard (1884:1).
[46] California State Board of Agricultural (1912:199).
[47] Hilgard (1884:4–5).
[48] Carosso (1951:134).

the 1900s. As supply exceeded demand in the major markets, merchants cut back on their wine purchases, so winemakers required fewer grapes. Although wine production had undoubtedly improved in California, a significant amount of cheap and adulterated wines were sold in the major markets, and the price differential between quality and quantity wines was reduced. Charles Wetmore noted in 1894 that "the man who gets ten tons of grapes to the acre gets 10 cents for wine; the man, who on a steep hillside, gets two tons and a half, gets 12 cents; and the 12-cent wine is mixed with the 10-cent." Some growers responded to low prices by grafting high-yield, low-quality varieties to their better vines, while others carried out only the minimum operations necessary, leading to disease running "unchecked in vineyards." Over 12,000 hectares of vines were uprooted in the early 1890s.[49] Yet as George Husmann noted in the 1896 edition of his book, the asset-specific nature of wine-making installations made wine owners even more vulnerable to downturns: "It was very clear to them that their immense storage houses, casks and machinery, and all the capital it had cost to build up their trade, would be wholly unremunerative if the vineyards died out, and they would be in worse plight than the producers, who had at least their lands, to cultivate in some other crops."[50]

California's wine crisis of the early 1890s affected both growers and winemakers, with adulterated wines driving down the price of genuine ones. The large number of growers that had no wine-making facilities (around four-fifths in 1891) were particularly vulnerable, and the *Pacific Rural Press* in 1894 claimed they depended on "three or four firms of San Francisco" for their livelihood. This grower's newspaper argued that a large percentage of the fifty million dollars invested in the wine industry belonged to the growers and producers, so the merchants' offer of five or six cents per gallon after the 1893 harvest gave a return on capital of just 2.5 percent, leaving growers nothing for labor and expenses." Despite these low prices, the newspaper claimed there was "no overproduction of wine in this State, unless it be in the 'brick vineyards' of San Francisco."[51]

To resolve the problems of poor quality and artificial wines, Professor Hilgard in 1889, along with others, proposed district cooperatives for labeling, bottling, and marketing wine and criticized the winemakers and merchants for emphasis on "pretty bottles and beautiful labels" rather than attempting to improve quality. As the established banks refused to lend on wine, an attempt was made to create a grape growers' and wine merchants' bank, but this failed because of its inability to secure financing, and the growing distrust between the two sides of the industry.[52]

A combine was proposed that would allow both growers and winemakers to

[49] Wetmore (1894), cited in Pinney (1989:355–56).
[50] Husmann (1896:260).
[51] *Pacific Rural Press*, June 9, 1894, p. 438.
[52] Carosso (1951:136–37).

pool their production and give them absolute control over the wine making and marketing. The original idea was for a growers' contract lasting several years, with prices increasing over time, but with the contracts only becoming binding when 75 percent of the total acreage of wine grapes had been secured.[53] Several hundred of the largest producers signed up, including Pietro Rossi and Andrea Sbarboro of the Italian-Swiss Colony, and the California Wine-Makers Corporation (CWMC) was established in 1894. Within a few months the price of ordinary wine increased from 6 cents to 12½ cents per gallon, in part because the CWMC was allowed good credit facilities, and in part because of the small harvest of 1894.[54] To sell the wine, the CWMC entered into a five year agreement with the California Wine Association, an organization also created in 1894 by the leading West Coast wine dealers (see below). Postcontractual opportunistic behavior quickly appeared when harvests recovered and lower prices followed. By the summer of 1897 a "wine war" had broken out between the two rival organizations, caused by the CWA's refusal to pay the prices demanded by the CWMC, and the CWMC, in violation of the contract, selling a million gallons (3.8 million liters) of wine and an option on another 1.5 million gallons directly to the New York house of A. Marshall & Co. , thereby allowing a powerful East coast merchant direct access to producers.[55] Although the CWMC controlled perhaps 80 percent of the state's wine producers by 1897, the CWA had 80 percent of the wine produced,[56] and the legal defeat for failing to honor the initial agreement saw the CWMC effectively disappear in 1899.[57] Almost immediately Henry J. Crocker, ex-president of the CWMC, attempted to start a new organization. He circulated over 1,400 letters to the state's grape growers in which he offered contracts for seven years and demanded a minimum of 80 percent of the

[53] The price of grapes and wine was expected to increase by 80–100 percent, depending on the grape variety and whether production was in the plain or hills. See *Pacific Rural Press*, July 28, 1894, p. 50; June 23, 1894, p. 406.

[54] Ibid., March 30, 1895, p. 199. The newspaper noted in May 1895 (p. 324):

the first pro rate distribution for the deliveries during the month of April has just been made, and makes our wine makers feel once more that they are not owned, body and soul, by a few wholesale dealers, who paid about what they pleased to those who were in sore need of money, and ruined the prices outside, but cutting each other's throats, regardless of what became of the 'goose that laid the golden egg.'

The 1895 harvest was also small, "but the quality of the product was the best for many years." Ibid., October 26, 1895, p. 271. See also Adams (1899:517–24).

[55] The response of the CWA was to reduce its own wine prices, with the object of discouraging Marshall and other eastern dealers from competing directly with the large California wine houses (*Pacific Rural Press*, June 5, 1897, p. 353).

[56] Peninou and Unzelman (2000:79).

[57] The CWMC was required to pay $101,000 to the CWA, although the final settlement was $8,000 (*Pacific Rural Press*, May 13, 1899, p. 289). Sales of wine and other assets continued for several years.

state's wine grapes.[58] The *Pacific Rural Press* noted that these grape prices could be achieved only if merchants paid artificially high prices for their wine, and that, with the exception of raisins, no other combine had been able to attract 80 percent of the crop.[59] Crocker's project failed because of the diversity of the growers' interests. One difficulty was that of assessing grape quality, especially as many of the smaller producers were beginning to plant shy-bearing European varieties on the hills, which produced small quantities of quality wine rather than the heavy bearers found on the plains. Another problem was that any organization that monopolized grape production would naturally want to use the best wine-making facilities available, and those small growers that already possessed older wine-making facilities feared that their investment would become worthless if these facilities were no longer used and they ceased to be winemakers.[60] Higher grape prices would also encourage new plantations and therefore depress future wine prices.[61] Finally, many growers without wine-making facilities resented being locked into long-term contracts, especially as they were benefiting from the competition from three major buyers: the CWA, Lachman & Jacobi, and the Italian-Swiss Company.[62]

The possibility for growers to have wineries competing for their grapes proved short-lived, however, as the CWA took control of half of both Lachman & Jacobi and the Italian-Swiss Company.[63] Yet there were limits to how far the CWA could dictate grape prices to the growers, as Percy Morgan noted in a letter to the company's stockholders in 1903:

> The policy of your management is to encourage Grape Growers to feel that they have a friend and partner in the Wine Merchant who is willing to share with them whatever prosperity the business affords. With this view prices were paid for grapes during the vintage which caused the largest part of the profit which has hitherto been derived from wine making to inure to the growers. This policy should meet the approval of all well-wishers of the industry, for unless grape growing is remunerative the business would languish and there would be no incentive to plant the new acreage necessary to off-set the damage by phylloxera and to supply the increasing demand for Dry Wines.[64]

The increase in production and wine prices in real terms from the late nineteenth century suggests that CWA successfully regulated the sector, especially as other fruit farmers in California were experiencing a decline in prices at this

[58] Ibid., July 8, 1899, p. 30.

[59] Ibid., August 12, 1899, p. 111; August 19, 1899, p.114.

[60] Crocker assured producers that the grapes would be sent to the nearest winery. Ibid., August 19, 1899, p. 114.

[61] This problem is discussed in chapter 11.

[62] Ibid., August 18, 1900, p.102.

[63] Peninou (2000:82, 89). In 1911 the Italian-Swiss Colony had an annual output of 225,000 hectoliters from eight different wineries.

[64] February 26, 1903, letter to stockholders (CWA minutes, vol. 2, February 26, 1903:240).

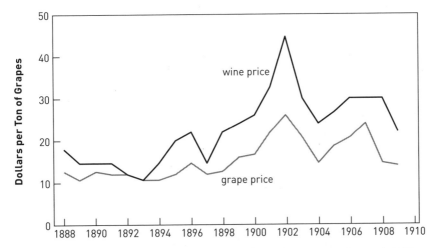

Figure 9.2. California grape prices and wine prices (equivalent per ton of grapes), 1888–1907. Source: Shear and Pearce (1934: table 26). The authors note that the "data are based on somewhat fragmentary and, at times, conflicting trade quotations." The wine price has been converted into an equivalent per ton of grapes.

time.[65] Furthermore, the fact that grape prices in general moved in line with both dry and dessert wine prices suggests the grape growers also shared that prosperity (figs. 9.2 and 9.3). Although no information is available on how many grapes the wineries outsourced, the CWA owned over 3,000 hectares, or less than 5 percent of California's wine grape acreage, but was responsible for 84 percent of wine output.[66]

THE CALIFORNIA WINE ASSOCIATION AND THE MARKET FOR CALIFORNIA'S WINES

In January 1904 Percy T. Morgan noted that the "quality and general excellence" of California's wines were "no longer in question," but that per capita wine consumption in the United States was less than a fiftieth of that in "wine drinking countries like France and Italy." He claimed that "almost every rolling hill and fertile valley of California could be profitably covered with vines," so that the future of the industry was "principally, if not entirely, dependent upon an expanding market."[67] Morgan did not have to worry about foreign competition, as the rapid

[65] Olmstead and Rhode (2010:285) note that rising wine prices would be incompatible with "persistent overproduction." For other fruit prices, see Rhode (1995:780–84).

[66] Peninou (2000:125, 134). It is not clear whether the CWA acreage includes that of other companies associated with it, such as the Italian-Swiss Colony.

[67] Morgan (1904:36).

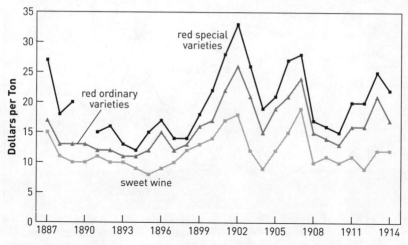

Figure 9.3. California prices for dry red and dessert wines, 1891–1913. Source: Shear and Pearce (1934: table 25)

growth in the California wine industry between the late 1870s and 1880s had taken place against a background of wine shortages in France because of phylloxera and high U.S. import duties on foreign wines in 1874 and again in 1883. Exports of French wines to the United States declined from 236,000 hectoliters in 1873 to 23,000 in 1889, while California's production increased from 95,000 to 587,000 hectoliters, and out-of-state exports rose from 19,000 to 310,000 hectoliters, between the two dates.[68] As Carlo Cipolla noted, the character of the wine industry changed dramatically, from selling two-thirds of California wine production in the state at the end of the 1870s to only 10 percent by the end of the 1890s.[69] In about three decades California increased its output tenfold, and its share of U.S. wine consumption rose from one-fifth to four-fifths (table 9.2).

For contemporaries, a major restriction to the growth of the industry was adulteration and fraud, and in 1906, despite the supremacy of the state's wines in the national market, California growers and merchants complained that their best wines were sold under foreign labels and their poorer ones, together with those from every other wine-producing state in the country, under the California brand.[70] The motives, as one writer at the end of the century noted, were

---

[68] Cipolla (1975: 299) for French imports. Calculations refer to wine imports in casks; for California wine production and for out-of-state sales, *Pacific Rural Press* (April 21, 1900:246).

[69] Cipolla (1975:303).

[70] The *Pacific Wine & Spirits Review* (December 1906:20), writes

Heretofore the greater part of California's best wines have gone to the Eastern markets in bulk and they have been sold to hotels, cafes and high-class bars, only to appear under French labels, while the commoner wines have gone into consumption as the products of California.... In other words...if California wines of the higher class had not largely masqueraded under foreign labels, the industry would have had a far greater development.

TABLE 9.2
Origin of Wines Consumed in the United States

	California (hectoliters)	California (% of total)	Other states (% of total)	Imports minus exports (% of total)	Per capita consumption (liters)
1870–73	130,686	19.6	23.7	56.6	1.6
1874–78	157,202	20.2	54.7	25.1	1.7
1879–83	323,457	38.0	38.2	23.7	1.6
1884–88	575,776	59.0	24.1	16.9	1.7
1889–93	670,097	64.9	19.1	16.1	1.6
1894–98	651,953	81.3	7.4	11.2	1.2
1899–1903	1,079,466	84.5	3.9	11.6	1.6
1904–8	1,333,528	78.7	6.6	14.6	2.0
1909–13	1,651,379	78.5	10.0	11.6	2.2

Sources: My calculations; Shear and Pearce (1934), table 41.

obvious: "A bottle of wine, which as domestic, could be sold at a good profit for fifty cents, with no extra expense except the affixing of a new label, may bring, as imported, a dollar and a half, and the customer be just as well pleased and as well served. Labels of all known brands, with imitations of special corks, bottles, or other peculiarities, are kept constantly in stock in all cities, at trifling cost, to be used for this purpose."[71] Adulteration was also the consequence of poor wine-making techniques, and in California as elsewhere, some producers turned to chemists not just to improve wine quality, but to cover their errors.[72] Complaints of adulteration and the fraudulent use of the "California brand" inevitably coincided with the low wine prices and glutted markets such as the late 1880s and early 1890s, but the nature of market demand led to very different solutions in California from those in Europe. The problems associated with wine adulteration were solved by three interconnected strategies: horizontal consolidation and vertical integration; the Pure Food and Drugs Act of 1906; and the tax reform for brandy in 1891.

Agoston Haraszthy as early as 1862 had proposed that the California wine industry fund an agent in San Francisco to organize the sale of the state's wines and vouch for their authenticity.[73] In 1872 a joint-stock association, the Napa Valley Wine Company, was formed to promote the region's wine in eastern cities.[74] In San Francisco itself the Board of State Viticultural Commissioners spon-

[71] Adams (1899:519).

[72] The commissioner of agriculture in 1887, noted that wine adulteration "has increased in amount and in the skillfulness of its practitioners until at the present day it requires for its detection all the knowledge and resources which chemical science can bring to bear upon it, and even then a large part doubtless escapes detection." Crampton (1888:207).

[73] Haraszthy (1862:xxi), cited in Carosso (1951:54).

[74] Carosso (1951:93); Heintz (1999:97–98).

sored in 1888 a permanent exhibition that allowed consumers to purchase locally branded wines.

However, it was a private initiative that led to the California Wine Association being created in August 1894 when seven of the state's largest wine firms formed a trust with a capital stock of $10 million.[75] All assets of these seven wine houses were turned over to the CWA, although wine would still be sold under some of their labels and trademarks. In 1900 the CWA acquired half interests in the three largest independent wine houses still remaining and continued to buy up other wineries. It recovered quickly from its considerable losses in the San Francisco fire of 1906, and in the same year work started on a huge wine complex called Winehaven on a 19-hectare site at Richmond, which had storage for some 378,500 hectoliters, or about 25 percent of California's output, and wine-manufacturing facilities for 140,000 hectoliters.[76] This was an experiment in horizontal consolidation and vertical integration—from grape growing to distribution—on a massive scale.

A major objective of the CWA was to improve wine quality, and in 1896 Husmann claimed that there had been rapid advances, in part because of the planting of better grape varieties in adequate locations, but also because wine making was being conducted more scientifically.[77] In that year Bioletti argued that the high temperatures found during fermentation explained "nine tenths of the trouble to which our wines are subject."[78] Wineries were quickly modernized, and Lachman noted in 1903 that "in the last eight years rapid progress has been made in the manufacture and maturing of wines, wine making having been conducted on more scientific lines," and the *Pacific Wine and Spirit Review* (hereafter *PWSR*) also noted the improvements in wine quality with the separation of wine making from grape production.[79] Neither comment is completely objective, as Lachman was a CWA director and its chief taster and the *PWSR* was also dependent on the same organization, but it has been said that the CWA never produced a bad wine, or an outstanding one.[80]

Although vertical integration helped the CWA to control the quality of its own wines, large quantities of California wines were still reportedly being sold under foreign labels. In addition, poor-quality wines of dubious origin, such as those that the *PSWR* dubbed as being manufactured chemically in Ohio's "brick vineyards," were sold as originating from California, thereby damaging the state's reputation. The CWA particularly welcomed the Federal Pure Food and Drugs Act because previous regulations had covered only interstate trade, and it had

[75] Carpy & Co., Kohler & Van Bergen, Dreyfus & Co., Kohler & Frohling, Lachmann Co., Napa Valley Wine Co., and Arpad Haraszthy & Co. Haraszthy withdrew shortly afterward.
[76] Peninou and Unzelman (2000:104).
[77] Husmann (1896:258).
[78] Bioletti (1896:39). See also Hilgard (1886).
[79] Lachman (1903:25); *PWSR* (December 1906:43).
[80] Amerine and Singleton (1977:286).

been possible to legally ship a barrel labeled "poison" between two states and then sell it as wine if the destination state had not passed its own pure food law.[81]

However, as in other wine regions, what was considered fraud and what legitimate was often very parochial, and in the same issue that *PWSR* criticized the brick vineyards, highly favorable comments were made concerning the "more than 2,000,000 bottles of *genuine champagne*" annually produced in the United States.[82] California wines that were described as a particular French type, such as "Burgundy" or "Champagne," were also rejected for the Paris Exposition of 1900. A couple of months later the same journal reproduced a cartoon (fig. 9.4) from the *Chicago Tribune* that neatly combines both the East Coast prejudice against genuine fine California wines and the widespread adulteration of French wines and the Midi protests, which was widely reported in the international press during that year.[83]

The third major factor in improving quality was legislation that allowed California's winemakers to fortify their wines with domestic grape brandy and not pay the $1.10 a gallon tax on brandy. This made dessert wine the cheapest form of alcohol on the market and allowed producers to profitably distil their unstable wines and produce sweet fortified wines.[84] It also implied that they did not have to resolve the technical problems associated with making wines in a hot climate and therefore was especially attractive to producers in the Central Valley. California's total wine production increased by a factor of four between 1891–93 and 1911–14, but for dry wines the increase was just a third compared with dessert wines, which saw their production multiply ninefold (fig. 9.5). Fortified wines kept longer when open, deteriorated less through transportation, and were easier to brand because quality could be kept constant from one year to the next, all factors that encouraged the CWA to increase their production. By 1913 dessert wines accounted for 45 percent of all California's production, of which 46 percent was classified as port, 31 percent as sherry, 12 percent as muscatel, and 9 percent as angelica, a wine named after Los Angeles.[85] There was also a major growth in the production of brandy, and the economies of scale found in both dessert wine production and distilling[86] and the legal requirements of having to have an excise officer present at all times benefited a large producer such as the CWA. On the eve of the First World War brandy production used 40.8 percent of wine grapes, compared with 35.5 percent for dry wines and 23.7 percent for dessert wines (fig. 9.6).[87]

During its first decade, the CWA distributed wines in bulk to wholesalers, but

[81] *PWSR* (May 1907:14).

[82] Ibid., 39. My emphasis.

[83] Ibid. (July 1907:14).

[84] Seff and Cooney (1984:418).

[85] Angelica was fortified grape juice rather than a wine. Calculated from California State Agricultural Society (1915:140).

[86] See Amerine and Singleton (1977:171).

[87] Period refers to 1909–13. Shear and Pearce (1934, table 42).

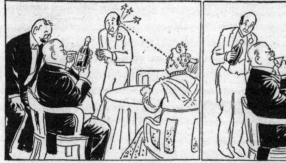

Figure 9.4. Cartoon, "What the French Wine Strike Is Proving," *Pacific Wine and Spirit*, July 1907, p. 14. Attributed to *Chicago Tribune*, June 12, 1907. Courtesy of Special Collections, University of California Library, Davis.

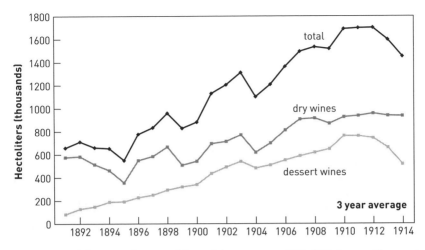

Figure 9.5. California production of dry and dessert wines, 1891–1913. Source: Shear and Pearce (1934: table 10)

Morgan talked of a change in company strategy in 1905 when he informed stockholders that he wanted to create brands of bottled wines and advertise them "in the hope of educating the public to a greater appreciation of the excellence of the better varieties of Californian wines."[88] A decade later, in 1917, he wrote:

> Until the coming of the California Wine Association only a few wineries tried to deliver their original packages direct to the consumer and build up a following for their label. The large dealers almost always sold California wines in bulk to distant jobbers who either bottled them with a domestic or foreign label known in their particular localities, or sold them to retailers who pursued a similar course. Moreover, these distributors and retailers had neither the knowledge not the facilities to age and handle wines properly. Only a large firm with capital could select from millions of gallons, blend to standards, market under labels that could gain the confidence of the public, and stand back of the label wherever sold.[89]

A number of reasons can be cited for why horizontal consolidation and vertical integration on this massive scale were chosen as a means of controlling quality in the U.S. market. In the first instance, the distance between California and its major markets (New York, New Orleans, and Chicago) was significant. It could be argued that the Midi or Algerian producers also faced problems of distance in France when they sold their wines in Paris and the industrial North, but in this case they were selling to consumers already accustomed to wine drinking. California, by contrast, had to create new markets, so the com-

---

[88] CWA minutes, vol. 2, February 26, 1903, p. 265.
[89] Percy Morgan (1917), cited in Peninou and Unzelman (2000:125). See also p. 94.

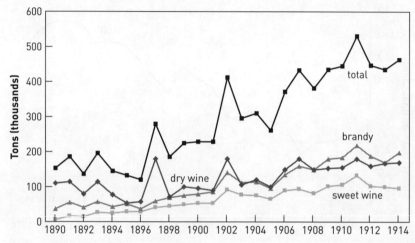

Figure 9.6. Grapes used in wine and brandy production. Source: Shear and Pearce (1934: table 42)

parison should be with selling French or Spanish wine to the reluctant wine drinkers in London rather than Paris. A second point was the limited political voice of the sector, and in particular of the growers. A final one is the heavy capital investment and asset-specific nature of this technology that led to wine production being just one of many sectors that participated in the "great merger movement." According to Naomi Lamoreaux, consolidations became common to escape price competition, especially in capital-intensive, mass-production industries where no single firm had a clear-cut advantage, and where expansion had been rapid in the years leading up to the depression of 1893.[90] This description matched clearly that of the California wine industry at this time. Most of the large, capital-intensive wineries had been constructed only a few years earlier, and the possibilities of collusion in the wine market were very difficult. One important consideration in the case of the CWA was the need to establish its own distribution networks. Although information for the United States is limited, retail wine prices fell much less in the major centers of consumption, such as Paris or Buenos Aires, than in areas of wine production.[91] By integrating forward and controlling distribution, winemakers could both obtain higher prices and guarantee a market for their wines. This helped provide the market stability for the group to invest in brand names.

Finally the improvement in the reputation of California wines also allowed a few independent wineries to begin to specialize in fine wines for niche markets, with the smaller Napa wineries such as Inglenook, Charles Krug, Beringer, and

[90] Lamoreaux (1985:45, 87).
[91] For Buenos Aires, see table 11.7.

Salmina (Larkmead) consistently taking awards and bottling their best wines with labels displaying "Napa" as their appellation.[92] By 1901 the *Pacific Rural Press* claimed that fine wines accounted for 5–10 percent of the state's output.[93]

Wine was a small player in the U.S. alcohol industry, and while over the half century prior to the First World War annual per capita consumption increased from 1.2 to 2.2 liters, beer consumption jumped from 20.1 to 79.2 liters, although spirits fell from 9.5 to 5.7 liters. Beer consumption increased not just because prices remained stable, but also because quality improved significantly as the average size of breweries increased twenty-four-fold between 1873 and 1914.[94] The success of the California wine industry was limited, and at the very best wine consumption in the state increased from around 20–25 liters per capita in the 1880s and 1890s to 36 liters in the mid-1900s, a figure that was little more than half the 60 liters found in Argentina or Chile.[95] However, the industry made major strides in improving both quality and market organization. At the College of Agriculture the work of individuals such as Eugene Hilgard and Frederic Bioletti was important to help growers tackle phylloxera and improve wine making. By 1914, California was part of the international scientific network that included research institutes in the Midi, Algeria, and Australia. The difficulties associated with making dry wine in hot countries were only beginning to be resolved on the eve of the First World War, but by this date a winery with adequate facilities and the correct expertise was able to produce an acceptable *vin ordinaire*. Prohibition ended much of this research, and the considerable human capital that had accumulated in the state's vineyards and wineries over the previous quarter century was lost, contributing to the delay, not just in California but in the New World in general, in the development of suitable wines for the European markets.[96]

What made California very different from other wine regions in this period, however, was the consolidation of the leading firms to create the CWA, and the degree of vertical integration of wine making and distribution that occurred. This allowed the CWA to sell branded wines in bottles directly to retailers and guarantee their purity. The fact that these were ordinary table or dessert wines rather than fine wines increased, rather than diminished the achievement. However, while the CWA appears to have had considerable success in producing a drinkable wine to sell among consumers who had originated from Europe's

---

[92] Lapsley (1996:67–68); Heintz (1990)

[93] *Pacific Rural Press* (December 14, 1901:372).

[94] Shear and Pearce (1934, table 2); Siebel and Schwarz (1933:78).

[95] The figure for 1878–82 was 24 liters, and for 1888–92, 22 liters. My calculations, from Cipolla (1975, table 2) for 1907; and *PWSR*, January 1909, p. 6, for 1906–1908. However, this source includes wines that were exported out of state by rail but not boat. The 36 liters also includes those wines used for distilling for brandy.

[96] Grapes continued to be produced for grape juice, raisin, and table grapes, but different varieties were used.

wine-producing regions it was less successful in creating new markets, especially outside the major centers of New York, Chicago, and New Orleans. There is little doubt that the CWA used its market power to fix grape and wine prices, but the rapid growth in the industry between 1894 and 1914 also suggests that all sectors of the industry benefited to some degree. The CWA brought greater stability to the market, and California probably suffered less during the turbulent 1900s than any other-wine growing region in the world, with the years of greatest difficulties being in the late 1880s and early 1890s, immediately prior to the founding of the CWA.[97]

Growers had limited political voice to influence legislation as their French counterparts did, but the CWA could manipulate prices which would have been unacceptable in countries where wine was an important item of popular consumption, such as Argentina. The industry benefited significantly from the high import tariffs and the elimination of the excise tax used to fortify domestic wines, which allowed a rapid growth in dessert wines and brandy, two commodities that were well suited to the company's large-scale industrial wineries. Unfortunately, while wine was of only modest importance to the U.S. economy, the same was not true for the alcohol industry as a whole. The latter was opposed by an increasing number of citizens, who succeeded in getting Prohibition approved by the federal government. Prohibition became effective in January 1920 and lasted until it was repealed in 1933. According to the wine historian Thomas Pinney, "even though the industry was not absolutely finished off, it was seriously diminished, obstructed, and distorted."[98] However, Prohibition made the sale of alcohol illegal, not its manufacture or consumption. Between 1920 and 1933 the area of vines actually doubled, and while some of the grapes were used for the table, to make fresh grape juice and raisins, a considerable, but an unknown quantity was turned into wine in private homes. The demands of home producers led to growers grafting their vines with varieties that produced poorer wines, but that had grapes with thicker skins that could withstand better transportation.[99] One estimate by the Wickersham Commission suggested that 4.2 million hectoliters of wine were produced annually between 1922 and 1929, slightly more than double what had been commercially produced in the final year be-fore Prohibition.[100] Warburton, by contrast, believed that consumption of all wines—sacramental, medicinal, and homemade—increased by 65 percent between 1911–14 and 1927–30, while beer consumption declined by 70 percent

[97] The CWA failured to buy grapes in 1908 because of the drop in demand because of Prohibition.

[98] Pinney (2005:11). It was repealed in part because of the need to increase taxes and create employment.

[99] This was not new, as Arpad Haraszthy noted in 1888 that "considerable quantities of wines" were made in San Francisco by the Italian, French, Spanish, and Portuguese population. These were for home consumption, or sold in a "small way to their neighbours and friends." California Board of State Viticultural Commissioners (1888:13).

[100] The figures are 111 and 55 million gallons, cited in Pinney (2005:20).

and spirits rose by 10 percent. Wine consumption increased more because it was the easiest form of alcohol to produce at home.[101] For the CWA, however, Prohibition made redundant its vast network of wineries and cellars, while scientific progress was halted as early as 1916 when the Regents of the University of California prohibited research on alcoholic fermentation and all applied research in wine and wine grapes.[102]

Californians had to relearn their wine after the end of Prohibition in 1933. Large areas of premium grape varieties had been either uprooted or grafted with poor-quality heavy bearers to produce table grapes. Wines were made by untrained winemakers in unsanitary conditions, leading W. V. Cruess to note that "after Repeal, the outstanding characteristic of our wines was instability."[103] The challenges following repeal were not dissimilar to those facing the industry in the 1880s, and it is fascinating to see how differently the sector responded. The University of California provided important scientific work, but a significant part of this research was directed toward understanding basic viticulture and viniculture practices associated with producing fine wines, rather than helping those producing for the initially vastly more important market for ordinary table and dessert wines. Local growers' associations such as the Napa Valley Wine Technical Group helped information circulate. It was the fine wine producers that first used varietal and geographical labeling to separate their better-quality products from the rest of the industry that still used European place names ("California claret," etc.). The CWA was not re-created, but the problems of having to vertically coordinate grape growing, wine making, and marketing remained. Instead of a highly integrated Californian firm that dominated the national wine market, out-of-state bottlers and distilleries moved into the area in early 1940s to purchase local wine-making businesses to guarantee supplies for the East Coast as the industry changed from the producer of "a bulk commodity" to a business that was "predominantly brand oriented."[104] Finally, winemakers succeeded in educating their drinkers, and in the words of James Lapsley, between the 1930s and 1980s "the public's expectation of 'wine' shifted from a fortified, often oxidized or spoilt beverage, produced from indistinct grape varieties, to a table wine possessing distinct flavor attributes derived from varietal grapes and from processing."[105]

---

[101] Warburton (1932:260); Mendelson (2009:50–51). Mendelson argues that illegal producers were more attracted to spirits because of the high ratio of alcohol to volume.

[102] Lapsley (1996:47).

[103] Creuss (1937:12), cited in Lapsley (1996:51).

[104] Lapsley (1996:95, 110)..

[105] Ibid., 1.

CHAPTER 10

Australia: The Tyranny of Distance and Domestic Beer Drinkers

> The production of wine in Australia has not increased as rapidly as
> the suitability of soil and climate would appear to warrant. The cause
> of this is probably twofold. . . . Australians are not a wine-drinking
> people and consequently do not provide a local market for the prod-
> uct, and . . . the new and comparatively unknown wines of Australia
> find it difficult to establish a footing in the markets of the old world,
> owing to the competition of well-known brands.
>
> —*Australian Official Year Book 1901–7*, 328,
> cited in Osmond and Anderson, 1998: 1.

THE AUSTRALIAN WINE INDUSTRY dates from the end of the eighteenth century, but as in California and Argentina, it was only in the two or three decades
prior to the First World War that it became commercially important. The early
settlers and government authorities were attracted to viticulture because it was
a labor-intensive crop, allowing larger settlements than found with extensive
livestock or cereal farming. From a social perspective, it was believed that wine-
drinking Europeans became less drunk than spirit-drinking Australians, while
the government saw wine as a potential export commodity and source of taxation. However, Australia suffered from a small, fragmented national market,
and producers were forced to look to the British market. Despite a distance of
20,000 kilometers and receiving no preferential tariffs, exports of unfortified
"full-bodied dry red wine" grew from 2,700 hectoliters in 1885 to peak at
45,000 hectoliters in 1902, equivalent to a fifth of those from France and representing 6 percent of Britain's total imports.[1] This market required Australians
to be at the forefront of technological change, and Alexander Kelly attempted
to cool wines during fermentation as early as the 1860s.[2] By contrast, Australia's
own domestic consumption remained at little over 2 liters per capita, often of
poor quality, although the production of brandy was becoming an important
sideline for large wine producers by 1913.

The first two sections of this chapter look at how Australians learned to grow

[1] Laffer (1949:124–25).
[2] Kelly (1861:115–20).

grapes and make wine, and the advances in wine-making technologies linked to dry table wine production. The next section considers the problems of vertical coordination, or "cooperation" as contemporaries called it, between grape growing and wine making. The final one shows that the Australian commodity chain differed from California in that it was market driven from Britain and explains why attempts to create an alternative Australian distribution network failed.

LEARNING GRAPE GROWING AND WINE MAKING

Vines were brought on the first boat arriving with settlers in 1788, but there were still only around 800 hectares in 1856, when *The Times* reported that Australian wine production was the pursuit of a few wealthy landlords and "more a fancy than an industry."[3] By 1866 the figure had jumped to 5,500 hectares, as the vine spread from New South Wales to other states where growing conditions were easier, with South Australia accounting for roughly half the area, Victoria three-tenths, and New South Wales a sixth. A half century later, and including only those grapes used for wine, South Australia (57 percent) continued to dominate, followed by Victoria (22 percent), whose production was decimated at this moment because of phylloxera, and New South Wales (18 percent).[4]

Demand for wine remained limited because of the large distances between settlements, high duties facing interstate trade before federation in 1901, and small size of the major cities, with Melbourne only reaching 100,000 inhabitants in 1861, and Sydney in 1871. Yet despite these problems, contemporaries initially worried more about the lack of grape-growing and wine-making skills so obviously absent among the early British settlers, as well as the fact that many of them also had no appreciation of how wine should taste.

To help resolve these problems, a number of settlers, including John Macarthur (1815 and 1816) and James Busby (before 1824 and in 1831), traveled to France to learn grape-growing and wine-making skills and collect vine stock. Busby had already lived in Cadillac (southwestern France) prior to his arrival in Sydney, and his *Treatise on the Culture of the Vine*, published in Australia in 1825, was inspired by the country's need to produce export crops. His next book, *A Manuel of Plain Directions for Planting and Cultivating Vineyards and Making Wine in New South Wales* (1830), was the first to be written on the subject with Australian conditions in mind and was widely distributed within the colony. Busby published his classic *Journal of a Tour through Some of the Vineyards of Spain and France* after visiting these countries in late 1831, and he made detailed

[3] Cited in Dunstan (1994:26) and Osmond and Anderson (1998:4).
[4] Figures for 1910, *Year Book 1912*, 396–97. When all vines are considered, including grapes for wine, the table, and raisins, South Australia and Victoria both had two-fifths of the total, and New South Wales a sixth.

observations on grape varieties, cultivation techniques, and wine making and suggested that the secret of producing fine wines was high levels of capital investment.[5] Busby collected 543 different varieties of vines, of which 437 came from Montpellier in southern France, and those that survived the long journey were planted in Sydney's botanical gardens and circulated among other growers. Just like Arthur Young, another great traveler whom Busby much admired, there is little evidence that he was a particularly successful grower himself.[6]

The wine boom of the early 1860s led to other important publications, including Dr. Alexander Kelly's *The Vine in Australia* (1861) and *Wine Growing in Australia* (1867), while Ebenezer Ward published a series of newspaper articles on the leading vineyards in Victoria and South Australia. In 1892 George Sutherland's *South Australian Vinegrower's Manual* became the first of many official publications to appear, while the *Australian Vigneron and Fruit-Growers Journal*, first issued in 1890, allowed state viticulturalists such as Arthur Perkins, Malcolm Burney, and Francois de Castella a forum to inform readers of European advances. Despite the title, this journal represented more the interests of the large winemakers and merchants than the humble grower.[7]

The shortage of skilled labor was partly eased by immigration. When Charles La Trobe arrived in 1839, Melbourne had only a couple thousand inhabitants, and he encouraged over a hundred vignerons and agricultural laborers from Neuchâtel, one of the few Swiss wine-growing regions, to settle. According to the historian John Beeston, within "ten years of its foundation there were not only numerous small vineyards in Melbourne and nearby areas, but also in Geelong and the Yarra Valley."[8] The Barossa Valley was initially farmed by settlers from Silesia in Prussia, although French experts arrived in the 1880s, including Edmond Mazure and Charles Gelly, who managed the Auldana Vineyard and was considered one of the best local winemakers.[9]

It soon became apparent that despite the importance of these immigrants in establishing the early industry, many brought with them knowledge and skills that were highly localized and often inappropriate to Australian conditions. Winemakers therefore traveled themselves, so much so that by the early 1890s Thomas Hardy of South Australia argued that they had become more knowledgeable than Europeans:

[5] See chapter 1.

[6] Busby (1825/1979:xix). Busby described Young as "that acute and accurate observer." Busby had an estate in the Hunter Valley, although there is no evidence of him visiting it. The contribution of Busby to Australian viticulture is told in Beeston (2001) and Faith (2003). For Young's failure as a farmer, see Mingay (1975:7–8).

[7] From July 1906 it became known as the *Wine and Spirit News and Australian Vigneron*, claiming that "the policy of this paper is to cater for the needs of the wine-maker and the wine and spirit merchant (Australian and import), and to energetically support the claims of the wine-grower" (255).

[8] Faith (2003:26) and Beeston (2001:38).

[9] Bouvet and Roberts (2004).

TABLE 10.1
Cycles of Growth and Depression in the Australian Wine Industry, 1854–1916

	Number of years	Annual increases / decreases			% wine production exported
		Area of vines	Wine production	Wine exports	
1854–71	17	15.5	18.4	14.1	1.8
1871–81	10	−1.1	−0.6	−5.2	1.6
1881–96	15	9.7	7.5	23.0	9.8
1896–1915	19	−0.1	−0.4	0.4	16.5

Source: Osmond and Anderson (1998), table 1.

It is thought by many that the vignerons of Australia have had very little experience, but that is a great mistake; as a rule we know more about wine-making and vine-growing than nineteen out of every twenty that come from Europe: their knowledge is almost always confined to the practices of their own immediate neighborhood, whilst many of us, in addition to 30 or 40 years' experience here, have had the advantage of the ideas of men from nearly all the wine-growing countries of the world, and also of travel in them ourselves.[10]

Hardy himself traveled extensively, most notably in the early 1880s when he wrote *Notes on Vineyards in America and Europe*, which contains detailed descriptions of the industry in California, Portugal, and Spain. Other Australians who studied European wine making included the Basedow brothers, Frank Penfold Hyland, Leo Buring, Charles Morris, and Oscar Seppelt.[11]

The gold rushes hampered the early industry by creating labor shortages, but they also brought benefits in the form of capital for investment, an increasing number of non-British immigrants, and a potential market for wine. From a very low base, and with not very trustworthy figures, the area of vines increased by an annual 15 percent between 1854 and 1871, wine production by 18 percent, and exports by 14 percent (table 10.1). However, as in California, the excessive production of poor-quality wine restricted demand and lowered prices, and during the following decade output fell from 806,000 to 760,000 hectoliters, and the area under vines from 6,976 hectares in 1871 to a low of 6,227 hectares in 1881.[12] The recovery can be dated from the late 1880s as wine production virtually doubled between 1886 and 1891 and coincides with a period when contemporaries were aware of the need to improve quality. While better wine-making facilities began to appear, the rush to plant in the late nineteenth century, especially in Victoria, brought with it a flood of the cheaper grape varieties.

[10] *Australian Vigneron*, February 1891, p. 177.

[11] Beeston (2001:128, 130, 133); Bishop (1998:65, 72).

[12] Figures from Osmond and Anderson (1998:40). The area of vines includes figures for table grapes and dried fruit.

The need to improve wine making is evident in the discussion by the South Australian Vinegrower's Association concerning the appointment of the state's first viticulturist. A speech by Thomas Hardy was reported in the *Australian Vigneron*:

> The selection of soil and sites for vineyards; the best available kinds to plant, the best modes of preparing the land and planting, the training, and pruning, and general cultivation, were generally well-known already, and beginners could always easily get the advice necessary for all these operations from those who had a long experience. Therefore he did not consider that we wanted or could be taught much more than we knew already by anyone coming from Europe in those matters. . . . A man who, having had a good training in the chemistry of wine, would, after some experience of our climate and general conditions, enable him to give advice in the all-important matter of fermentation, how it might be regulated and controlled in our variable weather during the vintage, how certain defects in our wines, arising from imperfect fermentation, could be overcome or prevented, and how wines which might get into bad condition could be rectified or improved.[13]

The problem, as has been noted before, was that the high temperatures found in many areas led to cessation of the vinous fermentation and acetic or lactic fermentation taking place, producing an unstable and "sweet sour wine."[14] The subsequent appointment of Arthur Perkins in 1892, a twenty-one-year-old English speaker who was a graduate of Montpellier's L'École Nationale d'Agriculture and had practical experience in Tunisia and knowledge of the rapid advances taking place in the production of dry table wines in hot climates, proved to be inspirational. In addition, the entrepreneurial leadership provided by Thomas Hardy and others resulted not just in the appointment of professionals such as Perkins, but also in the establishment of Roseworthy College, just west of the Barossa Valley.[15] South Australia enjoyed one other important advantage over the other colonies: it has remained phylloxera free even until this day.

Government assistance in the state of Victoria was very different. Phylloxera first appeared in Geelong in 1878, and the diseased vines were uprooted and destroyed with growers receiving compensation. Victoria actively encouraged the planting of labor-intensive crops to increase the number of farmers, with a bonus of £2 per acre being offered for each new acre of vines (£3 for fruit cultivation), which led to around 4,500 hectares of vines being planted by 1,400 vignerons between 1889 and the end of 1893.[16] Most vignerons had little or no experience, and they lacked capital and the skills to make sound wine. These planting bonuses produced considerable opposition from established growers because the

[13] *Australian Vigneron*, November 1891, p. 137.
[14] Ibid., November 1890, p. 116.
[15] Faith (2003:54). See also Bishop (1980).
[16] Some 900 of them planted less than 4 hectares each of vines (Pope 1971:28, 29).

increase in poor-quality wines led to a fall in grape prices,[17] and the problem was resolved only with the reappearance of phylloxera in Rutherglen after 1897.[18]

By 1898 Victoria already had two of the five agricultural colleges open in Australia (Dookie in 1896 and Longeronong in 1889), a viticulture college established at Rutherglen, and a horticultural school at Burnley. In 1898 Raymond Dubois was appointed principal of Rutherglen, and shortly after he was given the task of leading the fight against phylloxera, which was again devastating the district.[19] The viticulture college, however, failed initially to live up to expectations, and when the Victoria's minister of agriculture visited it in February 1901 he found new dormitories but no students.[20] Dubois was now instructed to visit wine producers and teach modern wine making, and with Percy Wilkinson and Edmund Twight he also translated from French a number of important books, including those by Roos (*Wine Making in Hot Climates*) and Pierre Viala and Louis Ravaz (*American Vines: Their Adaptation, Culture, Grafting, and Propagation*).[21]

Finally, although New South Wales benefited from Sydney being Australia's largest protected urban market, viticulture grew only slowly, and growers complained on the eve of federation of the lack of state support, compared to that received in Victoria: "no bonuses to assist them; secondly, only recently have they have been offered expert advice; thirdly, they have no experimental stations or places of instruction; fourthly, the dissemination of information etc. has been conducted on such spasmodic and unsatisfactory lines as to prove little or no practical use."[22]

Organization of Wine Production

The abundance of land and the weakness of demand led to most growers planting other crops among their vines. In the Barossa Valley, for example, the "three

---

[17] *Australian Vigneron*, June 1894, p. 38. Critics claimed that the bonus led to the production of poor-quality wines because vines were planted on unsuitable land or unsuitable grape varieties were planted on suitable land, and because of the "ignorant handling in fermentation by unskilled persons who rushed into the business" (ibid., June 1898, p. 21). One established grower complained that "we can thank the Victorian Government for bringing down the price of wine by giving the planting bonus. If it had spent the same amount of money in advertising our wines, we would have done the planting" (Victoria, William O'Brien, 1900:45).

[18] Phylloxera's reappearance in the Rutherford district in the final years of the century brought strong condemnations from growers and led to the sacking of Bragato, the government official in charge.

[19] *Australian Vigneron*, February 1898, p. 155; July 1898, p. 59; and October 1899, p. 117. For the demand for agricultural research in Australia at this time, see especially McLean (1982).

[20] *Australian Vigneron*, February 1901, p. 204.

[21] The Viticultural Station at Rutherglen distributed 350 copies each of these first two translations. Other studies translated included Dubois and Wilkinson (1901a), Dubois and Wilkinson (1901b), Gayon (1901), Mazade (1900), and Foex (1902).

[22] *Australian Vigneron*, July 1898, p. 41.

primary industries"—mining, grazing, and viticulture—proceeded side by side during the 1860s, while Ebenezer Ward noted that at one Victorian vineyard the owner had alternated walnut trees with vines with the intention of uprooting the least successful.[23] Most vineyards were also small. In Victoria in 1890 before the planting bonus there were 850 growers with more than 0.8 hectare, but 511 of these had only between 0.8 and 4 hectares. A further 244 farmed between 4 and 12 hectares; 68 between 12 and 24 hectares; and 13 between 24 and 36 hectares. There were only 14 vignerons with more than 40 hectares of vines, with the largest being 144 hectares.[24] A significant number of growers held less than 0.8 hectare. The total number of Victorian growers peaked in 1896, with 2,975 vignerons cultivating an average of just over 4 hectares each, before declining to 1,776 in 1914 (with an average holding of 5 hectares).[25] However, these figures are distorted by the inclusion of producers who grew grapes for uses other than wine. In 1914, for example, about two-fifths (700) of growers were found in the Mildura district, where no grapes were used for wine making. By contrast, Rutherglen's 116 growers produced 16,711 hectoliters, or 40 percent of the state's production in that year, giving an average of 144 hectoliters per grower.[26] The small scale of grape-growing activities was also found in New South Wales, while in the Adelaide region in 1892 three quarters of the 753 growers worked between 0.4 and 4 hectares of vines; 175 between 4 and 12 hectares; 46 between 12 and 24; 17 between 24 and 36, with just 7 with more than 36 hectares.[27]

Even the largest vineyards were relatively modest affairs compared with California. According to Whitington, Fowler's Kalimna vineyard in the Barossa Valley was the biggest in South Australia in 1903 with 132 hectares of vines, followed by Kelly's Tatachilla vineyard (124 hectares) in the McLaren valley.[28] The Kalimna vineyard bought large quantities of grapes produced in the neighboring district, but the Tatachilla vineyard started making wine only in 1903, after losing a contract to supply grapes to Thomas Hardy. By contrast, by far the biggest winemaker in the Barossa (and in Australia) was Seppeltsfield, producing an estimated 20,000 hectoliters, or five times Kalimna's production in 1903. This winery had less than 50 hectares of vines of its own and bought grapes from 165 private growers within a radius of 25 kilometers of the estate.

If vineyards remained generally small family concerns, the increasing economies of scale associated with the new wine-making technologies in hot climates

---

[23] Beeston (2001:82); Ward (1864/1980:43).

[24] *Australian Vigneron*, July 1890.

[25] *Victoria Year Book 1903*, 391, and 1914–15, 713.

[26] *Australian Vigneron*, June 1914, p. 293. No dried fruit was produced at Rutherglen. Many of the growers sold their grapes, and the wine was produced by firms such as Burgoyne. In the next three largest areas, average production was, for Ararat, 106 hectoliters; for Stawell, 76.5 hectoliters; and for Shepparton, 56.9 hectoliters. My calculations.

[27] Ibid., April 1892, p. 219.

[28] This paragraph is based on Whitington (1903). By contrast, Thomas Hardy in Victoria (1900:38) claimed to have 200 hectares of vines, presumably in a variety of different locations purchased grapes from 30–40 part-time growers, equivalent to a further 200–240 hectares.

TABLE 10.2
Leading Winemakers in South Australia, 1868, 1876, and 1903

Vineyards	1868		1876		1903	
	Quantity produced (hectoliters)	% of SA production	Quantity produced (hectoliters)	% of SA production	Quantity produced (hectoliters)	% of SA production
Largest three	1,637	4.4	4,546	20.3	42,960	37.0
Largest five	2,546	6.9	6,428	28.7	57,962	49.9
Largest ten	5,231	14.2	9,966	44.4	78,191	67.4
Total state vintage	36,919		22,422		116,082	

Sources: Bell (1993), table 3; and Whitington (1903:71).

such as Rutherglen or the Barossa eroded the competitive position of the small winemaker.[29] As early as 1882 Thomas Hardy noted that "the manufacture of wine is now almost wholly gone into the hands of those who make a business of it, and do not follow it merely as a secondary pursuit," leading to a concentration in the industry (table 10.2).[30] By 1900 the new wine-making technologies included refrigerators, continuous presses, aero-crushing turbines, sterilizers, and pasteurizers, and these helped create economies of scale in four important areas. First, considerable skills were required in wine making if growers were going to produce a dry table wine that would keep. By the 1890s the leading wineries were attempting to control the temperature of the must during fermentation and using cultivated yeasts. Hardy, for example, claimed to "keep a highly-paid man as an expert in the chemical and bacteriological department, and we have highly-paid skilled men in the manufacture of the wine, and it is that part of the business which is the primary business, not a secondary one."[31] Second, new wine-making technologies and cellar designs helped to cut labor costs, an important consideration in a high-wage economy such as Australia's. Third, merchants demanded large quantities of wines of a uniform style that could be repeated each year, which was impossible for small producers to achieve. Finally, large-scale wine production encouraged distilling and the development of the brandy and fortified wine industries, which after the 1906 legislation was possible only on a large scale. By 1914 a number of houses, such as Penfold or Seppelt, were reinforcing their corporate brands through the sale of brandy and dessert wines. Brandy was a highly profitable business but open only to the large producers, as a government excise officer had to have permanent offices at the distilleries, allowing Penfold to claim that it was a "Government guarantee" that all its brandy

[29] This retreat from wine making by growers was one factor in the decline in support for both the *Australian Vigneron* in the 1900s and the South Australian Vigneron Association, which by 1911 had a membership of only forty-five. *Australian Vigneron*, June 1911, p. 241.
[30] Cited in Aeuckens (1988:148).
[31] Victoria (1900:38).

was made from the wine of grapes, and that the state laws on maturing the brandy were properly respected.[32]

The new gravity-flow Seppeltsfield winery that opened in 1888 was one of the most advanced of the period. To keep labor costs to a minimum, the winery was built down a hillside, with the sixty fermenting tanks (each with a capacity of 90 hectoliters) arranged on three terraces. Wooden chutes allowed the winemaker to direct the must and wine to the desired vats, and in 1890 tin-platted copper cooling coils were installed to regulate the temperature.[33] Center pumps were used to mix the different parts of the liquid, to keep the temperature inside the vat uniform, and to obtain the maximum coloring and extractive matter by causing the juice to repeatedly pass through them. Seppelt started using cultivated yeasts imported from Europe in the early 1890s, but by 1899 its laboratory was propagating its own.[34] The winery had two steam-powered crushers and hydraulic presses that could each process 100 tons of grapes a day. Even by international standards the Seppelt winery was impressive, and B. W. Bagenal, a student for three years at Montpellier and representative of the London importers in Adelaide for over two years, noted that he had personally visited seven of the ten best French and Algerian vineyards cited in a recent book, and "he had confidence in saying that there was no place he knew of where the industry was better carried on than at Seppeltsfield."[35]

The high level of vertical specialization in grape growing and wine making found in South Australia was less common elsewhere in Australia.[36] The advantages of specialization were also hotly debated within the profession. In particular, Arthur Perkins spoke of the "two separate classes with antagonistic interests—vignerons on the one hand, winemakers on the other."[37] He blamed winemakers for the low prices being paid in 1899 and argued that the creation of state-sponsored regional depots would save growers the cost of storing wine before shipment and allow wines to be blended, which would remove the problem of limited production from family vineyards. Dubois in Victoria argued that small growers should form wine-making cooperatives and obtain

[32] Mills (1908:17).

[33] *Australian Vigneron*, May 1897, p. 11; Bishop (1998:71–72, 79). The journal stated that coolers had been in use "for the past 18 years," but an earlier article (November 1890, pp.116–17) failed to mention this despite dealing specifically with the problems of controlling temperature during fermentation and the Seppelt winery. Salter could not use the coils until 1895 because of a lack of water (Bishop 1988:79).

[34] Bishop 1988:74–75.

[35] *Australian Vigneron*, March 1898, p. 188. The book referred to most probably was Ferrouillat and Charvet, *Les celliers* (1896).

[36] *Australian Vigneron*, January 1891, pp. 155–56 and March 1902:236 for New South Wales; and Victoria (1900:33).

[37] *Australian Vigneron* January 1899, p. 169. Elsewhere Perkins wrote that "to-day the grower should be wine-maker, his perishable crop should not be under the whip hand of the wine-maker" (Victoria 1900:14).

modern equipment and, "more important than anything, and cannot be too much emphasised," benefit from technical education.[38] By contrast, and not unnaturally, the large wine producers and their trade paper, *Australian Vigneron*, as well as London importers believed that large, privately owned wineries buying grapes from specialist growers were more efficient. The presence of independent growers did allow however for a rapid response to upswings in demand. The £2 bonus offered to Victorian growers, which saw the area of vines increase by 50 percent between 1890 and 1894, would not have occurred if growers also had to invest in new cellars.[39] Between 1889–91 and 1911–13 Australian wine production increased from 147,000 to 256,000 hectoliters, and while not all of this expansion took place in hot climate areas and the introduction of new technologies was inevitably slow in some wineries, contemporaries linked the ability to produce better wines with the growth in the area of vines. As W. & A. Gilbey wrote in their annual letter to *The Times* on the state of the wine market in 1908, "Australian wines have improved in quality to a remarkable extent in the last few years. The inferior vintages are now distilled into brandy, and the wines remaining are, with exceptions, excellent."[40]

Burgoyne and Seppeltsfield illustrate how winemakers coordinated with their growers to adapt to changes in demand. The British importer Burgoyne bought the Mount Ophir vineyard in 1893, and the investment in new wine-making facilities led one contemporary to note that "it would be difficult to find in any part of the world a winery in which California labor saving appliances and the most approved European methods of securing the best treatment of the wine are so completely adopted." Burgoyne contracted out for most of his grapes and bought wine from other producers, blending it all in cellars that had a capacity of 34,000 hectoliters (750,000 gallons). Burgoyne was interested only in "full-bodied red wines" of less than 15 degrees, which were exported after nineteen months. The expansion of trade saw grape prices rise from about £2 to £5 per ton by 1909, but this specialization for a niche market had its risks. In early 1909, in the midst of the local devastation caused by phylloxera, the company demanded price cuts to remain competitive in the British market.[41] D. B. Smith, chairman of the Vinegrower's Committee, argued that growers would not replant but turn their land to "other purposes" if prices were reduced. He told growers that they were too dependent on the London market for dry red wines and should consider growing grapes for sweet wines, noting that "there were only three or four

---

[38] *Australian Vigneron*, August 1900, pp. 71–73.

[39] Perkins, for example, complained in the 1890s that in the old areas of production in South Australia the increase in vines was not always accompanied by a growth in storage capacity. In 1896 over 350 tons of grapes (the crop of 100 hectares) in Tanuda (Barossa) remained unpicked because of the lack of cellar facilities. Ibid., February 1898, p. 162.

[40] Cited in ibid., January 1908, p. 27.

[41] The company denied growers' claims that prices had been cut. Prices were fixed each year in accordance with demand and the estimated size of the future harvest.

buyers of wine for the London market, while there are upwards of 20 for the Australian sweet wines." Burgoyne backed down, and a full page advertisement appeared in the *Rutherglen Sun and Chiltern Valley Advertiser* under the title "For the Sake of the Industry," offering £7 per ton of grapes for "Carbinet, Shiraz and Malbec" and £6 10s per ton for other varieties for the next (1910) vintage. This offer was subject to "certain technical conditions," namely, that growers re-plant their phylloxera-devastated vines.[42] While the dry red wine market was very profitable for Burgoyne between 1885 and 1914, it became much less so in the interwar period when demand switched to fortified wines. Being located in an area where conditions were optimal for dry red wines perhaps explains the firm's failure to spot the shift in demand in Britain to dessert wines in the inter-war period, as it required the company to look for supplies in other, more suit-able areas such as Milawa.[43]

The Seppelt winery also adapted to changing market conditions by changing the grape prices it offered, with the price for cabernet sauvignon, for example, being cut from £7 per ton in 1899 to £4 17s. 6d. in 1911, while the price of Shiraz was increased from £4 to £4 17s. 6d., and white hermitage and riesling from £3 5s. to £4 5s. between the same years.[44]

As Burgoyne discovered in 1909, growers were not totally dependent on an individual wine producer. Furthermore, they could leave the industry at times of low prices as their investment was limited to vines, which threatened to leave the capital-intensive wine-making firms without grapes. Growers could, and at times did, simply exit, with perhaps the most famous being the Yarra Valley (Victoria), where in the 1920s, in the words of Beeston, "the advent of the Jersey cow was equally devastating as phylloxera had been fifty years before in Geelong."[45]

In Search of Markets

A major problem facing the new industry in Australia was the lack of markets. Not only did Australia have a much smaller population compared to Argentina or the United States, but the dominance of British people among the settlers discouraged wine consumption. As late as 1890 it was claimed that "more than half of the Australian wine drunk here is consumed by Italians, French and Germans."[46] Sales outside individual states were also restricted because of high

[42] This paragraph is based on reports in *Australian Vigneron*, April 1904 and February 1909; and the *Rutherglen Sun and Chiltern Valley Advertiser*, July 9, 1909.

[43] Faith (2003:71) notes that the "previous overwhelming influence" of the Burgoyne family had started to decline because Peter Burgoyne's son Cuthbert mistakenly believed that the future was for light wines, not fortified ports.

[44] Bishop (1988:62).

[45] Beeston (2001:176).

[46] *Australian Vigneron*, December 1890, p. 151.

TABLE 10.3
Australian Interstate Trade and Wine Consumption

	1909				1898–1900
	Imports (hectoliters)	Exports (hectoliters)	Net trade (hectoliters)	Exports as % of production	Per capita wine consumption (liters)
New South Wales	14693	2596	−12097	7.1	3.1
Victoria	6023	7142	+1129	15.8	5.9
Queensland	5051	50	−5001	1.2	1.3
South Australia	882	22462	+21580	19.2	24.7
Western Australia	4596	82	+4514	1.3	4.5
Tasmania	1145	45	−1100	—	0.8

Source: Australia (1911:402–4).
Notes: Production taken as 1909–10. Per capita consumption is very approximate as some wine was distilled.

tariffs and remained limited even with free trade after October 1901. In 1909, for example, the last year that figures are available, only 7.1 percent of New South Wales's production, 15.8 percent of Victoria's, and 19.2 percent of South Australia's was sold to other states (table 10.3).[47]

Wine quality and industrial organization were significantly influenced by the nature of market demand, although there were important differences between the domestic and international markets. The early South Australian industry, as elsewhere, was limited to local rural markets and on-farm consumption, and wines were often adulterated before being drunk.[48] As Mücke noted in 1866, "hundreds of wine cellars, belonging to men of the middle and poorer classes, are filled with wines spoiled by unskillful treatment, which have swallowed up all their capital and destroyed all their hopes."[49] High state tariffs allowed producers in New South Wales and Victoria access to growing urban markets (table 10.4), but Richard Twopenny wrote in his *Town Life in Australia*, first published in 1883, that Australian wines were "too heady, and for the most part wanting in bouquet, whilst their distinctive character repels the palate, which is accustomed to European growths." Yet he claimed that some were much better than the expensive, imported wines, and that despite a duty of 10 shillings a dozen, "large

[47] The figures would be larger if the wine used for brandy production and sold to other states were included.

[48] One report talks of a mixture of newly pressed (or partly fermented) grape juice and raw spirits, which, when added with heavy metal contamination contained in the spirit and the presence of undesirable fusel oils, made a concoction that was highly addictive (Bell 1993:151).

[49] *Adelaide Observer*, July 28, 1866, cited in Bell (1993:153).

TABLE 10.4
Select Urban Centers in the New World, 1881–1911 (thousands)

	Sydney	Melbourne	Adelaide	Buenos Aires[1]	San Francisco[2]
1881	221	262	38	433	234
1901	481	484	39	664	343
1911	630	586	189	1,576	417

Sources: Vamplew (1987:41); Mitchell (2007:48, 52, 54, 56).
[1] 1887, 1895, and 1914
[2] 1880, 1900, and 1910

quantities of Adelaide wine are drunk in Melbourne," although its chief characteristic was its "sweetness and heaviness."[50] By the turn of the century Raymond Dubois lamented that internal tariffs allowed Victorian growers high profits, "peddling their inferior wines in small quantities to local markets," and he believed that one-third of Victoria's wine was unfit for consumption as wine and had to be distilled, "a positive proof that there is something fundamentally wrong" in the manufacturing methods adopted in the industry.[51] A study of the chemical composition of a sample of 103 wines in the Melbourne area showed that a third had been adulterated with salicylic acid.[52]

In the 1890s large numbers of small producers in Victoria created poor-quality wines that inevitably lacked uniformity.[53] There were also problems for the casual drinker: "Try and get a glass of Colonial wine in a Melbourne hotel, and in nine cases out of ten you will be served at an outlay of 6d. with a small glass, of which measure it takes from 8 to 14 to fill a common wine bottle, and such a bottle is taken from a shelf in a bar, where its contents have been stewing in the hot atmosphere, after having been uncorked and left half full, perhaps, a fortnight before!"[54]

If Victoria and New South Wales had the country's two largest cities as captive markets, this was not the case with producers in South Australia. Attempts to produce wines for export in the 1870s failed,[55] but twenty years later the South Australian industry had been transformed by a handful of winemakers to one where important economies of scale were achieved in production and the major markets

[50] Twopenny (1883/1973:67–68).
[51] Victoria Department of Agriculture (1900:8). Thomas Hardy disagreed, arguing that distilling was a response to the demand for alcohol to produce sweet wines. *Australian Vigneron*, September 1900, p. 101.
[52] Victoria (1900:13).
[53] *Australian Vigneron*, September 1893, p. 90. The same article noted the need for an organization that prevented the "placing on the market say 50 or 100 small parcels of wine of various qualities, differing individually more or less each recurring vintage."
[54] Ibid., December 1890, p. 151.
[55] The East Torrens Winemaking and Distillation Company, the South Auldana Vineyard Association, and the Tintara Vineyard Company. See below.

Figure 10.1. Australian production and UK imports. Sources: Osmond and Anderson (1998: table 2) and Laffer (1949:124–25)

were Britain and, after federation, the urban centers of Melbourne and Sydney. Between 1885 and 1913 Australian wine imports in Britain increased from 0.27 to 2.84 million liters, peaking in 1902 at 4.5 million (fig. 10.1). In 1902 British imports from Australia were equivalent to 19 percent of those from France, 24 percent from Spain, and 25 percent from Portugal, and they were 250 times greater than those from the Cape.[56] The British market accounted for virtually all of Australia's wine exports and the equivalent of a fifth of the country's harvest in that year. The rapid growth in exports was achieved by producers in South Australia and Victoria, who each accounted for roughly half of the British market.[57]

The British demand was for dry red table wines, often confusingly referred to as a "Burgundy type" and described as being somewhere "between the heavily fortified wines on the one hand, and light wines, such as clarets and hocks on the other."[58] A British study in the early 1870s claimed that approximately 60 percent of South Australian wines found in two samples of fifty-six and seventy wines, respectively, had been "fortified," but this was based on the assumption that no "natural" wine could be stronger than 14 or 15 degrees alcohol, a claimed disputed by the South Australian government.[59] Virtually all Australia's exports,

---

[56] British imports of Australian wines in 1913 were still equivalent to 20 percent of imports from France or Spain, or 17 percent of those from Portugal (Wilson 1940:362–63).

[57] Britain imports between 1893 and 1899 comprised 50.7 percent from South Australia, 46.4 percent from Victoria, and just 2.8 percent from New South Wales (*Australian Vigneron*, May 1901, p. 21.

[58] Ibid., March 1902, p. 236.

[59] South Australia claimed that natural wines could contain in excess of 14.8 degrees (Bell 1994:34). Fortifying also occurred in Britain. Thudichum and Dupre (1872:744) wrote that "to

some 99 percent in 1907, paid the minimum shilling duty per gallon, although by this date it had increased to wines under 30 degrees proof.[60] By contrast, no Australian port or sherry type wines were exported before 1914, as producers could not compete with Portugal and Spain.

Australian producers of nonfortified table wines enjoyed a number of advantages over those from Europe to offset the significant transport and transaction costs that distance created. One factor was that Australia's wines, if properly treated, had "naturally far greater keeping qualities than many European wines."[61] Quality also fluctuated much less from one harvest to the next, although this did not automatically imply that a homogenous wine was produced, as *The Times* noted in 1893: "At present one of the principal complaints which are made by buyers of Australian wine is the absence of uniformity, but this inequality is an inequality of manufacture, which is purely accidental, and to be distinguished from the fundamental inequality produced by the uncertain climate of France and Germany."[62]

Quality differences between vineyards and grape varieties were resolved by blending, as the minimum quantity of wine accepted by shippers was usually considerably beyond that produced by most growers. Thus Blandy Brothers suggested that "it is almost useless to consign small parcels of wine, 10 or 20 hogsheads, which they are not in a position to repeat each year," while Pownall offered to buy red Australian wines "suitable for the English market in parcels of 1000 to 10,000 gallons."[63]

Yet these advantages would have been insufficient in the absence of an efficiently organized commodity chain. Significant barriers to trade existed between Britain and Australia. The first was caused by the 20,000 or so kilometers between buyer and seller. In the late nineteenth century this required a shipping time of six weeks, instead of the three days from France, and implied freight costs of approximately three times more.[64] Another major problem was that the long sea voyage and the extremes in temperature caused by the change in seasons as the ships crossed the equator resulted in all wines, "especially the young wines and wines that have not been thoroughly well made" to undergo a "very consid-

---

many of these wines, which had been carelessly shipped, brandy had been added after their arrival in England."

[60] *Ridley's*, May 1907, p. 368. Thomas Hardy noted between 1880 and 1892 that he had shipped "many thousands of gallons of pure natural wines to England," and only one shipment went above the 26 percent proof limit, requiring him to pay 2s. 6d. duty per gallon instead of 1s (*Australian Vigneron*, August 1892, p. 69).

[61] *Australian Vigneron*, January 1910, p. 29.

[62] *The Times*, May 24, 1893, cited in ibid. August 1893, p. 74.

[63] A hogshead was approximately 275 liters, and 1,000 gallons some 4,546 liters.

[64] *Australian Vigneron*, July 1892, p. 48. E. Burney Young gives these costs as 5s. 6d. per hogshead between Bordeaux and London, and 15–20s. from Australia. Young was not an entirely independent observer, as he would become the manager of the South Australian Government Bonded Depot in London.

erable and detrimental change."[65] Wines required several months rest on arrival, and faulty or young ones were often permanently ruined. The problems of asymmetric information created by distance and the possibility of wine undergoing significant changes in quality encouraged opportunistic behavior on the part of both Australian exporters and British importers. Australian producers might believe that they exported a good wine, but British merchants could claim that it arrived in poor condition and they were only willing to pay low prices. One prominent London West End merchant who imported 60 hogsheads of South Australian wine because of its growing popularity found it so "ill-fermented" that it resolved never to import any more directly from Australia.[66] The result was that even the leading Australian producers found it difficult to find agents for their wines, as the *Pall Mall Gazette* noted in 1884:

> In the European wine trade there is seldom a change of agencies, but no man holds an agency for Australian wines long, or cares to invest money in the venture. The Auldana agents have been many, but all have retired from their speculation. Fells no longer pushes Irvine's wines. Why is this? There surely is a good profit to be made in the trade. The reason must be found in the absence of friendly cohesion, or other unexplained cause between the wine-grower and his representative. Penfold & Co. are moving heaven and earth to secure an agent, but men fight shy of laying out their capital and energy for another's advantage, for there is a feeling that no abiding arrangements of a mutual nature can be made with any grower. Mr. Pownall is very vexed with his shipments recently received from Adelaide; some arrived sour, and very many are mousy, and he puts it all down to the extreme poorness of the grapes from which it was made.[67]

Without agents, Australian producers would have had little alternative but to use the spot market (auctions), where prices rarely covered production costs of cheap, young French wines, let alone one that had been matured a year and a half on the other side of the world. For this particular late nineteenth-century market to work, therefore, trusted agents were required at both ends of the chain: in Australia to check that only acceptable wines were shipped; and in London to determine the appropriate remedies to correct the wines after the journey. One of the first serious attempts at promoting Australian wine in Britain was in 1871 by the author A. C. Kelly, who started selling his wine from the Tintara Vineyard Company (McLaren valley) through an English wine merchant, Peter Burgoyne. The business failed and the company's assets passed to Thomas Hardy, who retained Burgoyne as his agent for the British market, "thus cementing Burgoyne's connections with the Australian wine industry."[68] Another early attempt

[65] Ibid., July 1892, p. 47.
[66] Ibid., October 2, 1893, p. 114.
[67] Ibid., January 1894, pp. 171–72. Penfold & Co. had established an import house in London by the turn of the century. *Pall Mall Gazette*, December 3, 1893, cited in ibid., March 1903, p. 202.
[68] Aeuckens (1988:157). The Emu brand was registered in 1883.

was that of Patrick Auld, who established the Auldana Vineyards at Magill (Adelaide) and a company called Auld, Burton & Company in London in 1871, under the registered name of "The Australian Wine Company." This too failed, in 1885, and was bought by Walter Pownall, who continued to use the Emu brand.[69]

By the final decade of the nineteenth century it was these two British companies that controlled Australian wine imports in Britain, and in particular Peter Burgoyne, who claimed in 1900 that over the previous thirty years "fully 70 per cent of the wine exported from Australia to England had passed through his hands."[70] In 1893, as noted above, Burgoyne bought the Mount Ophir vineyard (Victoria), and according to one report in the *Wine and Spirit Trade Record* of 1912, "the maintenance of the special characteristics of each brand is effected by the careful blending of the produce of various vineyards, according to soil, and variety of grape, whereby their quality is increased."[71] Burgoyne had agents who selected wines in Australia, while in London the firm had around 4,500 hogsheads of wine in bond and a similar quantity at the Pelham Street premises in 1912.[72] The newly arrived wines were treated in London:

> After the necessary rest . . . the wines are racked bright from the lees, and as a rule receive a preliminary fining. It is from the wines thus treated that the blends are made up. After a rest of three weeks or more in the vats the blends are drawn off into hogsheads and again fined. When absolutely brilliant—that is, after a rest of three or four weeks—the wines are again racked bright off the finings into clean casks. They are then allowed to rest again before bottling. The object of this treatment is to ensure the smallest possible crust or deposit in the bottle.[73]

The wines were then sold either by the cask unbranded or by the bottle, using the Ophir (Burgoyne) brand. Burgoyne claimed to have invested £300,000 in advertising them in Britain. "Burgoyne's Australian Wine" placards were found "on every railway station in England," but he also resorted to other, more ingenious methods to sell his wines. Edmond Mazure, manager of Auldana Vineyard in Adelaide, noted how his clerk picked out the birth notices from the "leading London papers" and sent each mother a bottle of wine with the note that although it was good, "he trusts the lady will not take it until she has consulted her medical adviser." In the words of Mazure, this achieved "a double advertisement, the first with the recipient, and the other, which is probably even more valuable, with the doctor."[74] Finally, in February 1905, Burgoyne entered the local Austra-

---

[69] The company changed its name to W. W. Pownall in 1895 and remained the second largest exporter of Australian wines before the First World War.

[70] *Australian Vigneron*, September 1900, p. 115.

[71] Cited in ibid., May 1912, p. 203.

[72] Ibid., May 1912, p. 202.

[73] Ibid., May 1912, p. 203.

[74] Ibid., December 1900, p. 164, and February 1901, p.72.

lian wholesale and retail trade, establishing an agency for a number of leading European brands of alcohol, selling under Burgoyne's Own Brand (B.O.B.), and holding large stocks of very fine wines—hocks, ports, sherries, and so on.[75]

The Australian producers inevitably resented the control exercised by the two British importers and lobbied their state governments to create an alternative marketing system. The South Australian Government Central Wine Depot was originally established in London in 1894 with the objective of providing either a place for growers' wines to recover before they were sold to their agents or consigned to the depot's manager, who would find an agent for them.[76] Opposition by Burgoyne and the London trade press to the depot was immediate and vitriolic, and it appears to have been boycotted by the London merchants.[77] The London correspondent of the *Australian Vigneron*, a writer close to Burgoyne, published a monthly condemnation of the depot's activities for Australian readers. In Australia the depot was criticized for losing large amounts of government money, for failing to make wine producers independent of Burgoyne, and for refusing to communicate to British merchants the name of the wine producer or to the producer the name of its customer. The depot was at a major commercial disadvantage because, although wine quality was checked by Perkins in Adelaide before shipping, it was often sent in small parcels, which made it difficult to sell. In addition, the depot was not the legal owner of the wines and thus could not blend them to compete directly with Burgoyne. Instead, and to the annoyance of other London merchants, the manager E. B. Young signed an exclusive agreement with the Blandy Brothers wine firm, who bought the wine to sell under the depot's Orion brand. By 1900 Young claimed that he had sold South Australian wine to over four thousand customers, "including most of the best wholesale houses in London and the provinces, large wine merchants, and grocers, as well as a number of high-class restaurants and clubs, who are now retailing the depot wines," and sales had increased from about 1,650 to 4,565 hectoliters between 1896 and 1900. The last figure was just a fifth of what Burgoyne claimed to have handled.[78]

[75] Ibid., December 1905, p. 344.

[76] By contrast, the depot trade with other commodities, such as butter, fruit, and frozen meat, was profitable and uncontroversial.

[77] South Australia (1901, no. 24, p. 68, 2297). Thomas Hardy noted that "the large buyers would not touch the depot." The 1901 Select Committee also questioned whether, if the depot in the future concentrated just on the wholesale trade, this would provoke a boycott by the London merchants (ibid., 54). Burgoyne also had a particularly fiery temper, which led to a number of important and sometimes very public arguments. For his disputes with Dr. Alexander Kelly, see Beeston (2001:75); and with Pr. Perkins of Roseworthy and the winemaker John Christison, see the *Australian Vigneron*, March and April 1902; February 1903.

[78] South Australia (1901, no. 24, p. iii). Burgoyne claimed to have been responsible for 25,000 of the 36,000 hectoliters of Australian wine imports. *Australian Vigneron*, March 1902, p. 235. The figures shipped under "Government Certificate" from Adelaide were much smaller—2,320 hectoliters in 1899–1900 and 3,860 hectoliters the following year (South Australia. Parliamentary Papers no. 43. 1915:67).

At the turn of the century the South Australian and Victorian governments both held parliamentary enquiries into the benefits of central wine depots in the organization of the wine trade. Opposition to them came from Burgoyne, who threatened to end buying wines in Australia itself,[79] but as the Victorian commissioners noted, there was also a "great conflict of testimony amongst expert witnesses as to the best means for the state to promote the industry."[80] The debate over the best use of public funds led to four very different proposals, of which only one involved a London depot. One alternative was to subsidize a British retail chain to stock and sell Australian wines. In December 1898 the Australian Vigneron reported that Thomas Lipton had offered to open a depot in Melbourne for Victorian wines and to advertise them in Great Britain in exchange for £5,000 for eight years, while a similar offer was apparently received from W. & A. Gilbey.[81] Neither was accepted, and in 1911 Burgoyne (2,700,000 liters) and Pownall (750,000) still dominated exports from South Australia and Victoria, but the British retailer Gilbey (625,000) now came a close third.[82]

Other alternatives included local wineries and state depots located in the major ports to benefit small producers. The problem was one of creating the correct incentives because while many in the industry wanted government funds, it was widely believed that subsidies should be paid only for producing "sound and unadulterated wines" and not "bad wines," which then had to be distilled.[83] Public-funded cooperatives or depots had to be able to reject grapes and wines that were of poor quality, but, as Perkins noted, while it was possible to determine that a wine was "sound and unadulterated," on "the question of quality none will agree."[84] The fear was that if growers of poor-quality grapes received support, this would lead to an expansion in their production given the ease to plant a vineyard, while there was a "grave danger of the wineries being made repositories for large quantities of wines of inferior quality—in fact a dumping ground for the rubbish left after the sound, marketable article has been disposed of in the ordinary way."[85] Many established producers therefore feared that an ill-conceived state intervention would lead to the growth in output of poor-quality wines, dragging prices down, so the demand for intervention by the small

[79] As early as 1894 Burgoyne had complained that when his representative was away from Australia for some months, shipments were far from satisfactory as "native wood casks were used by many shippers, and complaint was made of the carelessness in the execution of the orders" (Australian Vigneron, September 1894, p. 83). He repeated these complaints in a letter in March 1903 to Benno Seppelt, which he asked to be read at the Winegrowers Association, stating that after the forthcoming harvest he would end buying wines in Australia itself (ibid., March 1903, pp. 202–3).

[80] Victoria (1900:10).

[81] Australian Vigneron, October 1896, p. 97; December 1898, p. 150; January 1899, p. 176.

[82] Ibid. January 1912, p. 25; February 1912, p. 71. Figures are approximate.

[83] South Australia (1901), Perkins, 109, and Paul de Castella, 31.

[84] Ibid., Perkins, 109.

[85] Victoria (1900), Bragato, 21. Australian Vigneron, March 1898, p. 129.

growers came to nothing. In Victoria, many abandoned the sector, especially with phylloxera ravaging in the state, thereby removing large amounts of poor-quality wines from the market. The support for a London depot among the large producers also remained limited, especially as Burgoyne's threat of abandoning the market would have left producers with the risk of sending their wines to London at their own cost and accepting the market price there.[86]

The Australian experience was very different from that of Argentina and California because Australia's domestic market was so small and was heavily fragmented, a result of both distances and internal customs duties before 1901. The last quarter century prior to the First World War saw a major increase in the scale of production of the leading wineries. The specialist skills and equipment required to make dry table wines in hot climates were not profitable in small wineries. By the last decade of the nineteenth century, dry red table wines were being exported to the British market despite the difficulties for a commodity chain that stretched around the world. Strict control was required on the quality of both the wines leaving Australia and those entering the Britain. Attempts at creating public institutions, such as the South Australian Government Depot, broke down, in part because of the opposition of competitors like Burgoyne and Pownall, but also because the semipublic nature of a depot made it difficult to reject poor-quality wines, and management's lack of experience and capital made it hard to build up stocks and establish new markets.

Export markets demanded large quantities of wines of similar styles, while the domestic market encouraged winemakers to produce brandy and dessert wines. Brandy producers benefited not only from economies of scale in the production, but also from the possibilities of import substitution behind high tariffs and limited competition because of the excise tax. Brandy and dessert wines were also considerably easier to brand than table wines. In this respect, companies such as Seppelt had much in common with the California Wine Association or Gonzalez Byass in that they were able to take advantage of growing economies of scale in production and marketing behind government protection.

[86] Ibid., March 1896, p. 388; April 1897, p. xvi; and February 1898, p. 167. In South Australia, it was argued that "the half-dozen larger capitalist growers" were "quite independent of any central cellar."

CHAPTER 11

Argentina: New World Producers and Old World Consumers

> To sell what the couple of dozen wineries do here (Mendoza) would need in France a hundred merchants . . . and need the output of several thousand producers.
>
> —José Trianes, 1908:26

ARGENTINA between 1869 and 1914 embarked on a period of exceptional growth, with per capita income increasing by an average of 5 percent a year and population jumping from 1.74 million to 7.89 million. By 1914 Argentina had a higher per capita income and real wages than in most European countries (table 11.1).[1] Economic growth was caused by exogenous factors as falling transport costs and external demand gave farmers a comparative advantage in the production of raw materials (hides and wool) and later foodstuffs (wheat and frozen meat). High wages attracted European migrants and were accompanied by large flows of capital to construct railways, port facilities and urban amenities. Between 1869 and 1914 the population of Buenos Aires grew from 182,000 to 1,576,000 inhabitants at an average rate of 6.5 percent a year; it was the second most populous on the Atlantic seaboard, after New York, and three times larger than Madrid or Rome.[2]

Grape-growing conditions in Mendoza were exceptional, and once this region was connected by rail with Buenos Aires (1884) and producers learned the art of controlling fermentation in the hot climate, the possibilities of selling large quantities of cheap branded wines in the rapidly growing market appeared endless. With the exception of Algeria, Argentina's industry grew faster than anywhere else in the world between 1885 and 1914, making it the seventh largest producer, with an annual per capita consumption of 60–70 liters, considerably more than either the United States (2 liters) or Australia (5 liters). Yet contemporaries found little to praise because the industry appeared incapable of eliminating the frequent deep recessions into which it was plunged, and despite high standard of living per capita wine consumption remained less than half that of France.

[1] Williamson (1995).

[2] Rock (1993:114); Gallo (1993:83–84). The population of Madrid in 1910 was 600,000; Rome, 543,000; and Paris, 2.9 million (Mitchell 1992:73–74).

TABLE 11.1
Population, GDP, and GDP per Capita in Various Countries, 1870–1914

	Argentina	Australia	Chile	United States	Italy	France	Spain	United Kingdom
Population (millions)								
1870	1.8	1.6	1.9	40.1	27.9	38.4	16.2	29.3
1913	7.7	4.8	3.5	97.6	37.2	41.5	20.3	42.6
GDP (millions of 1990 Geary-Khamis dollars)								
1870	2.4	6.2	Na	98.4	40.9	71.4	22.3	95.7
1913	29.1	26.5	9.3	518.0	93.4	143.1	45.7	214.5
GDP per capita (1990 International dollars)								
1870	1,311	3,801	na	2,457	1,467	1,858	1,376	3,263
1890	2,152	4,775	na	3,396	1,631	2,354	1,847	4,099
1913	3,797	5,505	2,653	5,307	2,507	3,452	2,255	5,032

Source: Maddison (1995).

Factor endowments in Argentina were similar to those in Australia and California, as labor and capital were scarce and nonirrigated land was abundant. One major difference was that most immigrants were already wine consumers, and some had firsthand knowledge of the industry, although frequently this had been acquired in regions that showed few of the characteristics of their new country, especially as Argentina's vines depended on irrigation. By the final decade of the nineteenth century, a dozen or more large wineries dominated the industry, supplied by large numbers of independent specialist family grape producers. Just as in California, a relatively small region (Cuyo) accounted for over nine-tenths of the country's output but had just 5 percent of the nation's population in 1914.[3] While big business and "trustification" were as much features of the Argentine economy as they were in the United States, attempts at collusion by Mendoza's leading wineries had only limited success. This was because if the large wineries could usually count on the active support of the local provincial government in Mendoza, the federal government in Buenos Aires was unwilling to accept blatant price-rigging for an item that was part of the country's staple diet.

This chapter first looks at the growth of the industry after 1885 and its organization at the turn of the century. The second section examines the response of different groups to the major slump in 1901–3 and shows that although attempts to self-regulate the industry were generally successful, wine quality remained poor because consumer choice was determined by price. Finally, the chapter considers the response of different sectors to the collapse in wine prices after 1913 and shows the difficulties leading producers faced in passing the costs of adjustment on to growers and restricting supply.

ESTABLISHING THE INDUSTRY

The Cuyo region, which contains the provinces of Mendoza and San Juan, is located in the extreme west of the country, at the foot of the Andes. In this arid region, with an annual rainfall of just 200 millimeters, crops can grow only with irrigation fed by the melting snow during the hot summer months. The *criolla* vine was introduced into the region from what is present-day Chile as early as the sixteenth century, but in the early 1880s the region still specialized in fattening cattle for the Chilean market, and the population of Mendoza was just 9,900 inhabitants and that of San Juan, 10,600.[4] The victory of the centralist state and enforcement of the constitution, the creation of a national cur-

---

[3] By 1914 Mendoza and neighboring San Juan accounted for 98 percent of the country's wine, of which 95 percent was consumed outside these two provinces. *Boletín del Centro Vitivinícola Nacional*, June 1915, p. 156; Centro Vitivinícola Nacional (1910:17). For population, Mitchell (2007:38).

[4] Population figures are for 1869. Small quantities of wine were sold to the coastal region in the

rency (1881), and the opening of a rail link between Mendoza and Buenos Aires changed the local economy and allowed commercial wine production to become a serious proposition.

A revival of interest in viticulture occurred a decade or two before the railways. In 1870 Eusebio Blanco translated Henry Marchard's *Traité de la Vinification*, adding notes of his own, and in 1887 Emilio Civit wrote *Los Viñedos de France y los de Mendoza* and advised Tiburio Benegas, who was the provincial governor (and also his brother-in-law), on the future of the local industry. In both cases, the recommendation was to produce fine wines, copying the best French practices. In part this was because of the French influence on the local wine industry in neighboring Chile, but also because poorly made wines would not survive the long trip to Buenos Aires, and only fine wines selling at high prices would be profitable for producers.[5]

The works of Blanco and Civit influenced state policy toward the sector, although interest in fine wine production remained limited. A local law in 1881 exempted growers from taxes on new vineyards, a new irrigation law was passed in 1884, attempts were made to attract European settlers, and the Banco de la Provincia de Mendoza was created in 1888.[6] However, the rapid growth in viticulture owed more to market incentives than government backing, and in particular the creation of a rail link between Mendoza and Buenos Aires. Wine output increased fourfold between 1883 and 1899 (table 11.2), driven by population growth, import substitution, and growing per capita income. On the eve of the railway link, Mendoza had just 2,000 hectares of vines, but fifteen years later this had increased to 21,500, with an additional 13,000 in neighboring San Juan.[7]

As in other regions of the New World, natural conditions were ideal: long hours of sunshine created grapes full of sucrose, and there was no phylloxera or other vine disease, with the exception of powdery mildew. Irrigation allowed yields that were considerably greater than those found in California or Australia (table 11.3). However, the need for irrigation implied that before 1885 most land was already owned by the provincial elite, who used sharecropping arrangements to plant and cultivate their vines.[8] Because the local Creole labor force showed little interest in full-time employment in the vineyards, much of the labor was supplied by European immigrants, who planted the vines "in the same way as was

---

seventeenth and eighteenth century, but trade slowed with independence and the civil wars and anarchy that followed. See Rivera Medina (2006) and Coria López (2006).

[5] Blanco (1870:15), cited in Richard Jorba (2008:7).

[6] Richard Jorba (2006:77–81). Tax relief was given only if there was a minimum of 1,260 vines per hectare. Official attempts to recruit labor had limited success but were unnecessary after 1885.

[7] Together these two provinces had 76 percent of the country's vines in 1899 and produced 85 percent of the wine in terms of volume and 87 percent in terms of value (Galanti 1900, appendix). The two provinces had 83 percent of the capital invested in vineyards and 93 percent in bodegas.

[8] Mendoza and San Juan are oases in the desert, with little more than 80,000 hectares cultivated (or around 3 percent of the region) in Mendoza, and 70,000 hectares in San Juan in the late nineteenth century (Richard Jorba 2006:24).

TABLE 11.2

Wine Production and Imports, Argentina, 1883–1912 (thousands of hectoliters)

	1883	1888	1893	1899	1902	1912
Mendoza	19.1	58.9	181.5	850	1051	3452
San Juan	Nd	Nd	Nd	550	235	637
Other provinces	Nd	Nd	Nd	231	74	323
Argentina	Nd	Nd	Nd	1,631	1360	4402
Imports	507	713	709	460	306	468
Total	—	—	—	2,091	1666	4870
Population					5060	7370

Sources: 1883, 1888, 1893—Richard Jorba (1994:80); 1899—Galanti (1900); 1902—Arata (1903:219, 253); 1912—*Boletín del Centro Vitivinícola Nacional,* July 1912, p. 2209; and for Mendoza, Barrio de Villanueva (2009:10).

done in their home country."[9] The obligations and rights stipulated in these contracts varied over time depending on the relative scarcity of labor and price of wine. Two contracts in particular were used: the *contratistas de plantación* (planting contracts) and the *contratistas de viñas* (cultivation contracts). With the planting contracts, the landowner provided the cuttings, work animals, and farm tools, and the sharecropper received a cash payment for each surviving vine that he had planted, together with the produce of the last harvest or two. Planting was relatively easy given the soft soil and irrigation and could be done with cheap plows and without the need for trenching.[10] The lack of viticulture experience initially found among the landowners contrasted with that of the immigrants, and the latter often made the strategic decisions as to which grape varieties to plant or the spacing between vines.[11]

By contrast, the cultivation contracts were annual contracts used in mature vineyards. They included a fixed cash payment that was paid monthly and a small share of the harvest, often no more than 5 percent. Jules Huret noted in 1913 that Domingo Tomba's 200-hectare estate was worked in plots of 10–12 hectares per family, often of Chilean origin.[12] With both contracts, families supplemented household income with a few sheep, pigs, chickens, and a small vegetable patch. By 1936, 35 percent of Mendoza's vineyards and 68 percent of the vine area were cultivated using this type of contract.[13]

[9] Pacottet (1911:viii).

[10] Simois and Lavenir (1903:117). Richard Jorba (2007), in a study of twenty-six contracts, found that 65 percent of tenants were Italian, 19 percent French, and the rest probably Spanish.

[11] Richard Jorba (2007:180).

[12] Huret (1913:229). This seems large. Suárez (1922:xi) argued that it has been demonstrated that a worker could cultivate 6 hectares a year. However, the area obviously also reflects family size.

[13] Salvatore (1986:232).

TABLE 11.3
Production and Yields in the New World

	Area of vines (hectares)	Wine production (thousands of hectoliters)	Yields (hectoliters per hectare)
Australia			
1880	6,400	80	13
1890	14,820	139	9
1900	25,620	195	8
1910	24,360	231	10
Argentina			
1899	45,196	1,600	35
1912	66,400	4,400	66
Chile			
1900	30,000	1,062	33
1910	60,400	2,017	33
California			
1880	14,575	318	22
1890	36,539	669	18
1900	34,818	882	25
1910	58,704	1,689	29

Sources: Figures are very approximate. Australia—Osmond and Anderson (1998); Argentina—Galanti (1900:19) and table 11.2; Mitchell (2007); United States—Shear and Pearce (1934), table 10; area of vines—Sullivan (1998:48).

Note: Barrio de Villanueva (2009:2) gives 57,764 hectares for Mendoza in 1912, which has been assumed to represent 85 percent of the total area of Argentina.

Sharecropping contracts allowed the traditional provincial elites to own large areas of vines, and the thirty leading landowners in Mendoza saw their holdings increase from 714 hectares in 1883 to 6,317 in 1900, equivalent to a third of the total area at both dates. However, as the contracts established before 1900 were made under conditions of labor shortages and high grape prices, immigrants enjoyed very favorable terms, and many saved enough to become owners themselves. Just as in the Midi, the creation of wine estates was therefore accompanied by a major increase in the number of small vineyards, and a total of 1,486 plots of vines (82 percent of the total) planted between 1886 and 1895 had less than 10 hectares, allowing a skilled part-time workforce to be available for employment on the large plantations and in the wineries.[14]

[14] Richard Jorba (1994, cuadro 3). A total of 886 plots had less than 2 hectares, and 1,211 had less than 5 hectares. In 1919 there were 2,940 properties of less than 5 hectares; 1,078 between 5 and 10; 745 between 10 and 20; 535 between 20 and 50; 171 between 50 and 100; and 84 of more than 100 hectares (Suárez 1922:x).

The average winery in Mendoza in 1900 received grapes from only 13 hectares of vines, although this increases to an average of 21 hectares when just the major producer regions are considered, and 47 hectares in the department of Luján (table 11.4). By contrast, 18 of the 1,082 wineries produced more than 10,000 hectoliters; in San Juan the figure was 7 of the province's 612 wineries. In reality, the number of wineries that operated in any particular year fluctuated significantly, as growers preferred to sell their grapes to large wineries when prices were high but crushed them themselves when they had no buyers. Despite the presence of small wineries, a dozen or more dominated the industry, although the wine they produced was of poor quality and virtually all sold within the year.[15] In 1899 the largest sixteen wineries (1.7 percent of the total), accounted for 39 percent of all the capital invested in Mendoza's wineries, while in San Juan the extremes were even greater, with seven wineries (1.1 percent) accounting for 41 percent of the capital stock.[16] Three years later the Tomba winery produced 91,250 hectoliters of wine (about 12 percent of Mendoza's total, equivalent to approximately 1,500 hectares of vines in full production), the Giol y Gargantini winery produced 58,255 hectoliters, Arizu 57,022, and Barranquero, with two wineries, produced a total of 54,754 hectoliters. The leading ten wineries, eight of them were foreign owned, produced 445,137 hectoliters or a third of the province's total.[17] In the period 1908–10 winemakers bought almost three-fifths of their grapes from specialist growers, and by 1914 it was claimed that several winemakers produced 160,000 hectoliters, four times more than the largest in France.[18] Finally, foreign-born owners of vineyards increased from 29 percent of the total to 52 percent between 1895 and 1914, and from 28 percent of the wineries to 69 percent.[19]

Despite the favorable conditions for wine production, a rapidly growing labor supply, access to capital markets, and the presence of a dozen or more modern wineries, contemporaries were unanimous in their opinion of the poor quality of the work carried out in both vineyards and wineries. According to Pedro Arata, head of the 1903 commission that studied the industry: "The great majority of growers and winery owners believed that planting vines and making wine was like buying cows and bulls, turning them out into a field and selling immediately the calves, and leaving it all to Nature as is done in the Province of Buenos Aires."[20]

Arata's negative comments have been widely quoted by historians, but one

---

[15] Kaerger (1901), cited in Barrio de Villanueva (2009:16). See also Simois and Lavenir (1903: 127) and Arata (1903:202).

[16] Galanti (1900:92, 136), my calculations. These wineries had a minimal capacity of 10,000 hectoliters. The average in Mendoza was 463,000 pesos, and in San Juan, 346,000 pesos.

[17] Barrio de Villanueva (2008b, cuadro 1). They used electricity or steam power.

[18] Centro Vitivinícola Nacional (1910:17); *La Prensa*, April 24, 1914, p. 12.

[19] Salvatore (1986:233). The number of foreigners in Argentina was 30 percent of the total population in 1910.

[20] Arata (1903:192). For the commission, see below.

TABLE 11.4
Size of Wineries in Mendoza, circa 1900

Department	Mendoza vines (%)	Mendoza growers (%)	Mendoza wineries (%)	Number of modern wineries*	Growers with winery (%)	Area of vines per winery
Godoy Cruz	7.4	6.4	8.1	6	38	17
Maipú	29.0	13.1	15.6	4	36	35
Guaymallén	16.2	27.3	19.3	2	21	16
Ciudad	2.4	5.4	6.5	1	36	7
San Martín	8.6	10.0	13.2	2	40	12
Luján	14.7	8.5	6.0	1	21	47
Las Heras	5.2	5.5	6.1	0	33	16
Subtotal	83.5	76.2	74.9	16	29	21
Others	16.5	23.8	25.1	2	31	13
Total	100	100	100	18	30	19

Sources: Galanti (1900:91), and Richard Jorba (1998:304). My calculations
*Wineries with a capacity of over 10,000 hectoliters. Figures are distorted by growers being able to use facilities in neighboring departments. All figures refer to 1899, except the area of vines (1900).

obvious explanation for his bitter criticism is that his report was made at the depths of a major recession. Grape prices fell to levels that barely covered growers' variable production costs, so that only the most important tasks were carried out in the vineyard, and yields were kept high by excessive irrigation. The 1903 commission also noted that the short-term nature of sharecropping contracts encouraged the *contratistas* to maximize output, without considering the long-term consequences for the vine.[21] In addition, workers were often slow to switch from planting and pruning methods that were suitable to their native wine-producing regions to those needed for the rich, irrigated soils of Mendoza and San Juan.[22] Growers planted "French varieties," in particular malbec, cabernet (sauvignon and gros cabernet), sémillon, and pinots, so that by 1903 only 26 percent of the area of vines comprised the old *criollas*. However, the canes brought from Chile because of the ban on vines from phylloxera-affected Europe led to large numbers of sterile plants and low yields.[23]

Harvesting also left much to be desired because unskilled labor was paid by piecework, resulting in no attempt to select the most suitable fruit or separate the grapes properly from the stems and leaves. There were also other problems

[21] Simois and Lavenir (1903:120). For a wider discussion on the incentive structures and contracts in viticulture, see Carmona and Simpson (1999:292–94) and Carmona and Simpson (forthcoming).
[22] Richard Jorba (2006:114–15); Simois and Lavenir (1903:120); and Arata (1904:147).
[23] Simois and Lavenir (1903:118, 123–24).

peculiar to the region. The French geographer Pierre Denis noted that the dryness of the atmosphere caused ripe grapes to remain longer on the vine without any harm, and a longer harvest reduced the number of migrant workers required and allowed growers to cultivate a larger area.[24] Indeed, the huge size of some of the wineries can only be explained by the possibility of stretching the harvest over three or four months.[25] Grapes that remained on the vine for this length of time, rather than for a month or six weeks, lost moisture and weight but gained in sugar content. This encouraged growers to overirrigate and benefited the winery owners, who bought the grapes by weight and added water to the must to reduce the alcoholic strength from 16 to 12 degrees.[26] It did little to improve wine quality.

Arata's report criticized the unscientific nature of wine making that often took place in unhygienic conditions. The addition of tartaric acid to the must was limited by its cost, and despite abundant supplies of cold water very few wineries had invested in cooling equipment.[27] Yet the government enquiry of 1903 also noted that large amounts of capital invested in winery equipment had been wasted because producers "forgot" the basics of production and conservation of sound wines.[28] Most contemporaries were unanimous in their condemnation of the poor quality of the vast majority of wines at the turn of the twentieth century, but they were also divided as to whether this was caused by the lack of winemaking skills or by the nature of market demand. Consumers drank red wines almost exclusively—wines that were "very dense, with lots of color, a high alcoholic strength and rich in dry extract." Water was almost always added; the question was who along the commodity chain added it.

REDEFINING THE INDUSTRY

Elías Villanueva, Mendoza's provincial governor and eleventh largest wine producer, argued in 1902 that the crisis occurred not because wine quality had sud-

[24] It also encouraged a trade in grapes. Denis (1922:86–87) noted that in Mendoza there was "a division of labor which seems to the European visitor as strange as the climate which partly explains it."

[25] The 1908 harvest, for example, stretched from mid-February to the end of May. The largest winery, Tomba, crushed some grapes in vats of 50 hectoliters, which were brought by rail to the winery with the wine fermenting (Pacottet 1911:78, 82).

[26] Simois and Lavenir (1903:126); *La Prensa*, April 8, 1914, p. 12.

[27] As in other regions with hot climates, the major difficulties were the lack of acidity in the must and the high temperatures produced in the fermenting vats, caused not just by climatic conditions, but also by the high sugar content of the grapes. A number of wineries had purchased Müntz and Rousseaux cooling systems, but these were considered expensive and were designed for small wineries where water was scarce (Simois and Lavenir 1903:130–35).

[28] Ibid., 128. A major problem was that equipment was often imported and more appropriate to Europe's wine regions than Mendoza's.

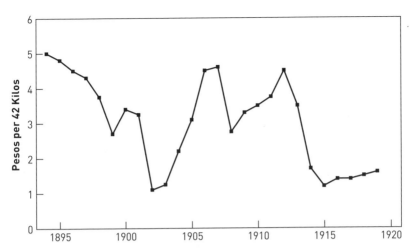

Figure 11.1. Grape prices in Mendoza, 1893–1920. Sources: Suárez (1914: *Estadística*) and (1922:viii)

denly deteriorated, but because merchants lacked money to make purchases, and wine was an expenditure that many consumers reduced at times of financial difficulties.[29] The high profits of the 1890s had encouraged growers and winery owners to go heavily into debt, and after 1901 they faced a credit squeeze brought about by a sharp contraction of foreign investment and outflow of gold to cover the trade deficit. Grapes, which had sold for around 3.4 pesos per 46 kilos in 1900 and 3.25 in 1901, fell to 1.1 pesos in 1902 and 1.25 in 1903 (fig. 11.1). Wine prices also declined, though by less.[30] Among the numerous casualties of the crisis was the family firm Benegas e Hijos, perhaps Mendoza's best-known wine producer, which was forced to offer shares in the company to its main creditors in order to remain in business.[31]

Although short-lived, the crisis highlighted a number of structural problems facing the industry. As in other countries, the poor quality of many wines made adulteration easy and especially attractive when prices were high. Wine producers demanded a national law to protect their product and which would be enforced. An additional problem in Mendoza was that the large wineries reacted to the decline in demand by refusing to buy the grapes of small producers. These were therefore forced to open their old cellars and sell the wine immediately after fermentation because of their lack of storage capacity and limited access to credit, all of which contributed to the "discredit of the product" and falling prices because of the excess of supply. Weather conditions also made wine mak-

[29] *El Comercio*, October 22, 1902, p. 2, cited in Barrio de Villanueva (2008a:339). GDP shrunk 2 percent in 1902 but recovered quickly the following year (Sturzenegger and Moya 2003:114).
[30] Arata (1903:253).
[31] Barrio de Villanueva (2005:46–56).

ing especially difficult in 1902, and the lack of buyers led to large amounts of wine being left to spoil in the large wineries' cellars.[32]

Four groups were deeply affected by the crisis: the Mendoza provincial government, which depended heavily on the sector for taxes; the banks, which had large quantities of nonperforming loans; the large wineries, with their unsold stocks; and the growers, who risked being left with grapes that they could not sell. On the eve of the 1903 harvest, an attempt at an agreement between the large producers, bankers, and "a few growers" to rent wine-making facilities and cellar space to the small producers so they would not dump poor quality wines on a saturated market, as had been done in 1902, was ridiculed in the local press for shifting all the adjustment costs onto the grape growers. The leading regional newspaper, *Los Andes*, called on all groups to make sacrifices, but this was limited to controlling adulteration.[33]

As in other wine districts at this time, many contemporaries believed that the real cause of the crisis was underconsumption, which was blamed on the poor quality of wine and widespread adulteration found along the entire commodity chain. A national law of 1893 permitted the sale of wines made from raisins, fermented pomace (*petiot*), or other products, but these had to be clearly labeled as such and paid higher taxes than "natural" wines.[34] Mendoza's provincial law of 1897, by contrast, prohibited the production of all wines made from substances other than fresh grapes, while that of 1902 doubled taxes on red wine with less than 26 per 1000 dry extract to stop producers watering down their wines. The real challenge, as everywhere, was enforcement, which implied creating some independent entity that could legally check what was happening in winemakers' and merchants' cellars. The collapse in wine prices in 1902, the abysmal quality of the wines that year, and the threat by some politicians in Buenos Aires to reduce tariffs on Chilean wine imports pushed the large producers and local government into action. Villanueva was successful in getting the Ministerio de Agricultura de la Nación to create the previously mentioned national commission headed by Pedro Arata to investigate the industry's problems, and this included provisions for inspecting wine cellars as well as destroying diseased and adulterated wines. The commission soon faced local opposition to its work. The newspaper *Los Andes* complained that the large winery owners were quick to applaud the activities of the commission when it removed poor-quality wines from small producers' cellars but placed legal obstacles to it entering their own.[35] In particular, the manager of Governor Villanueva's own winery refused permission for the destruction of 1,000 hectoliters of spoiled wine, and it required a second visit by the commission, when some 500 hectoliters were finally removed.[36]

---

[32] Simois and Lavenir (1903:127, 143) noted that "mannite affected wines were in abundance in 1902."

[33] *Los Andes*, January 17, 1903, p. 4.

[34] Balán (1978:76–77).

[35] *Los Andes*, March 17, 1903, p. 4).

[36] See especially Barrio de Villanueva (2008a:336).

A new national wine law was passed in 1904 (*Ley Nacional de Vinos*, no. 4363) which defined wine as being made only from fresh grape juice, with certain enological exceptions, such as the addition of tartaric or citric acid. Red wines had to contain between 24 and 35 per 1,000 dry extract (white wines less than 17 percent), and imported wines had to be sold in their original casks, with certificates proving the place of origin. The only activities permitted in a winery now were the manufacture of wine, distilling of wines and pomace, and refining of spirits. The making of artificial wines from raisins and pomace had to take place in establishments other than wineries and was strictly controlled. Finally, wines that had become spoiled and diseased had to be destroyed.[37]

As in other countries, the law reflected local conditions and the interests of certain groups within the industry. Imported table wines were strong in alcohol and high in extract as they were mainly blended with domestic wines, and the new law discriminated against them. The maximum 35 per 1,000 in extract permitted in domestically produced red wines was to discourage the watering down of wines before sale, and the minimum level was to encourage the production of wines that could compete with imports.

The Centro Vitivinícola de Mendoza was created in 1904, in part as a response to fraud, and it merged the following year with the Sociedad de defensa vitivinícola nacional, a similar organization of wine merchants in Buenos Aires, to form the Centro vitivinícola nacional (CVN), based in the capital. This geographic shift on the part of Mendoza's largest producers was caused by a convergence of interests between themselves and the capital's leading wine merchants, as imports declined from around a quarter of consumption in 1900 to a tenth by 1912.[38] To control fraud the CVN paid the federal government for five new subinspectors, which led to a number of prosecutions.[39]

Unlike European countries, a high share of Argentina's production was located in a small geographical area, and industrial concentration was exceptionally high. Mendoza accounted for four-fifths of the nation's production in 1902–3, and just twelve producers (who enjoyed close links with the provincial government) produced a third of the total.[40] The industry contributed as much as a half of the province's tax revenue and 60 percent of Mendoza's gross industrial output as late as 1914.[41] The problems of 1901–3 led the provincial government to consider the need to control wine adulteration to protect its revenue base as well as to use government funds to diversify the local economy and re-

---

[37] Barrio de Villanueva (2007:7–9). Fine wines that had been bottled were exempt from these restrictions on *extracto seco*.

[38] Market control was fragmented in Buenos Aires between those merchants who dealt with imported wines (which were blended with domestic or artificial wines) and those who depended exclusively on domestic supplies.

[39] Barrio de Villanueva (2006:200).

[40] Table 11.2 and Barrio de Villanueva (2008b:89).

[41] Richard Jorba (2008:7), who writes that viticulture allowed the province "an important degree of independence" with respect to the federal government; and Coria López (2008, anexo viii). Local industries were developed to supply the sector. See especially Pérez Romagnoli (2006).

duce its dependence on the sector. The arrival of Emilio Civit as governor in 1907 marked a turning point and led to conflicts between politicians and the leading wineries.[42] In 1907 the Dirección de Industria was created to provide technical assistance to the sector, and within this institution the Oficina Química had the responsibility for inspecting wineries and the capacity to levy fines and destroy wine that was deemed as illegal. All wine-making facilities had to be registered (those unregistered were considered illegal), and every building where wine was made, stored, or sold was liable to surprise inspections. The first director was Enrique Taulis, who as a Chilean was supposed to provide an element of impartiality. Fines, which totaled $166,177 in 1908, rose to $388,883 two years later, provoking fierce opposition from some of the large producers.[43]

The control of fraud was highly controversial and depended on the continued vigilance of government inspectors, paid for by an often reluctant industry. *Los Andes* in October 1909, for example, criticized "the almost complete lack of vigilance" of the retail trade in Mendoza, in part because of the inept government controls, but also because of the "favors" shown toward some owners who were friends of "politicians."[44] Writers closer to the large wineries thought otherwise, and one considered government enforcement as bordering on being excessive.[45] A tribunal consisting of the leading winery owners was established when Taulis was forced to resign after fierce opposition from the sector, and fines fell in 1912 and 1913 to 5 percent of the 1908 level.[46] Although some artificial wines inevitably continued to be produced, they had ceased to be the industry's major concern by 1910.

The extreme difficulties facing the industry between 1901 and 1903 encouraged large producers to propose changes that went beyond just controlling adulteration. The first initiative was in November 1901 when Tiburcio Benegas suggested the creation of a trust whereby each member placed 5 percent of their wine in a new company in exchange for shares. By this system, the industry's leaders hoped not just to limit supply, but also to preserve their control of the industry. While the Benegas project originated from within the sector, that of Elías Villaverde the following year came from the provincial government and involved greater state involvement. The proposal consisted of using funds raised by new production and sales taxes on wine to improve quality by investing in better equipment and scientific skills. Finally, Horacio Falco, a highly influential member among the leading producers, returned from Europe with a "project to

[42] Barrio de Villanueva (2006:210–12).

[43] Bottaro (1917).

[44] *Los Andes*, October 7, 1909.

[45] Alazraqui (1911:90). Taulis, who was responsible for control in Mendoza, suggested that fraud accounted for only 15 percent of consumption, with most occurring in Buenos Aires (*Boletín del Centro Vitivinícola Nacional*, June 1909, p. 1155).

[46] Bottaro (1917); *La Prensa*, April 6, 1914, p. 8.

save the industry" in 1903, which turned out to be little more than another attempt by the large wineries to create a company that they themselves controlled, to restrict supplies by buying up surplus wine when necessary.[47]

There were several major obstacles to the creation of a wine trust such as the California Wine Association, and the Mendocino newspaper *Los Andes* described the local attempts as both "ridiculous" and "impossible."[48] Unlike California (or Australia), wine was the national drink, and therefore any suspicion of a Mendoza cartel creating monopoly prices led to politicians in Buenos Aires demanding imports from Chile or elsewhere.[49] The low quality of much of the wine also implied that a trust that did not include all the major wineries would suffer from problems associated with free riders—independent producers who did not contribute to the costs of regulating the market but increased their output to take advantage of higher prices. By the autumn of 1903 the promise of better grape and wine prices following the February harvest ended the debates among major producers who had been proposing collusion or a trust.

The low prices and problems of finding wineries willing to purchase their grapes also left small growers reflecting on the need to create cooperatives. In 1903, at the depth of the depression, a number of growers proposed a "sindicato" to tackle specific problems facing the industry, including transport, marketing, and the shortage of wine-making equipment. It was neither a genuine producer cooperative nor a trust, and interest quickly faded. A short-lived cooperative, the Helvecia winery in Maipú, appeared in 1903.[50] When Leopold Mabilleau, director of the Social Museum in Paris, spoke in Mendoza in 1912, he explained the need to overcome excessive individualism and advocated a collectivist approach to problems, noting the success of French farmers in creating cooperatives.[51] However, in France, as shown in chapter 3, the state responded positively to demands by small farmers, especially by providing access to capital markets, which favored the creation of producer cooperatives. In Mendoza it was the interests of the large wine producers that the provincial government responded to, and these were reluctant to see small growers form cooperatives that might threaten their supplies of grapes.

The industry's problems were resolved by the renewal of massive capital imports and immigration, leading to the Argentina's economy growing at an annual 7 percent between 1903 and 1913.[52] Demand for wine also increased dramatically. The area of vines grew from 22,875 hectares in 1904 to 57,764 hectares in

[47] These projects are described in greater detail in Barrio de Villanueva (2006:190–98) and Mateu (2007:14–15).

[48] *Los Andes*, August 1901, p. 4.

[49] The Buenos Aires newspaper *La Nación* warned of the dangers of a cooperative organized by speculators among wine producers, which claimed to protect the interests of the industry, consumers, and government. Cited in *Los Andes*, January 15, 1903.

[50] Mateu (2007:15).

[51] *La Tarde*, September 18, 1912, cited in ibid., 13.

[52] Sturzenegger and Moya (2003:114).

1912; wine output jumped from 1.1 million hectoliters to 3.5 million; imports declined from a million hectoliters to half a million; and annual per capita consumption increased from 41 liters to 72 liters (1913).[53] Capital investment was used to modernize and raise winery capacity, rather than increase the number of firms, so that while the 1,064 bodegas producing less than 10,000 hectoliters in Mendoza in 1899 fell to 956 in 1910–14, those above this figure increased from 18 wineries in 1899 to 60 in 1910, 76 in 1912, and 96 in 1914, by which time the three largest bodegas produced a total of 350,000 hectoliters and required grapes from around 6,000 hectares.[54]

As elsewhere in the New World, major winery investment reduced production costs by saving expensive labor, as well as capturing the economies of scale that could be achieved by building wineries on green-field sites. The hot, dry climates produced grapes of high quality, and scientific wine-making processes and selective blending in theory now allowed a wine to be produced in large quantities and of a consistent and better quality from one year to the next that could be branded. A handful of firms, including Benegas and López, had begun to do just this in the final year or two of the nineteenth century, and the number increased to thirty-six brand names during the first decade of the new century.[55] The result was that a much greater percentage of ordinary commodity wine was sold under producer brand names in Buenos Aires than in Paris or Madrid (table 11.5). While contemporaries had been highly critical of the diversity of wine produced in Mendoza because of the poor wine-making conditions at the turn of the century, a decade later they bemoaned the fact that there was little to distinguish between the output of the largest producers, in part because of the widespread use of the malbec variety.[56] This uniformity, and perhaps collusion between the leading producers, led to minimal price differences for their wines (table 11.5). As most consumers were of Mediterranean extraction, the demand was for a strong, alcoholic wine, which was then watered down, for both taste and economy. The only guarantee that a brand offered in these conditions was that it would be the consumer, rather than an intermediary, who added the water. Migrant labor (*golondrinas*) did not drink branded wine in Italy, and they were unwilling to pay a premium for it in Argentina. This uniformity and low quality of wine is also shown by the fact that it sold for only a third more than the *vinos de traslado*, wines produced by independent growers in small wineries and sold immediately after fermentation to be blended with the produce of the

[53] Barrio de Villanueva (2009:1).

[54] Figures refer to 1910, 1912, and 1914. As noted above, the number of wineries that operated in each year varied. The provincial governor in 1910 noted that 910 wineries used modern methods, and only 250 traditional ones. See Barrio de Villanueva (2009:5–7).

[55] Barrio de Villanueva (2003:39, 42); Richard Jorba (1998:306). One of the functions of the CVN was to facilitate registration of brands by producers.

[56] Galanti (1900:26–27); Simois and Lavenir (1903:200); Galanti (1915:34); and *Boletín del Centro Vitivinícola Nacional*, December 1910, p. 1671. For the 1930s, see Trianes (1935:37).

TABLE 11.5
Selected Wine Prices and Brand Names, June 1910

		Price—centavos per liter	
Winery	Brand	Mendoza	Buenos Aires
Domingo Tomba	Tomba	0.27*	0.32*
Giol & Gargantini	Toro	0.27	
Sociedad Bodegas Arizú	Arizú	0.275	0.32
Honorio Barraquero	Baco	0.25*	
Sociedad Germania	Cigüena	0.24	
B.y C. del Bono		0.23	
Herwig & Cia.	Perdiz	0.25	
Red wine in general "sobre wagon"		23–27	
Vinos de "traslado" from winery		15.0–17.0	

Source: Boletín del Centro Vitivinícola Nacional, June 1910, p. 1516.
*New barrels

big houses to meet their orders. In this way, the large wineries were able to adjust their supply to meet demand for their branded wines in the final market. However, when the big producers stopped buying as prices fell after 1914, these wines were sold directly to dealers in Buenos Aires, forcing down the prices of all wines, branded and unbranded.

By the first decade of the twentieth century contemporaries such as José Trianes and José Alazraqui were aware of the important differences between the Argentine and European wine industry.[57] Yet while the production and commerce of cheap wines were more vertically integrated in Argentina than in Europe, they were less so than in California and probably in Australia. Wine quality undoubtedly improved in the decade after 1901, but because consumers were primarily interested in price, there were no financial incentives to improve quality to the levels found in either Australia or California, where producers had to create new markets among consumers unaccustomed to drinking wines. The attempts to establish competitions and adjudicate prizes for the best wines, for example, came much later and appear to have created little consumer interest.[58] Mendoza's wine industry before 1914 lacked figures of the stature of Eugene Hilgard and George Husmann in California, or Thomas Hardy and Arthur Perkins in South Australia. Highly observant commentators such as Pedro Arata and Amerino Galanti were not involved in wine making, while the influence of people such as Aaron Pavlovsky, Paul Pacottet, or Leopoldo Suárez was limited, at

---

[57] Trianes (1908:26) quote at the beginning of this chapter; and Alazraqui (1911:85), who noted that winemakers assumed many of the merchants' functions.

[58] Barrio de Villanueva (2003).

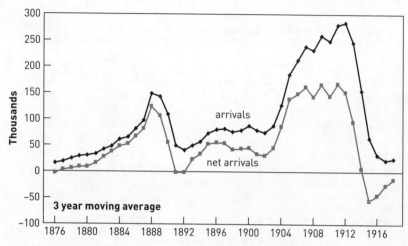

Figure 11.2. Argentine immigration, 1875–77 to 1917–19. Source: Mitchell (2007:105–6)

least prior to 1914.[59] The wine-making experiments, especially those of Algeria, Australia, and California, were rarely mentioned in the local wine press. This can be explained by the fact that immigrants still looked to their origins for solutions: Italy (and especially the Scuola di Viticoltura e di Enologia in Conegliano in Treviso), France (essentially Bordeaux), and Spain. In addition, market incentives and local investment devoted to scientific research were simply not considered necessary by the industry. Wineries sold all they could produce in times of high demand, and when prices collapsed they simply reduced their grape (and wine) purchases from small producers and looked to the provincial government for a solution.

THE LIMITS TO GROWTH AND THE RETURN TO CRISIS

The rapid growth of the Argentine economy was brought to a sudden halt when GDP collapsed by 10 percent in 1914 as financial problems in London provoked a repatriation of capital, creating an economic depression that lasted until the end of 1917.[60] Growers who had sold their grapes for 4.5 and 3.5 pesos a quintal (42 kilos) in 1912 and 1913 had to accept just 1.7 and 1.2 pesos in 1914 and 1915 (fig. 11.1). Against a background of credit shortages, contemporaries once more reflected on the industry's shortcomings. As the *Boletín del Centro Vitivinícola*

[59] Pacottet was an agronomist and head of viticultural research at the Institut National de France and author of a number of works on grape and wine production; Pavlovsky directed Mendoza's Escuela de Agricultura and was a winery owner.

[60] Sturzenegger and Moya (2003:93); Rock (1993:141–42).

Table 11.6
Availability and Consumption of Wines in Various Countries, 1909–13 (liters per capita)

	Domestic production	Net Trade (– exp + imp)	Total consumption
Argentina	60	+6	66
Australia	6	–1	5
Chile	60		60
United States	2		2
Algeria	142	–110	32
France	118	+16	134
UK	0	+1	1

Sources: Argentina—table 11.2; Australia—Osmond and Anderson (1998); Chile—Mitchell (2007); United States—table 9.2; Algeria, France, and the United Kingdom—table 4.5.

Nacional noted in January 1914: "only at critical moments ... does the sector worry about its collective interests and then, as if woken from a dream concerning its own grandeur, it agitates madly to find a way to return to tranquility ... and sleep once more."[61]

One explanation given by some contemporaries for the crisis was the lack of demand for wines. Movements in consumption and immigration were closely linked, and the sharp decline in arrivals after 1914 affected sales (fig. 11.2). In 1869, when the foreign population numbered 12 percent, per capita consumption was 23 liters per year; by 1895 the foreign population was 21.5 percent and per capita consumption had risen to 32 liters; finally, by 1914, 30 percent of the Argentina's population was foreign born and wine consumption reached 62 liters per head (table 11.6).[62] The fact that this was less than half the figure in France, despite Argentina's high wages, led some to argue that the problem was not overproduction but rather underconsumption, and that the difference could be attributed to the presence of greater adulteration.[63]

The income elasticity of wine is difficult to measure, but there was a tendency for French per capita consumption to increase with living standards, at least until the First World War. Yet a simple link between higher wages and wine consumption is deceptive, as wine was traditionally drunk much more in regions of production than in areas where the vine was absent. In France wine consumption was common over comparatively large areas of the country, but some regions, such as Brittany, produced and consumed very little wine. Per capita consumption in both Spain and Italy was lower than in France, in part because the vine was absent in a greater number of regions where there was no wine-drinking

[61] *Boletín del Centro Vitivinícola Nacional*, January 1914, p. 2757.

[62] Bunge (1929:128). Per capita consumption refers to the years 1895–99 and 1910–14.

[63] *La Prensa*, April 17, 1914, p. 12, for example, claimed, but without supporting evidence, that annual consumption was actually 140 liters per person, including adulterated wines.

tradition. The fact that in the period 1885–95 almost two-thirds of Spanish emigrants to Argentina left from Galicia and Asturias—regions that accounted for 14 percent of Spain's population but less than 2 percent of its vines and wine production—inevitably restricted demand.[64] This also helps to explain why, with a few notable exceptions, such as Balbino Arizu, Miguel Escorihuela, and José Gregorio López, Spaniards in Mendoza's wine industry were significantly less important than Italians.[65] Yet even in Italy, the major source of immigrants before 1900 was Veneto, a region responsible for 4 percent of Italy's wine, although by 1900 Sicily, a major southern wine producer, had become equally important.[66] Immigrants from regions such as Galicia or Veneto inevitably limited the potential wine market in Argentina.[67]

Market volatility was common in all wine regions because high prices encouraged the planting of vineyards that would become productive only three or four years later, when market conditions might be very different. The rapid increase in output in Argentina after 1910 can thus be partly explained by the high grape prices of 1906 and 1907, while the high prices between 1909 and 1912 increased supply after 1913 and 1914.[68] Volatility was greater in Argentina, however, in part because it operated under New World production conditions, with specialist growers and rapid population growth, and in part because it sold in a market with European characteristics, where wine was considered a basic consumption item, and therefore the Buenos Aires government would not tolerate price manipulation by producers.

The high fixed costs along the commodity chain implied that wine prices in Buenos Aires fluctuated much less than grape prices in Mendoza, which could double or halve from one harvest to the next.[69] Pedro Arata estimated that when the price of grapes in Mendoza was three and half centavos a quintal (46 kilos), wine cost 23.7 centavos a liter in Buenos Aires, but when grapes fell by 70 percent, wine prices fell by only 35 percent (table 11.7). Retailers purchased wine only at prices at which they could make a profit, grapes were bought by wineries

---

[64] Sánchez Alonso (1992:89; 1995:292–93); and Simpson (1985a:359–60). Figures for wine refer to 1889. Catalonia was the only region that supplied a significant number of emigrants (12 percent) and was a major wine region (22 percent of vines). In later years both Alicante and Almeria, traditionally wine-importing provinces, would have high levels of emigration, but to Algeria and France rather than Argentina.

[65] Arizu originated from Unzué (Navarra), Escorihuela from Tronchón (Teruel), and López from Algarobo (Malaga). See especially Mateu (2009); Bragoni (2008). For the presence of Spanish as *contratistas*, see note 10 above.

[66] Devoto (2004:100). Sicily produced 15 percent of Italy's wine in 1895–99 (Mondini 1900: 31).

[67] Ramon Muñoz (2009) makes a similar point for olive oil exports.

[68] An exact fit is difficult to establish, however, because annual production fluctuated for a variety of reasons. In Mendoza estimates of the area under vines are very inexact, and it is not possible to know to what extent the official statistics include young, unproductive vines. See especially Suárez (1922:56) and *La Prensa*, April 14, 1914, which complained that in Mendoza "it costs less to plant vines than count them."

[69] *La Prensa*, April 12, 1914, p. 11.

TABLE 11.7
Correlation between Grape Prices and Wine Prices

Grape price in Mendoza	Wine price in Buenos Aires
3.50 pesos @ 42 kilos	23.7 centavos per liter
3.00	22.0
2.50	20.4
2.00	18.7
1.50	17
1.00	15.4

Source: Arata (1903:204).

if producers believed they could sell the wine, but grapes were always collected and crushed in primitive wineries if the growers believed that the wine could be sold at a price above the immediate cost of harvesting and fermentation.[70]

In Mendoza the creation of new vineyards required relatively little capital, and the use of planting contracts allowed landowners to increase their area of vines without even having to pay wages. Small holders could plant vines on their own land by working during periods of low seasonal demand for wage labor. By contrast, the high capital cost of the new wineries implied that investment in new facilities was linked to the availability of credit and when producers could be sure of being able to sell their wines at a sufficiently high price in Buenos Aires to cover their production costs. This separate ownership of vines and wine-making facilities led to an unbalanced growth between the two sectors because when large wineries failed to purchase grapes from independent producers, aggregate supply was not diminished. As noted with the 1901–3 downturn, an important number of wineries reopened and producers dumped their young wines in the market.[71] The same was true in 1914, when an additional seven hundred wineries opened compared with the previous year, and when the large wineries refused to buy the *vino de traslado* the producers sold it directly in the major markets at important discounts in relation to the branded items.[72] Belated attempts at collective action on the part of the leading producers to purchase these wines failed because few were actually willing to commit themselves and buy, so prices remained a third lower than the previous year in the principal markets.

Specialist grape producers in both Australia and California had greater negotiating strength and market instability probably was less than in Argentina.[73] This was because of the better-quality table wines produced there, and conse-

[70] It was widely believed that prices were imposed on the grower by the large wineries. Ibid., April 6, 1914, p. 9; April 10, 1914, p. 9.

[71] There were reportedly 1,742 wineries in operation in 1902 compared with 1,082 in 1899 and 1,010 in 1907 (Barrio de Villanueva 2009, cuadro 6).

[72] *La Prensa*, April 10, 1914, p. 9.

[73] In California the CWA provided stability. The demand for wine depots suggests, however, that the problem was also present, although to a lesser extent, in South Australia and Victoria.

quently the need for winemakers to guarantee a supply of high-quality grapes when markets recovered. In addition, grape producers in general enjoyed more dynamic labor markets in the large, diverse local economies of San Francisco and Melbourne than growers did in Mendoza, which was a thousand kilometers from Buenos Aires. If winemakers refused to buy their grapes, growers could exit the industry. The absence of a large urban center nearby on the scale of San Francisco or Melbourne also made it harder for Mendoza's growers to sell their surplus grapes for the table, although small amounts began to be sent to Buenos Aires in 1903.[74] In their favor, although the size of family vineyards was similar to that of other New World producers, yields in Mendoza were significantly greater because of irrigation (table 11.3), and before 1914 market downturns were short, so that many growers carried low levels of debt.[75] Those growers in Mendoza and San Juan who were contracted to large producers were also partly protected because as *contratistas* they received a fixed salary, even though Pedro Arata reported labor leaving the region in 1903.[76]

The response by Mendoza's leading wineries to the cyclical downturns was clear: the costs of adjustment should be borne by other sectors. In Europe, producers reacted to overproduction by demanding that "surplus" wine be removed from the market to bring supply and demand back into equilibrium at previous price levels. However, in the New World this was harder to achieve because any recovery in wine prices would immediately lead to a renewed growth in planting vineyards by specialist growers, and an even greater overproduction a few years later. To resolve this problem, Mendoza's leading wineries colluded to keep grape prices artificially low, at a little more than a peso per 42 kilos between 1914 and 1919 (fig. 11.1). This was hardly enough to cover variable production costs in the vineyards, but it significantly reduced the incentives to increase the supply of grapes by new plantings.[77]

In the wider Argentine economy, as in the United States, industrial supply "moved toward 'trustification,' . . . capital concentration and big business" during the first decade of the twentieth century.[78] While there is plenty of evidence that the large wine producers colluded to restrict competition in their grape purchases, there is little to suggest that a trust such as the California Wine Association was ever seriously considered. As Mendoza's leading daily newspaper, *Los Andes*, noted in 1903, the large producers found it more attractive to capture tax revenue from the provincial government, which they could use to regulate sup-

[74] Barrio de Villanueva (2008a:334). Growers had successfully lobbied to abolish the tax on grape exports out of the province and to get the rail freight costs reduced in time for the 1903 harvest.

[75] *La Prensa*, April 10, 1914, p. 9, noted that the growers who suffered most were the new producers whose vines had just started producing.

[76] Arata (1903:196–97).

[77] The area of vines officially increased during this period, but this reflects the poor quality of official statistics rather than incentives for growers.

[78] Rocchi (2006:9).

ply, than to invest their own money in risky enterprises.[79] The Mendoza law of 1916 (*ley* 703) in particular created a cooperative that allowed the leading wineries to use public funds to remove from the market unwanted wine. Past attempts to create similar voluntary organizations had failed because insufficient producers were willing to participate, so a punitive tax was now levied on the production of those who remained outside the scheme.[80] The cooperative was run by the large wineries and from the start was highly controversial. Its life proved short, as Argentina's Supreme Court determined it to be unconstitutional, ruling that a provincial government could not create a private monopoly.

The failure of the Argentine industry to improve wine quality was criticized by contemporaries and historians alike.[81] Much of the wine was of poor quality, adulteration was common prior to 1903 if not later, and advances in wine making were slow compared with those found in other regions with hot climates, such as Algeria and Australia. Interest in improvements occurred only at times of major market disruptions, and even then more time and energy were given to discussing attempts at market regulation than scientific progress. Yet it was also equally true that not only had a dozen or more wine producers become very rich, but considerable wealth had also filtered down the industry, so much so that a grower in 1912 with 50 hectares of vines was considered sufficiently wealthy to enjoy a box seat at the Colón opera house in Buenos Aires.[82]

These two visions of the industry were perfectly compatible. Many regarded as fanciful the suggestions that the industry should start exporting wine during the downturn after 1914. According to the Centro vitivinícola nacional in its album to celebrate the country's centenary in 1910, about 95 percent of Argentina's wine production was *vin ordinaire*, and the remaining 5 percent of better quality was blended with imports and then sold under foreign labels.[83] Prosperity was found in making cheap wines in large quantities, not small quantities of fine wine, and Mendoza's spectacular growth was based on the demand for cheap, strong, red wines to be consumed with water. This fact alone explains the limited interest in state investment in specialist research institutions and the lack of curiosity by winemakers in the new technologies being developed elsewhere. While companies such as Penfold and Hardy were shipping unfortified table wines from Australia to England, producers in Mendoza were discussing whether it was possible for such wines to survive the journey to Buenos Aires.[84] Wine quality did improve over the 1900s, but producers still expected to sell the vast ma-

---

[79] *Los Andes*, September 20, 1903, p. 4.

[80] This was eight pesos per hectoliter of wine and six pesos per quintal of grapes, when market prices the previous year had been just six and three pesos, respectively (Mateu 2007:16).

[81] Mateu and Stein (2006); Stein (2007).

[82] *La Industria*, May 4, 1911, p. 5, cited in Barrio de Villanueva (2009:1).

[83] Centro Vitivinícola Nacional (1910:xx).

[84] Arata (1904 :111–35).

jority of their wines during the year of production, unlike in Australia, where wines spent about nineteen months maturing before export.

In Europe wineries were limited in size, partly by the need to crush large quantities of grapes quickly before they deteriorated. This particular restriction was absent in Mendoza as the harvest was allowed to stretch over three or four months, and the result was some of the world's largest wineries, who depended on specialist growers for large quantities of their grapes . Yet the coordination mechanisms between specialist growers and independent industrial wineries were far from perfect. Grape prices were strong most years before 1914, but the specialist growers were forced to maintain basic wine-making facilities for those times when the wineries failed to buy their grapes and wines, such as occurred in 1902 and 1914. The system benefited the large wineries as they did not have to buy unwanted grapes and wine at times of weak demand, but it failed to stop the small producers flooding the Buenos Aires market with cheap wines that had barely finished fermenting. Industrial wineries now found that consumers showed little brand loyalty, and the price of all wines was dragged down. By contrast, the small growers, despite frequent support by the local press, failed to get the necessary political backing before 1914 to create some form of cooperative or a "Mendoza" brand, as this was opposed by the large producers, who feared that it would threaten the flexibility they enjoyed in the grape and wine markets.

Finally, the high level of dependence of the Mendoza provincial government on the sector proved in time to be a double-edged sword. Although many of the governors were themselves wine producers, the importance of wine as a source of revenue encouraged the government to attempt to diversify away from the crop. Wine was often taxed, which raised costs and, between 1912–14 and 1930–34, contributed to the decline of Mendoza's share of national production from 82 percent to 70 percent.[85] Crucially the sector was also dependent on other provinces and especially Buenos Aires for its markets. National politicians threatened to reduce import taxes when they thought that Mendoza's producers were colluding or quality deteriorated because of fraud. Wine in Argentina was a basic necessity, and a cheap and plentiful supply for the urban middle and lower classes would be a centerpiece of Peronismo.[86]

[85] Liaudat (1934:18).
[86] Stein (2007:102).

THIS BOOK has followed the growth and development of wine production from mostly small-scale family operators in southern Europe to a worldwide concern, with large-scale industrial producers using scientific wine-making methods and modern marketing techniques to sell their wine. Change was not uniform, and by 1914 major differences were found in the organization of production and marketing of commodity wines in places as far-flung as France, California, South Australia, and Mendoza. Even within a country such as France, new and differing institutions had appeared that altered market incentives for growers, winemakers, and merchants in places such as Bordeaux, Reims, and Montpellier. The changes that occurred in the half century or so before the First World War can be fully understood only by taking a broad view of the sector. In particular, while the international transfer of scientific knowledge and production technologies was becoming increasingly common, most producers found themselves in two very distinct camps: those supplying markets where wine was the alcoholic beverage of choice, and those where new marketing systems had to be developed to allow wine to compete with other well-established drinks, such as beer and spirits. The divergence between the Old and New Worlds has also been explained here by the differences in resource endowments and the highly favorable growing conditions for grapes in the New World, and the political strength of growers to influence government policy in Europe. This part of the book looks briefly at the changes that took place among traditional producer countries in Europe and then offers some comments concerning the obstacles facing the producers in the New World. It finishes with reflections on the extent to which the organization of the wine industry today is the result of changes that took place before 1914.

OLD WORLD PRODUCERS AND CONSUMERS

In Europe's wine-producing countries, rising real wages, falling transport costs, and growing urbanization contributed to per capita consumption increasing significantly, and by 1914 the French (including their children) drank on average around 150 liters per person per year, while consumption reached 120 liters in Italy, 90 liters in Portugal, and 85 liters in Spain. These levels are impressive especially considering that large areas of Europe's vineyards were decimated by phylloxera and growers also had to fight other new vine diseases, such as powdery and downy mildew. In the case of France, the only country for which reasonably accurate figures exist, output collapsed from an annual average of 57 million hectoliters between 1866 and 1875 to 30 million between 1886 and 1895. In response to these domestic shortages, France switched from exporting the equivalent of 5 percent of domestic production to importing 19 percent of its needs.

Phylloxera-induced shortages and the resulting higher wine prices produced a variety of responses among growers and winemakers. First, in large areas of Spain, for example, there was a major increase in production as land and labor were diverted away from extensive livestock and cereal production to viticulture. Technical change was limited, and wines were made stable for export by adding spirits, often produced from cheaper vegetable matter such as sugar beet and potatoes rather than grapes. By the late 1880s perhaps a third of Spain's production was exported to France.

Elsewhere, technological change played a greater role, in part because the fight against phylloxera led to large amounts of scientific research, which allowed growers to choose grape varieties better suited to their vineyards. For most producers, profitability was positively correlated with yields rather than quality, encouraging them to increase output, so that while production was little different in France at the beginning and end of our period (1850–1914), the area under vines declined by about a third. At the extreme, growers in the Midi used irrigation and heavy inputs of chemicals and planted hybrid vines that produced yields of over 100 hectoliters per hectare. Regions specialized according to their comparative advantage, and the Midi's thin, low-alcohol wine was blended with Algeria's wines for their good color and high alcohol content, thereby creating a cheap, drinkable alcoholic beverage for the Parisian and industrial consumers.

Changes in wine-making technologies were potentially as important as anything that took place in the vineyard, and by the 1900s the leading producers were controlling the temperature of the must during fermentation, correcting its acidity, and using cultivated yeasts. The new wine-making technologies and cellar designs increased the amount of wine produced from a ton of grapes and cut labor costs, an important consideration as in many areas farm wages were increasing. Winemakers were also now able to consistently produce drinkable commodity wines in hot climates. There was a much lower incidence of vine disease in these regions; land was usually considerably cheaper than in northern Europe; and large areas of vines could be easily cultivated using plows. There followed a major relocation of the industry, and France's four departments in the Midi and Algeria saw their production increase from the equivalent of a fifth of France's domestic consumption in 1852 to a half by the 1900s. In the rest of Europe, the shift to hotter regions occurred more slowly, in part because phylloxera took longer to destroy traditional areas of viticulture, but also because of the much slower diffusion of new wine-making technologies and the weaker integration of the national market.

The high prices of the 1880s led to adulterated wines becoming common in both producer and nonproducer countries, and their presence continued when prices collapsed at the turn of the century. While fraud had always been present in food and beverage markets, the growing physical separation between producers and consumers and the development of new preservatives allowed manufacturers to mask food deterioration and lower costs, making food adulteration imperceptible to consumers. By 1900 consumers often had little guarantee of where the product had been made, what percentage of it was made from grapes, or indeed whether it was actually safe to drink. Furthermore, the French economist Charles Gide believed that the presence of cheap, spirit-based drinks had finally checked the growth in wine consumption.[1] The combination of stagnant demand, widespread adulteration, and the recovery in domestic production following the replanting with high-yielding vines after phylloxera led to overproduction. Markets failed to self-correct because growers and merchants had financial incentives to adulterate wines, especially after a poor harvest when prices increased; growers in the Midi sold their wines at a loss in five out of seven years between 1900 and 1906, provoking massive demonstrations.

The sector's success in increasing output and consumption in the face of destruction caused by vine disease makes its failure to develop export markets all the more important. Except when phylloxera devastated French domestic production, trade between producer countries was restricted by high tariffs. Among nonproducers, Britain imported over half its food and was potentially a major wine market. This urban and relatively rich country, with a long tradition of importing fine wines, passed legislation in the early 1860s that was specifically designed to encourage the import and consumption of cheap wines, especially among the growing urban middle classes. Wine consumption increased by 140 percent between 1856–60 and 1871–75, but although import duties remained stable and real wages continued to increase, per capita consumption then declined so that on the eve of the First World War it was similar to what it had been a hundred years earlier on the eve of Waterloo.

The attempts to convert a country of nonconsumers to become wine drinkers failed because private firms were unable to maintain or even guarantee wine quality. Unlike producer countries, where legislation was passed that defined wine as being made from crushed grapes and prohibited the sale of imitations, the law remained more lenient in nonproducer countries such as Britain. Alcoholic drinks such as Hamburg sherry, or Britain's infamous "basis wines," which were manufactured from imported grape juice and other substances and then mixed with French wines and sold as claret, even damaged the reputation of fine clarets and sherries. The important economies of scale found with some foods and beverages in processing, packaging, and distribution encouraged firms such as Cadbury's and Lipton's to spend heavily on branding and advertising were therefore absent with cheap wines whose quality could change radically from one harvest to the next. There was insufficient volumes of business to make mass publicity profitable for chains stores such as Gilbey's and the Victoria Wine Company.

---

[1] Gide (1907).

TABLE C.1
Negotiating Strength of Growers and Shippers in Domestic and British Markets in the Early Twentieth Century

		Domestic market	*United Kingdom*
Port	Growers	Strong	
	Shippers	Weak	Strong
Sherry	Growers	Weak	
	Shippers	Strong	Weak
Champagne	Growers	Strong	
	Shippers	Weak	Strong
Claret	Growers	Strong	
	Shippers	Weak	Weak

The problems associated with adulteration and fraud encouraged Europe's growers, winemakers, and merchants to try and redefine the nature of the industry. The speed at which new economic institutions such as regional appellations and cooperatives appeared reflected the political structure and relative negotiating strength of the growers and shippers in each country. In France Jules Ferry famously declared in 1884 that the Third Republic "will be a peasants' republic or it will cease to exist."[2] In exchange for votes, French politicians were happy to adopt policies that protected small family producers, but the rural elites in Spain and Portugal continued to enjoy considerably more political power. Consequently growers in Bordeaux and Champagne were successful in creating the framework for a regional appellation before 1914, but not those in Jerez. Growers in Porto also lacked electoral influence, but the Portuguese government backed their demands in the face of opposition from British shippers. The situation in the British market was different, however, as both Bordeaux and Jerez shippers faced competition from clarets and sherries produced in other countries, but the port wine shippers persuaded the British government to protect their product in the 1916 treaty. Champagne producers, by contrast, had much stronger brands, in part because the wines had to be bottled at source.

Technological change and market integration also produced tensions along the commodity chain for cheap table wines. In particular, problems of vertical coordination between grape producers and winemakers arose at the beginning of the twentieth century because of the growing economies of scale that were associated with the new wine-making technologies. Even under optimal conditions, with the new vineyards planted on the fertile plains to allow plows to operate between the long rows of vines and thereby save labor, a family farm was rarely more than 8 or 10 hectares. In reality the high concentration of vines and their haphazard planting, together with the fragmentation of property, led to

[2] Wright (1964:13).

most vineyards in the Old World being much smaller. By contrast, changes in wine-making technologies increased capital and skill requirements and threatened traditional family wine making, while improved transportation allowed merchants to obtain wines from much wider areas, permitting them to avoid particular regions after poor harvests, and making it harder for small producers to sell their surplus produce.

A number of influential French writers, such as Charles Gide and Michel Augé-Laribé, saw wine cooperatives as a solution as they allowed the efficient family vineyards to coexist alongside large, capital-intensive wineries. Cooperatives promised to produce better-quality wines at a lower cost than those made in the small "peasant" cellars and provide cheap storage space for wines to mature, rather than having to be sold immediately, and thereby allow growers to enjoy higher prices. They also offered the possibility for processing the wine lees—the remains of the grapes after they had been crushed—to make alcohol and tartaric acid. A typical French wine cooperative on the eve of the First World War had about 160 members, whose members each had an average production of just 50 hectoliters of wine, coming from little more than a hectare of vines. However, there were formidable organizational problems in creating cooperatives, including the lack of capital, the absence of experienced management, and the difficulties associated with measuring grape and wine quality. Government legislation eased credit restrictions and led to the number cooperatives in France jumping from thirteen in 1908 to seventy-nine in 1913, of which fifty were found in the South. Considerably interest was also shown in Italy and Spain, but in these countries the cheap credit needed to construct new wineries was lacking until at least the interwar period.

New World Producers and Consumers

The vine followed European settlement in the New World and North Africa, but commercial production became important only at the end of the nineteenth century, as Algerian production increased from a million hectoliters in 1885 to 8.4 million by 1910 on the back of its privileged access to French markets, while in the New World it went from virtually nothing in the 1870s to 3.8 million hectoliters in 1910 in Argentina, 2 million in Chile, 1.8 million in California, and 0.2 million in Australia. Much of the growth coincided with the introduction of new wine-making technologies that incorporated significant economies of scale and lowered unit production costs.

The very different factor endowments of the New World implied that the industry developed its own style and characteristics, and producers enjoyed a number of important advantages over those in northern Europe. Grape growing was relatively easy, as the warm conditions allowed the fruit to mature fully and there was a much lower incidence of disease and rot. Land was cheap and the vines started producing after just three years, rather than the four or five common in most of Europe. Vineyards were extensively cultivated, allowing plows to operate freely along the rows, reducing labor inputs to a minimum. A number of very large vineyards existed, but even in the labor-scarce New World, considerable quantities of grapes were still produced on small, family-operated vineyards rather than on large estates. Thus in Napa and Sonoma in California, only 37 percent of the vines were found on holdings greater than 25 hectares in 1891, and in Victoria, Australia, only 27 of the state's 850 growers had more than 24 hectares in 1890.

By contrast, and to a much greater extent than in either the Midi or Algeria, specialist grape growers supplied the new industrial wineries because the high and consistent grape quality made it much easier for winemakers to negotiate contracts with such growers, thereby reducing potential conflicts at harvest time. Wineries were purpose-built to reduce the costs of crushing large quantities of grapes in short periods of time and the labor-intensive processes of moving wine around the winery in an age before electric pumps. The new wineries' greater capacity permitted investments in laboratories, technicians, and fermentation cooling systems to be spread over larger quantities of wine, helping to erode the competitive position of the small wine maker.

A favorable terroir also allowed more homogenous wines to be produced. These were not necessary better than Europe's *vin ordinaire*, but they did allow merchants to accumulate large batches of wines with similar characteristics from one year to the next. Yet despite the hopes of some contemporaries that their countries would soon be able to supply Europe with cheap table wines, the reality was that the New World industry, with the exception of Australia, remained one of import substitution depending on tariffs and protected markets. Production was often located at considerable distances from major markets, and the structure of the commodity chain differed according to whether producers were able to sell to immigrants who originated in Europe's wine-producing regions such as in Argentina, or whether they had to create new markets as in Australia and California.

In California, the growth in demand, new wine-making technologies, and widespread adulteration of wines required new forms of business organization if the high levels of capital investment in vineyards and winery equipment were not left idle. Growers lacked political support, but a lenient regulatory environment toward trusts encouraged the formation in 1894 of the California Wine Association, a combine created by the leading San Francisco wine dealers. By 1897 the CWA controlled 80 percent of wine sales and integrated vertically from grape growing to distribution on a massive scale. By controlling distribution, it achieved the necessary market stability for it to invest in brand names, and growers in California probably suffered less during the turbulent 1900s than in any other wine-growing region in the world. Yet the vast majority of U.S. citizens were unaccustomed to drinking wine, and the economies of scale associated with the production and marketing of dessert wines, as well as a favorable tax regime, led to a much faster growth in their production than table wines, so that by 1913 they accounted for 45 percent of all Californian wines, of which 46 percent was classified as port, 31 percent as sherry, 12 percent as muscatel, and 9 percent angelica. However, although politicians considered wine to be sufficiently unimportant and permitted the CWA to act with relative freedom, they were unwilling to ignore the powerful Prohibition lobby, which made it illegal to sell wines and other forms of alcohol after 1919.

The high entry costs to marketing wines on the U.S. East Coast before Prohibition was overcome by a producer-led commodity chain. By contrast, in Australia it was a market-led chain that was created to sell dry, red table wines in London, some 20,000 kilometers away, despite freight costs being three times greater than those facing French exporters. Trade was dominated by two major London houses that specialized in Australian wines: Walter Pownall and, in particular, Peter Burgoyne, who claimed in 1900 to have exported fully 70 percent of the wine sent from Australia to England between 1870 and 1900 and to have invested £300,000 in advertising so that "Burgoyne's Australian Wine" placards were found "on every railway station in England." Australian producers resented the control exercised by the two British importers, but attempts to create an alternative marketing system failed. Winemakers such as Hardy or Penfold lacked the resources and scale to establish permanent representation in London, and attempts to create government-sponsored companies suffered from a lack of finance. In addition, any publicly owned institution, whether it was a regional cooperative in Australia or a depot in London, suffered from selection problems because of the need to be able to reject substandard grapes and wines. As Arthur Perkins noted in 1901, it was possible to detect whether a wine was sound or unadulterated, but on "the question of quality none will agree." The Australian industry remained divided in 1914, with approximately 20 percent being high-quality table wines destined for the British market and much of the rest being low-grade wines for domestic consumption. As in California, the leading wineries, such as Seppeltsfield and Penfold, turned increasingly to brandy and fortified wines, which were both easier to brand and market than table wines and popular among infrequent drinkers as they enjoyed a much longer shelf life once opened.

Finally, in the early twentieth century a dozen or so massive wineries dominated the Cuyo industry in Argentina. Wine quality was sufficiently homogenous to allow these wineries to brand their own wines, but production techniques remained primitive because the Mediterranean immigrants were unwilling to pay a premium for better-quality wines. During the periodic crises, when the major winemakers simply refused to buy the growers' grapes, disused wineries were reopened, allowing supplies to be maintained in a glutted market and keeping prices low. In Mendoza and California the political climate was relatively permissive to big business and trusts, and both regions with less than 5 percent of their nation's population produced around 95 percent of the domestic wine supply. Unlike in California, however, wine was considered an integral part of the Argentine diet, and the Buenos Aires government was unwilling to permit a trust or monopoly to manipulate prices, while regulators were much less concerned in the United States.

THE WINE INDUSTRY IN THE TWENTIETH CENTURY

Today's distinctive organizational structures in the Old World and the New World were clearly visible by the First World War. In Europe the industry was fragmented into hundreds of thousands of small, family farms, while in the New World it was dominated by a handful of large, highly integrated firms such as the CWA, Domingo Tomba, and Burgoyne. Producers in places such as Bordeaux, Burgundy, and Champagne had long provided small quantities of fine wines to a limited number of connoisseurs, but now the combination of new production technologies, institutional innovations, and organizational structures offered the possibilities of producing better-quality commodity wines everywhere. Yet progress was slow as the economic incentives to make the necessary investment remained weak for at least half a century after 1914, so there was no reason for producers to plant shy- instead of heavy-bearing vines, invest in the new winemaking equipment, or study the basic principles of viticulture and enology. As late as the 1960s, production and consumption in the alcoholic beverage industry was still "country and culture specific,"[3] and only with the major changes in consumption habits in the past thirty or forty years has it become profitable to improve the quality of cheap wines. But as consumers in countries such as Italy, France, and Spain steadily moved up the quality ladder, they not only stopped drinking large quantities of cheap wine, but also switched to other types of alcoholic beverages and soft drinks. High tariffs limited trade and kept production country specific, and this perhaps explains why even today it is difficult for European consumers to find foreign wines outside specialist stores. A combination of growers' reluctance to scrub up "inferior" vines and productivity improvements have led to large quantities of wine being produced that no one wants to drink. Today Bordeaux's best clarets may sell for astronomical prices, but vast quantities of this region's cheaper wines remain unsold and destined for the still.

Change was initially driven by shifts in demand in nonproducer countries, and in particular Britain and the United States. In the former, millions of tourists were introduced to wine during their package holidays on the Mediterranean coast. Rising living standards and the possibility of purchasing cheap wine in supermarkets at home led to a huge increase in sales, while a new generation of writers led by Hugh Johnson, who published his first *Wine Atlas* in 1971, helped to educate the more adventurous consumer.

[3] Lopes (2007:1).

In California commercial wine production had to effectively start from scratch after Prohibition. As James Lapsley notes, the work of Hilgard and Bioletti suggests that on the eve of Prohibition California had passed the first phase of avoiding bacterial spoilage and was firmly in the second phase of improving quality through the use of pure yeast cultures, sulfur dioxide, and cooling, and if Prohibition had come ten years later, a "scientific understanding of winemaking would have had time to spread and take hold throughout the industry."[4] Instead, by the 1930s, after Prohibition was repealed, most winemakers could not produce a drinkable, good dry wine and consumers were unable to appreciate one, so that about three quarters of production was dessert wines and approximately half of all dry wine was homemade from purchased grapes. Change was slow, and in the late 1940s almost all the quality producers in Napa were also still producing bulk and generic wines. However, by this time the region has achieved "a critical mass where social organization for regional promotion and interchange of technology was assured."[5]

A similar story is found in Australia, where Burgoyne's influenced disappeared after the First World War with the growth in the British demand for dessert rather than dry table wines. There was no prohibition in Australia, but the strong temperance movement, on the one hand, and a migrant population with a tradition of strong spirits and beer, on the other, limited the size of the domestic market for dry table wines. As in California, even in the mid-1960s more fortified wines were produced than table wines, but now there was a "giant leap in output" of the classical grape varieties, as well as important advances in winemaking technologies that allowed for a "remarkable improvement in the quality of light table wines made from Riesling, Semillon, Cabernet and Shiraz grapes grown in Australian irrigated vineyards."[6]

In Europe, if wine quality was slow to change, the set of economic institutions chosen on the eve of the First World War was a major factor in determining the future structure of the industry. Over the twentieth century the small family grower continued to enjoy the protection of the state, although the nature of institutions such as cooperatives and *appellations d'origine contrôlées* adapted over time to reflect the shifts in economic and political influence of producers. In France the creation of a wine monopoly, such as those proposed by Bartissol or Palazy in the 1900s, became unnecessary as growers looked directly to the state to resolve the problems of overproduction and low wine prices when they reappeared in the 1930s. Legislation was passed that regulated markets favoring small producers, and the transaction costs associated with complying with the new regulations were absorbed by the state. From the 1950s state intervention increased and spread to other European countries, helping the small family grower to remain in business today, but often at the expense of slowing produc-

[4] Lapsley (1996: 51).
[5] Ibid., 135.
[6] Simon (1966:xi, 3).

ers' response to shifts in market demand and leading to persistent structural problems of overproduction.

The switch from commodity to fine wine production has been slow over the last couple of centuries. In his recent book David Hancock talks of how British merchants in the eighteenth century happily exchanged information among themselves to perfect wine-making technologies, just as James Lapsley finds producers doing in the Napa Valley after the Second World War.[7] Collective invention and product innovation that led to better-quality wines improved the region's reputation and made it easier for all producers to sell their wines. There was also recognition of the need for the generic promotion of regional wines, as the potential for advertising was considerably greater than for what a single grower could achieve. Perhaps not so much has changed since the times of Busby, Civit, and Haraszthy, who, in their attempts to emulate Europe's great wine producers, discussed the extent that technology can substitute for terroir.

[7] Hancock (2009); Lapsley (1994:204).

Vineyards and Wineries

THE INFORMATION concerning vineyards and wineries is poor for most countries, and accurate figures become available only late in the twentieth century. One major problem is that very small holdings were common, and whether these figures are included or not significantly changes the overall picture of landownership. For example, for the Aude department in the Midi in 1913, 42 percent of growers had less than 1 hectare, but these represented only 6 percent of the total area and 5 percent of wine production. The average area was just 3.6 hectares and 197 hectoliters. However, this was not a region of small property as over two-thirds of the vines and 70 percent of output was found on holdings of over 5 hectares (and with almost three-tenths of total output produced on vineyards of more than 30 hectares). Unfortunately the statistics do not allow a distinction to be made between part-time and full-time producers.

Table A.1 summarizes the information for France using the 1924 figures for vineyards and the harvest declared in 1934. The large numbers of smallholders is apparent, but so too the larger holdings found in the Midi. In this region 27.5 percent of the vines were on holdings of 20 hectares or more, compared with just 4.6 percent in the rest of France, and 32.3 percent of production was from farmers that produced in excess of a thousand hectoliters, against 6.8 percent in other regions. In total there were thirty-two wineries that produced more than 10,000 hectoliters in the Midi (and just thirteen more in the rest of the country).

Table A.2 gives the area of vines by county in California in 1891. These figures appear to ignore many of the minuscule plots and are frequently rounded to the nearest acre.

Finally, table A.3 has been calculated using Gervais's large directory of southern France and Algeria. The figures appear to be winery capacity rather than the production figures of a particular year as in table A.1, which explains the considerably greater number of large wineries in 1903. Comparisons can be made with South Australia (table 10.2) and Argentina (table 11.4).

TABLE A.1
Area of Vines and Output per Winery in France, 1924 and 1934

	<1 hectares	1–5 hectares	5–10 hectares
surface in 1924			
number of farms	1000430	283935	21422
%	76,0	21,6	1,6
total acreage per size	360389	613672	159648
%	25,6	43,7	11,4
production in 1934	<100 hl	101–400 hls.	401–1000 hl
number recoltanta (declaration)	1411449	132842	14891
%	90,2	8,5	1,0
total recoltes per size	27950636	23883047	8736270
%	38,0	32,5	11,9
	MIDI		
surface in 1924			
number of farms	73355	50755	9859
%	52,1	36,0	7,0
total acreage per size	35086	159727	76902
%	7,6	34,4	16,6
production in 1934			
number of recoltants (declaration)	118785	54194	9759
%	63,5	29,0	5,2
total recoltes per size	4035768	11038817	5895664
%	13,0	35,6	19,0
	rest of france		
surface in 1924			
number of farms	927075	233180	11563
%	78,8	19,8	1,0
total acreage per size	325303	453945	82746
%	34,5	48,2	8,8
production in 1934			
number recoltants (declaration)	1292664	78648	5132
%	93,8	5,7	0,4
total recoltes per size	23914868	12844230	2840606
%	56,3	30,2	6,7

10–20 hectares	20–50 ha	50–100 ha	>100ha	total
6640	3362	1029	187	1317005
0,5	0,3	00,1	0,0	100
100935	97683	64758	8592	1405676
7,2	6,9	4,6	0,6	100
1001–3000 hl	3001–5000 hl	5001–10000 hl	>10000 hl	
4399	728	301	45	1564655
0,3	0,0	0,0	0,0	100
7073847	2726103	1244859	1850376	73465138
9,6	3,7	1,7	2,5	100
4185	1990	585	81	140810
3,0	1,4	0,4	0,1	100
64495	66817	53712	7165	463904
13,9	14,4	11,6	1,5	100
3284	636	248	32	186938
1,8	0,3	0,1	0,0	100
5262913	2394509	969524	1399206	30996401
17,0	7,7	3,1	4,5	100
2455	1372	444	106	1176195
0,2	0,1	0,0	0,0	100
36440	30866	11046	1427	941772
3,9	3,3	1,2	0,2	100
1115	92	53	13	1377717
0,1	0,0	0,0	0,0	100
1810934	331594	275335	451170	42468737
4,3	0,8	0,6	1,1	100

TABLE A.2

Number of Growers and Area of Vines by County in California, 1891

	3–10 acres		10–20 acres		20–40 acres	
	Growers	Area	Growers	Area	Growers	Area
Alameda	31	242	36	574	39	1,153
Amador	0	0	0	0	0	0
Butte	6	43	1	16	1	26
Calaveras	14	115	6	107	6	190
Colusa	54	396	36	574	16	492
El Dorado	44	376	12	227	8	238
Fresno	47	318	28	437	10	297
Inyo	1	5	0	0	0	0
Kern	0	0	0	0	0	0
Lake	24	160	13	211	7	215
Los Angeles	29	223	12	208	10	322
Marin	7	41	1	20	4	116
Mendocino	14	101	5	89	0	0
Merced	14	115	3	55	4	108
Monterey	0	0	0	0	0	0
Napa	124	967	178	2,887	137	4,203
Nevada	11	65	3	52	2	70
Orange	2	14	0	0	0	0
Placer	9	60	3	55	6	195
Sacramento	41	305	22	322	14	415
San Benito	0	0	0	0	1	30
San Bernardino	9	66	1	15	0	0
San Diego	7	47	2	31	0	0
San Joaquin	39	259	10	166	5	137
San Luis Obisbo	2	13	3	37	2	80
San Mateo	6	36	8	134	3	101
Santa Barbara	15	90	1	20	1	40
Santa Clara	126	1,072	84	1,366	86	2,651
Santa Cruz	20	145	20	336	12	344
Shasta	5	31	3	47	0	0
Solano	2	15	5	77	3	104
Sonoma	264	2,096	267	4,392	174	5,291
Stanislaus	0	0	0	0	0	0
Sutter	2	13	1	12	0	0
Tehama	0	0	0	0	0	0
Tulare	4	32	2	36	0	0
Tuoumne	0	0	0	0	0	0
Ventura	2	11	3	44	1	35
Yolo	2	20	3	55	7	215
Yuba	4	40	8	160	4	150
TOTAL	981	7,532	780	12,762	563	17,218

Source: California, Board of State Viticultural Commissioners. *Directory of the Grape Growers, Wine Makers and Distillers of California, and of the Principal Grape Growers and Wine Makers of the Eastern States* (Sacramento, 1891); my calculations.

40–62 acres		63+ acres		Total area of vines	% of vines in county
Growers	Area	Growers	Area		
13	668	20	3,634	6,270	7.3
0	0	0	0	0	0.0
0	0	0	0	85	0.1
0	0	0	0	412	0.5
6	322	4	350	2,134	2.5
3	170	0	0	1,011	1.2
7	351	17	3,964	5,367	6.2
0	0	0	0	5	0.0
0	0	0	0	0	0.0
1	60	2	400	1,046	1.2
7	360	13	3,160	4,273	5.0
0	0	2	300	477	0.6
0	0	0	0	190	0.2
0	0	1	122	400	0.5
0	0	0	0	0	0.0
55	2,891	55	6,664	17,612	20.4
0	0	0	0	187	0.2
1	56	1	70	140	0.2
1	75	0	0	385	0.4
1	60	5	1,710	2,812	3.3
0	0	1	110	140	0.2
0	0	7	933	1,014	1.2
1	45	0	0	123	0.1
3	155	2	160	877	1.0
0	0	0	0	130	0.2
1	55	4	407	733	0.9
1	60	0	0	210	0.2
27	1,425	29	3,394	9,908	11.5
7	374	1	90	1,289	1.5
0	0	0	0	78	0.1
0	0	1	250	446	0.5
45	2,381	65	8,239	22,399	26.0
0	0	1	80	80	0.1
0	0	0	0	25	0.0
0	0	1	3,705	3,705	4.3
0	0	0	0	68	0.1
0	0	0	0	0	0.0
0	0	0	0	90	0.1
1	45	6	1,160	1,495	1.7
2	120	1	100	570	0.7
183	9,673	239	39,002	86,185	100.0

TABLE A.3
Winery Size in the Midi and Algeria, 1903

	Production (millions of hectoliters)	Winery size (hectoliters)			
		10,000–14,999	15,000–24,999	25,000 +	Total
Aude	5.9	13	11	0	24
Gard	3.5	13	6	6	25
Hérault	11.4	56	13	3	72
Pyrénées-Orientales	2.6	6	3	0	9
Midi	23.4	88	33	9	130
Algeria	3.1	15	14	3	33
Constantine	0.9	4	2	3	8
Oran	3.7	7	5	0	12
Algeria	7.7	26	21	6	53

Source: Gervais (1903), my calculations.

Wine Prices

TABLE A.4
Farm and Paris Wine Prices, July 1910

	Paris (hectoliters)	Vineyard	Minimum Paris (hectoliters)
Ordinary wines			
Aramon	24–25	19–20	24
Rosé	24–25	19–20	24
Montagne	25–27	20–21	25
Gard	24–28	18–25	24
Jacquez		22–25	
Aude	26–30	18–24	26
Corbières	26–27	20	26
Roussillon	26–40	20–30	26
Alger	24–30	16–24	24
Oran	23–29	17–20	23
Blanc de Rouge	24–28	19–20	24
Blanc de Blanc	28–35	20–25	28
Vins du Centre	piece (228 hls)		
Auvergne	80–90	65–90	35
Cher	70–80	55–60	31
Chinon	80–110		35
Touraine	60–70	60–80	26
Anjou Blanc	95–160	75–250	42
Muscadet	95–110	100–120	42
Nantais	60–70	50–110	26
Saumur	100–150	90–130	44
Sologne	75–85	55–60	33
Vouvray	85–150	65–90	37
Vins de Bourgogne	piece (228 liters)		
Auxerre		90–130	
Beaujolais			
1er choix	100–150	110–130	44
2e	75–95	60–90	33
Bas-Beaujolais	60–80	50–60	26
Lyonnais	70–100		31
Mâconnais			
1er choix	85–100	60–85	37
2e	75–80	50–60	33

TABLE A.4 (*Continued*)

	Paris (hectoliters)	Vineyard	Minimum Paris (hectoliters)
Bourgogne			
ordinaire		100	
grand ordinaire		150	
passe-tout-grain		1,000	
· fin classé		1,200	
tête de curvée	1,600	1,500	
Chablis			
village	115–30		50
classé	150–250		66
Ordinaire blanc	90–110	100	39
Grand ordinaire blanc	110–320	300	48
Meursault	800	800	
Vins de Bordeaux	225 liters		
Blayais	120	110	53
Bourgeais	130	120	58
Côtes	100–115	70–130	44
Médoc			
Bourgeois	160	150	71
5° crus	210	200	93
Petites Graves	110	100–125	49
Grandes Graves		125–900	
Palus	75–100	75–80	33
Bas-Médoc blanc	85	75	38
Entre-Deux-Mers	70–90	75–100	31
Graves	110–320	100–300	49
Barsac	300–500	300–400	133
Sauternes	320–1,200	300–1,000	142
Vins du Sud-Ouest			
Dordogne	70–95		31
Haute-Garonne	65–75		29
lot	60–65	40–60	27
Lot-et-Garonne	60–70	50–50	27
Tarn-et-Garonne	65–75		29
Gaillac blanc	85–95		38
Gers	60–65		27

Source: Revue de Viticulture, July 21, 1910

Telephone :
1788 MAYFAIR.
Telegrams :
"BERRINCHE, LONDON."

May, 1909

PRICE LIST.

BERRY BROS. & CO.,

Established in the XVII Century, at

3 ST JAMES'S STREET,

LONDON, S.W.

CELLARS:-
1, 2, 3, & 4 Pickering Place, S.W.
Hay's Mews Vaults, Berkeley Sq., W.
26 Savile Row, W.

BONDED WAREHOUSES:-
LONDON,
GLASGOW,
and
SOUTHAMPTON.

Bankers :-BANK OF ENGLAND.

INDEX

TERMS - - CASH

FOR CREDIT 2/- in the £ extra is charged, and 5 per cent. Interest after 12 months.

Most of the Wines and Liqueurs can be supplied in Half-Bottles.

BERRY BROS. & CO. 3

SHERRY.

		Per Doz.	Quarter Cask.
Light, and Pale Golden	...	24/-	£14
Amoroso	...	30/-	£17 10s.
„ Older	...	36/-	£21
Vino de Pasto	...	36/- & 42/-	£21, £25
Amontillado ... 36/-, 42/-,	48/-	& £21, £25, £28
„ Old and Fine	...	72/-	
„ Exceptional "L.M."	...	84/-	
Isabelita Very Dry	...	38/-	
Pando, Exceedingly Dry	...	38/-	
Manzanilla	...	42/- & 48/-	
Dry Sack (Williams and Humbert)	...	48/-	
Old Oloroso	...	48/-	
Old Golden Very Old in Bottle	...	84/-	
„ Very Fine, 1847	...	84/-	
Old Pale, Dry, Full Bodied		66/- & 72/-	
„ Dry, Full Bodied Very Old Bottled	...	84/-	
East India	...	72/-, & 96/-	
Fine Old Brown	...	72/-, 84/-, 108/-	
Don Carlos, Pale Brown Very Old Wine		96/-	
Don Tadeo Perfection of Sherry, 100 years old	...	144/-	
From the Royal Cellars	...	66/- to 240/-	
Rota Tent	...	24/- ½-Bots.	

3, St James's Street, London, S.W.

4

BERRY BROS. & CO.

PORT.

				Per Doz.
Ordinary Good	...	...	...	24/6 & 30/6
Finer, "Service"	...	...	...	36/-
"Crown," 20 years in wood		...	...	44/- Dry 44/-
"Queen's," 30 years in wood		...	...	64/- Dry 60/-
"The Prince's," 50 years in wood. Pale, Soft, Fine			...	80/-
"Victoria," 1820	...	...	...	108/-
"The King's," 1805		...	...	144/-
Very Pale Old Tawny		...	...	72/-
Old Bottled	...	...	...	42/- & 48/-
Older and Finer		...	...	**54/- & 66/-**
Fine Dry	...	...	...	72/-
White Port	...	...	...	42/- & 54/-

Vintage.

1900	Taylor's	...	...	42/- ½-Bots. 23/-
	Warre's	...	...	42/-
	Fonseca's	...	...	38/-
1896	Taylor's	...	...	56/-
	Warre's	...	...	56/-
	Cockburn's	...	...	60/-
	Smith Woodhouse's	...	54/- ½-Bots. 29/-
	Kingston's	...	...	52/-
1890	Offley's	...	...	78/-
	Sandeman's	...	...	80/-
	Dow's	...	...	82/-
	Croft's	...	...	82/-

3; St James's Street, London, S.W.

5

BERRY BROS. & CO.

PORT—*continued*.

Vintage.					Per Doz.
1887	Martinez'	...	...	...	102/-
	Offley's	...	...	...	105/-
	Dow's	...	...	...	105/-
	Sandeman's	...	...	...	108/-
1884	Sandeman's	...	...	...	115/-
	Martinez'	...	...	...	115/-
	Cockburn's	...	...	...	120/-
	Croft's	...	...	...	120/-
1881	Croft's	...	...	...	132/-
	Dow's	...	...	...	126/-
	Martinez'	...	...	...	126/-
1881	Cockburn's	...	...	...	150/-

1904 | Port for laying down:—
Croft, Taylor, Martinez, Warre
Fonseca: 35/- per doz.
(Bottled 1906).

CHABLIS, &c.

Imperial Chablis	...	...	**18/- & 25/-**
Pouilly	...	...	...	36/-
Montrachet, 1899	...	...	60/-
Chablis Extra, 1889	...	...	66/-
„ Moutonne, 1884	...	...	84/-
Meursault, 1904	...	...	42/-

3; St James's Street, London, S.W.

6   BERRY BROS. & CO.

SAUTERNES.

	Per Doz.
Ch. D'Eyrans ...	16/6
Ch. La Roue, White Bordeaux ...	21/-
Domaine de Broustaret ...	**24/-**
Light and Dry.	
St Croix du Mont, 1893 ...	36/-
Preignac, 1900 ...	36/-
Ch. Ricaud, 1895 ...	40/-
Ch. Carbonnieux, 1er Graves ...	42/-
Ch. Rieussic ...	60/-
Ch. Filhot, 1864 ...	78/-
Ch. Yquem, 1893, Ch. Bottled ...	108/-
„ „ 1874, „ ...	275/-

CLARET.

	Per Doz.
1904 Ch. Constant Bages ...	21/-
Ch. Belgrave ...	24/-
Ch. Angludet ½ bots. only ...	15/-
Ch. Talbot ...	30/-
Ch. Pichon Longueville ...	33/-
Ch. Gruaud Larose Sarget ...	34/-
„ „ „ Ch. Bottled ...	38/-
Ch. Canon, 1er St Emilion ...	36/-
1905 Ch. La Mission Haut Brion Ch. bottled ...	42/-
Ch. Margaux, 1st growth „	54/-
Ch. Latour, 1st growth „	50/-
Ch. Mouton Rothschild, „	50/-
Ch. Haut Brion, 1st growth „	60/-

For Laying Down.

3, St James's Street, London, S.W.

BERRY BROS. & CO.   7

CLARET—*continued*

	Per Doz.
Vin Ordinaire ...	12/6, 14/-
Clos du Prieur, own growth ...	**16/6**
St Estèphe ...	18/6
Médoc Supérieur ...	21/-
1900 Ch. Rochemorin ...	30/-
Ch. Moulin de Calon Ségur, ½ Bots. only	18/-
Ch. d'Alesme Bekker ...	42/-
Ch. Chasse Spleen ...	42/-
Ch. Mouton Rothschild ...	60/-
1899 Ch. Lafon Rochet ...	36/-
Ch. Léoville, Poyferré ...	42/-
„ Lascases ...	45/-
Ch. Brown Cantenac ...	45/-
1896 Ch. Lafite ...	72/-
Ch. Lafite ...	78/-
Ch. La Mission Haut Brion ...	**72/-**
1893 Ch. Léoville, Lascases ...	60/-
Ch. Cheval Blanc ...	78/-
1888 Ch. Smith Haut Lafite ...	72/-
1878 Ch. Lafite Ch. Bottled ...	180/-
Ch. Palmer Margaux ...	96/-
1877 Ch. Léoville ...	108/-
1874 Ch. Larose, in Wine Bottles ...	66/-
1870 Ch. Margaux ...	90/-
1869 Ch. Lafite, Grand Vin, Ch. Bottled ...	200/-
Ch. Lafite, probably 1869 ...	108/-

3, St James's Street, London, S.W.

8

BERRY BROS. & CO.

BURGUNDY.

	Per Doz.
Beaujolais	18/6 & 24/-
Beaune	30/- & 36/-
Beaune Supérieur, 1899 ...	42/-
Aloxe Corton	42/-
Santenay, 1904 & 1900 ...	36/- & 42/-
Volnay, 1902 & 1900 ...	42/- & 48/-
Pommard, 1904 & 1902 ...	42/- & 45/-
,, 1899 ...	48/-
,, 1891 ...	72/-
Nuits St. Georges, 1899 ... ⅓-Bots. only	26/-
,, 1898 ...	60/-
,, 1902 ...	42/-
Corton, 1893, Very Fine ...	96/-
,, Clos Du Roi, 1881 ...	108/-
Chambertin, 1900 ...	54/-
,, 1892 ...	108/-
,, 1886 Very Fine ...	160/-
Richebourg, 1892 ...	108/-
Clos de Vougeot, 1896...	78/-
,, 1898 ... ⅓ Bots. only	45/-
,, 1887 ...	150/-
Château de Clos Vougeot, 1889	144/-
Romanée Conti, 1895, Estate Bottling	150/-
,, 1891 ,,	180/-
,, 1888 ,,	200/-
Chambertain, Clos St Jacques, 1904 Monopoly (Very Fine, for laying down)	60/-

9

BERRY BROS. & CO.

CHAMPAGNE.

	Per Doz.
Saumur	30/-
Epernay	42/-
"St Cyr," light, pure Rheims Wine	54/-
Clos St Jacques, Monopoly. 1899	78/-
,, for laying down ... 1904	70/-
,, ¼-Bots.	22/6
Jules Mumm, Extra Dry, 70/- less 5/- =	65/-
Dagonet, Ex. Quality, Brut or Dry	90/-
Giesler, 1st Quality ...	74/-
Pommery & Greno, Ex. Sec.	96/-
Vve. Clicquot, Dry ...	96/-
1892 Heidsieck, Dry Monopole ...	180/-
Pommery & Greno ...	210/-
Vve. Clicquot ...	240/-
1895 Dagonet, Ex. Sup. Dry ...	96/-
1898 ,, Cuvée Excepelle ⅓-Bots. only	48/-
Pol Roger ...	108/-
Louis Roederer ...	98/-
1899 Binet, Dry Elite ...	108/-
Louis Roederer ...	108/-
Perrier Jouët ...	100/-
1900 G. H. Mumm & Co., Cordon Rouge	96/-
,, ,, Ex. Dry	126/-
Pommery & Greno ...	90/-
Vve. Clicquot, Dry ...	126/-
Louis Roederer ...	150/-
George Goulet ...	90/-
Duc de Montebello ...	86/-
Perrier Jouët ...	84/-
...	86/-

Quotations for other Brands on application.
1904 Vintage on next page.

BERRY BROS. & CO. 10

CHAMPAGNE—continued

	Per Doz.
1904 G. H. Mumm & Co., Cordon Rouge	92/-
Pommery & Greno, Nature	92/-
Vve Clicquot, Dry ...	96/-
Bollinger	84/-
Krug, Private Cuvée ...	82/-
Binet, Dry Elite ...	80/-
Ruinart	82/-
Louis Roederer ...	86/-
Pol Roger	86/-
Duc de Montebello... ...	78/-
Lanson, Père et Fils ...	80/-

Quotations for other Brands on application.

HOCK.

St Jacobsberg	20/6
Niersteiner	30/-
Erbach	40/-
Eltville Sonnerberg 1904	60/-
Rudesheimer, Berg, 1900 ...	66/-
" " 1904 ...	60/-
" " Cabinet 1886	132/-
Steinberger, 1893 ...	96/-
Steinberger Cabinet, 1876	90/-
Marcobrunner, Cabinet 1886	190/-
Schloss Johannisberg Cabinet—	
Prince Metternich's, 1893 ...	200/-
" " 1889, 144/-; 1884, 180/-	
Red Hock	50/- & 60/-
Sparkling ditto	50/- & 66/-
Sparkling White Hock ...	54/- & 66/-

3, St James's Street, London, S.W.

These prices are subject to market fluctuations.

BERRY BROS. & CO. 11

HOCK—continued

Famous Auslese Wines from the Cabinet Cellar of the Duke of Nassau, at greatly reduced prices.

	Per Doz.
1786 Hochheimer	96/-
1865 Marcobrunner	100/-
1865 Steinberger	108/-
1868 Marcobrunner (Hellgrüner Lack)	144/-

MOSELLE.

Moselwein	18/-
Berncastler	21/6
Erdener	30/-
Trarbacher Schlossberg ...	36/-
Brauneberger	36/-
Zeltinger, 1900	42/-
Stephansberger, 1900 ...	48/-
"Berncastler Doctor," 1900 ...	54/-
" EX. QUALITY, 1900 ...	66/-
" " 1904 ...	60/-
Graacher Hummelreich, 1893 ...	66/-
Neumagener Hengelhofberg Auslese, 1904, Exceedingly Fine ...	108/-
Sparkling Moselle	50/-, 66/- & 78/-

3, St James's Street, London, S.W.

BERRY BROS. & CO. 12

MARSALA.

	Per Doz.
Ingham's L.P.	20/6
" Racalia	26/-
Woodhouse's L.P.	20/-
" Natural	24/-
" Virgin	26/-
" O.P., very old	30/-

Can be supplied in Octaves and Quarter Casks.

MADEIRA.

	Per Doz.
Good Sound Wine	36/-
1874 Bual	60/-
Choice Old East India	180/-
1826 Very Fine, 50 years in bottle ...	150/-
1834	150/-
"Coronation" as supplied to the Royal Cellars, 1816 Vintage	144/-
Crown Malmsey, old and choice ...	100/-
1865 Sercial	180/-
"Wellington" Madeira, laid down at Strathfieldsaye in 1861 ...	200/-

TOKAY.

		Each.
1857 Tokay	.50 litre	40/-
1863 " ...	"	35/-
1868 Korona Tokaji ...	"	30/-
1889 Herczeg Tokaji ...	"	21/-

These are the essence of Wine; invaluable in cases of severe illness and for restorative purposes.

3, St James's Street, London, S.W.

BERRY BROS. & CO. 13

TOKAY—*continued*

		Per Doz.
Vin Sec de Tokay ...	.65 litre	120/-
Very fine old, dry; recommended for after-dinner use.		
Muscat de Tokay ...	.65 litre	78/-
Fine bouquet and delicate muscat flavour.		
Hungaria Sec		72/-
Szamorod, Fine		78/-

BRANDY.

	Per Doz.
Young, good for cooking ...	56/-61/-
Good ordinary	67/-
12 years old, "Les Barrières" ...	73/-
20 Years old, fine quality ...	85/-
25 years old	103/-
30 years old, Liqueur ...	115/-
70 years old, Liqueur (1820) ...	157/-
1855 vintage	180/-
1847 do do. very fine ...	250/-
1867 do. do. Brut Absolu	282/-
1870 do. Hine's do. Brut	300/-
1858 do. Grande fine Champagne	300/-
1805 do. Grande fine Champagne	315/-
1848 do do.	360/-
1830 Grande fine Champagne, superb quality...	36gns.
1797 Vintage. A curiosity ...	35/- per Bot.

3, St James's Street, London, S.W.

Page 14

WHISKY.

	Per Gal.	Per Doz.
Scotch	23/-	48/-
" 7 years old	24/6	52/-
" "Glenavon"	26/6	55/-
" **Fine, 10 Years Old**	**28/6**	**61/-**
" 1897 Macallan Glenlivet	28/6	61/-
" Fine, 15 years old	37/6	79/-
" Very old Liqueur	43/6	91/-
" Very old Glenlivet	49/6	103/-
" 1885 Talisker		127/-
Irish, 7 Years Old	**24/6**	**52/-**
" Very fine and very old	31/6	67/-
In Sherry Casks of 7, 14, or 28 Gallons.		

American Rye Whisky.

J. Wagner & Sons, Philadelphia,	
No. 3; special. Finest procurable	103/-

SPIRITS (Various).

Hollands, Fine, old, in Cruchons	...	78/-
Gin, Dry	...	43/-
" Sweetened "Old Tom"	...	47/-
Rum	...	49/- & 61/-
" 30 years old (finest quality)	...	79/-
Aquavit, Lundgreen's	...	61/-
" **Kjeldsberg's,** Old ...	...	**67/-**

AMERICAN COCKTAIL.

"Independence," Dry Martini Style		57/-
J. Wagner & Sons, Philadelphia, speciality.		

Page 15

LIQUEURS.

		Per Doz.
Apricot Brandy, Special quality	...	101/-
Bitters, Angostura, Orange, Peach, &c.		48/- to 102/-
Pères Chartreux,		
made by the Monks } Green	...	136/-
in Spain		
" Yellow	...	103/6
Chartreuse, Original, made in France, *see page 17.*		
Cherry Brandy ...	...	70/-
" Amsterdam, very fine		100/-
" Special, Dry		100/-
Cherry Whisky (Old Joe), 15 years old		**66/-**
Crème de fine Champagne	...	84/-
Curaçao Berry, à la fine Champagne		**126/-**
" Brown or White	...	84/-
" Bols, Dry, Green or White	...	102/-
" Triple Sec, Guillot	...	78/-
" Grand Duc, Guillot	...	90/-
Ginger Brandy ...	...	52/-
" " **Special Liqueur**	...	70/-
Green Ginger Liqueur ...	...	66/-
Kirschwasser	...	103/- & 117/-

LIQUEURS—*continued*.

	Per Doz.
Kümmel, Riga	70/-
,, ,, Stockmannshöfer, Dry	72/-
,, ,, **Fleur de Cumin**	**90/-**
,, ,, Grüner, Finest	108/-
,, ,, Finest Eckauer "OO"	104/-
,, ,, Amsterdam, Fine Green	100/-
Mandarine	84/-
Maraschino	82/- & 100/-
Crême de Menthe, **True Mint,**	
Specially prepared	102/-
Milk Punch	58/-
Crême de Noyau	82/- & 124/-
Orange Brandy, Old and Dry	71/-
Orange Whisky	**71/-**
Peach Brandy (A 1), Old and Fine	**102/-**
Peach Whisky	78/-
Silos Benedictinos ...	**126/-**
Sloe Gin (Special)	58/-
,, ,, Extra Dry ...	64/-
Crême de Thé (Half-bottles) ...	63/-
Tangerina	82/-
Vermouth (French and Italian) ...	36/- & 42/-

3, St James's Street, London, S.W.

ORIGINAL AND GENUINE
LIQUEUR
DE GRANDE CHARTREUSE.

GUARANTEED made by the Chartreuse Monks at their Monastery previous to their expulsion from France.

GREEN.

Litres, Exceedingly Rare, raised letters on bottles ... ... per bottle, 3 guineas		
Litres ... ... per bottle	20/-
Half-Litres ... ... per ½-bottle	10/-
,, Made in 1885 ,,	12/6

YELLOW.

Litres ... ... per bottle	20/-
Half-Litres ... ... per ½-bottle	10/-
,, Made in 1885 ,,	12/6

3, St James's Street, London, S.W.

Glossary

This glossary of English words and their French and Spanish equivalents is based on Janis Robinson, *The Oxford Companion to Wine*, 3rd ed. (Oxford: Oxford University Press, 2006).

chaptalization (*sucrage* / *azucarado*) — The addition of sugar to grape juice or must before fermentation. Permitted in some areas when the glucose levels in the grapes are low because of insufficient sun.

direct producers (*producteurs-directs* / *hibridos productores directos*) — Hybrids, which do not need grafting.

downy mildew (*mildiou* / *mildiu*) — Introduced into Europe from America before 1878, causing havoc in France after 1882. "Bordeaux mixture" (*bouillie bordelaise*), a mixture of lime, copper sulfate, and water, was developed as an effective control.

first growth (*premier cru*) — A wine that is judged to be of the first rank according to a classification.

free-run wine (*vin d'égouttage* / *vino de lagrima*) — The juice obtained by crushing rather than pressing the grapes. It consists of 60–70 percent of the total juice available.

goblet (*taille en gobelet* /*poda en redondo*/ *en vaso*) — Free standing vine-training system, used extensively in low vigor vineyards in dry climates.

hybrids (*hybrids* / *hibridos*) — In viticulture, the offspring of two varieties of different species, rather than a cross between two of the same species. The American phylloxera-resistant vines were combined with the European vinifera varieties from the late nineteenth century.

layering (*marcottage* / *acodadura*) — System of vine propagation. A cane is trained from the mother vine, and buried to the normal planting depth, with the end bent upwards to where the new vine is required. The method was used frequently to replace vines in the pre-phylloxera vineyards.

plastering (*plâtrage* / *enyesado*) — Addition of powdered plaster to crush grapes to improve clarification, color, and conservation. The 1891 French law restricted its use, and winemakers turned to tartaric and citric acid. The system continued well into the twentieth century in Spain.

pomace (*marc* / *orujo*) — The debris of the grapes; with white wines it is unfermented, and with red wines fermented.

powdery mildew or oidium (*oïdium* / *oidium*) — The first of the vine fungal diseases, identified in 1834. Native to North America and noted in England in 1845. Caused major damage in Europe in the mid-1850s. Develops rapidly in warm temperatures (20–27°C) and little affected by humidity. It was quickly discovered that it could be controlled by dusting with sulfur.

sulfur dioxide, SO² (*anhydride sulfureux / anhidrido sulfuroso*) — Gas formed by burning sulfur in the air and "the chemical compound most widely used by the winemaker, principally as a preservative and a disinfectant."

varietal wine — Term used for a wine made after the dominant grape variety from which is made. First used in California in the 1950s.

Selected Grape Varieties

alicante bouschet — High-yielding grape producing deep red wines. Created by Henri Bouschet between 1865 and 1885, crossing the petit Bouschet with grenache (alicante).

aramon — A variety that was planted on the flat, fertile soils of the Midi to produce high yields of low-alcohol red wines. It was mixed first with Spanish wines and later with those from Algeria to provide a cheap drink for urban areas in northern France. The low sugar avoided the fermentation coming to a premature end because of excessive temperatures in the vat.

carignac — Late-ripening black grape with high acidity, tannins, and color.

Bibliography

CONTEMPORARY JOURNALS AND NEWSPAPERS

Los Andes. Mendoza, 1901–4.

Australian Vigneron and Fruit-Growers Journal (from July 1906, *Wine and Spirit News and Australian Vigneron*). Sydney, 1890–1915.

Boletín de Agricultura Técnica y Económica. Madrid, 1909–14.

Boletín del Centro Vitivinícola Nacional. Buenos Aires, 1905–14.

Bulletin de l'Office International du Vin. Paris, various years.

Feuille vinicole de la Gironde. Bordeaux, 1882–1908.

El Guadalete. Jerez de la Frontera, various years.

La Nacional. Buenos Aires, 1914.

Pacific Rural Press. San Francisco, 1894–1909.

Pacific Wine & Spirits Review. San Francisco, 1900–1910.

La Prensa. Mendoza, 1914.

Le Progrès agricole et viticole. Montpellier, various years.

El Progreso Agrícola y Pecuario. Madrid, 1899–1914.

Revista vinícola Jerezana. Jerez de la Frontera, January 1866–August 1867.

Revista vitícola y vinícola. Jerez de la Frontera, February 1884–February 1885.

Revue agricole, viticole, horticole illustrée, code, conseil rural et organe de la Fédération des sociétés horticoles et viticoles du Sud-Ouest. Bordeaux, 1905–11.

Revue de viticulture. Paris, various years.

Ridley's Wine and Spirit Trade Circular. London, 1856–1916.

Rutherglen Sun and Chiltern Valley Advertiser. Rutherglen, 1909.

Le Vigneron Champenois. Épernay, 1901–13.

BOOKS AND ARTICLES

Adams, Edward Francis. 1899. *The Modern Farmer in His Business Relations*. San Francisco: Stone.

Aeuckens, Annely. 1988. "Emergence of a New Industry, 1892–1918." In *Vineyard of the Empire: Early Barossa Vignerons 1842–1939*, edited by A. Aeuckens et al. Adelaide: Australian Industrial Publishers.

Akerlof, George A. 1970. "The Market for 'Lemons': Quality Uncertainty and the Market Mechanism." *Quarterly Journal of Economics* 84:488–500.

Alazraqui, José. 1911. "La viticulture en Argentine." In *International Congress of Viticulture*. Montpellier: Coulet.

Alsberg, Carl. 1931. "Economic Aspects of Adulteration and Imitation. *Quarterly Journal of Economics* 46:1–33.

Amerine, Maynard A. 1973. "Proceedings, Ohio Grape-Wine Short Course." Ohio Agricultural Research and Development Center.

———. 1981. "Development of the American Wine Industry to 1960." In *Wine Production Technology in the United States: A Symposium*, edited by M. A. Amerine. Washington, D.C.: American Chemical Society.

Amerine, Maynard A., and Vernon L. Singleton. 1977. *Wine: An Introduction*. 2nd ed. Berkeley: University of California Press.

Anderson, Kym, David Norman, and Glyn Wittwer. 2004. "The Global Picture." In *The World's Wine Markets: Globalization at Work*, edited by K. Anderson. Cheltenham: Edward Elgar.

Andrade Martins, Conceição. 1990. *Memória do Vinho do Porto*. Lisboa: Portuguesa.

———. 1991. "A filoxera na viticultura nacional." *Análise Social* 26:653–88.

Antúnez, L. 1887. *Informe sobre la crisis actual de las industrias pecuarias y vitivinícola*. Barcelona.

Arata, Pedro. 1903. "Industria Viti-vinícola." In *Investigación Vinícola*, edited by P. Arata et al. Buenos Aires: Talleres de la Oficina Meteorológica Argentina.

Arata, Pedro, et al. 1904. *Investigación vinícola complementaria de 1904. Trabajos presentados al Ministro de Agricultura por la comisión compuesta por . . .*

Arcari, Paola. 1936. "Le variazioni dei salari agricoli in Italia dalla fondazione del Regno al 1933." *Annali di Statistica* 36.

Archivo Ministerio de Agricultura, legajo 82.2. 1886. "Cultivos, Información vinícola. Resumen por provincia." Cádiz.

Atger, Frédéric. 1907. *La crise viticole et la viticulture méridionale (1900–1907)* Paris: Giard & Brère.

Audebert, Octave. 1916. *Les exportations des vins Français á l'Étranger et le transit des vins exotiques en France*. Bordeaux: Pech.

Augé-Laribé, Michel 1907. *Le problème agraire du socialisme: La viticulture industrielle du Midi de la France*. Paris: Giard & Brère.

———. 1926. *Syndicats et Coopératives agricoles*. Paris.

Australia. 1911. *Yearbook of the Commonwealth of Australia, 1911*. Melbourne: McCarron, Bird.

Australian Official Year Book 1901–7. 1908. Melbourne.

Balán, Jorge. 1978. "Una cuestión regional en la Argentina: burguesías provinciales y el mercado nacional en el desarrollo agroexportador." *Desarrollo Económico* 69.

Balcells, Albert. 1980. *El problema agrario en Cataluña: La cuestión Rabassaire*. Madrid: MAPA.

Bara, Paul. 1998. *Histoire de Bouzy*. Reims: Paysage.

Bardhan, Pranab K. 1984. *Land, Labor, and Rural Poverty: Essays in Development Economics*. New York: Columbia University Press.

Barrio de Villanueva, Patricia. 2003. "Hacia la consolidación del Mercado Nacional de Vinos. Modernización y desarrollo del sector vitivinícola de Mendoza (Argentina), 1900–1914." *Espacios-Historia* (26):33–60.

———. 2005. "Crisis económica y estrategias empresariales: Bodegueros mendocinos a principios de siglo xx." *Cuadernos de Historia, Cordoba* (7):31–69.

———. 2006. "Las asociaciones de empresarios vitivinícolas mendocinos en tiempos de crisis y de expansión económica (1900–1912)." In *La región vitivinícola argentina. Transformaciones del territorio, la economía y la sociedad, 1870–1914*, edited by R. A. Richard Jorba et al. Buenos Aires: Universidad Nacional de Quilmes.

———. 2007. "En busca del vino genuino: Origen y consecuencias de la Ley nacional de vinos de 1904." *Mundo Agrario. Revista de estudios rurales* 8 (15).

———. 2008a. "Una crisis de la vitivinicultura mendocina a principios del siglo XX (1901–1903)." In *El vino y sus revoluciones*, edited by A. M. Mateu and S. Stein. Mendoza: Universidad Nacional de Cuyo.

———. 2008b. "El empresariado vitivinícola de la provincia de Mendoza (Argentina) a principios del siglo XX." *Historia Agraria* (45):81–111.

———. 2009. "Caracterización del boom vitivinícola en Mendoza (Argentina), 1904–1912." *Mundo Agrario* (18):1–19.

Bayet, Alain. 1997. "Deux siècles d'évolution des salaires en France: Insée, Document de travail."

Béaur, Gérard. 1998. "Land Accumulation, Life Course, and Inequalities among Generations in Eighteenth-Century France: The Winegrowers from the Chartres Region." *The History of the Family* 3 (3):285–302.

Béchade, Edmond. 1910. *La marque "Bordeaux."* Bordeaux: Coopérativè.

Beeston, John. 2001. *Concise History of Australian Wine*. St. Leonards, NSW: Allen and Unwin.

Bell, George. 1993. "The South Australian Wine Industry 1858–1876." *Journal of Wine Research* 4 (3):147–63.

———. 1994. "The London Market for Australian Wines 1851–1901: A South Australian Perspective." *Journal of Wine Research* 5 (1):19–40.

Bennett, Norman. 1990. "The Golden Age of the Port Wine System." *International History Review* 12 (2):221–48.

———. 1994. "The Port Wine System in the 1890s." *International History Review* 16 (2):251–66.

———. 2005. *That Indispensable Article: Brandy and Port Wine, c.1650–1908*. Porto: SerSilito.

Berget, Adrien. 1908. "La viticulture septentrionale." *Revue de Viticulture*, xxix.

Biagioli, Giuliana. 2000. *Il modello del proprietario imprenditore nella Toscana dell'Ottocento: Bettino Ricasoli: il patrimonio, le fattorie*. Firenze: Olschki.

Bioletti, Frederic. 1896. "Fermentation." *University of California. Report of the Viticultural Work during the Seasons 1887–93, Sacramento*.

———. 1905. *Manufacture of Dry Wines in Hot Countries*. Vol. 167. Sacramento: University of California, College of Agriculture, Agricultural Experiment Station.

———. 1908. "An Experience in Deep Plowing. *Pacific Rural Press*: 53–54.

———. 1909. "Wine Making on a Small Scale." *Pacific Rural Press*, December 11.

Bishop, Geoffrey C. 1980. *Australian Winemaking. The Roseworthy Influence*. Hawthorndene: Investigator.

———. 1988. "Viticulture and Winemaking, 1842–1905." In *Vineyard of the Empire: Early Barossa Vignerons 1842–1939*, edited by Annely Aeuckens et al. Adelaide: Australian Industrial Publishers.

———. 1998. "Viticulture and Winemaking, 1842–1905." In *Vineyard of the Empire: Early Barossa Vignerons 1842–1939*, edited by Annely Aeuckens et al. Adelaide: Australian Industrial Publishers.

Bonal, François. 1994. *Le Livre d'or du champagne*. Lausanne: Editions du Grand-Pont.

Bottaro, Santiago. 1917. "La industria vitivinícola entre nosotros." Ph.D. thesis, Facultad de Ciencias Económicas, Buenos Aires.

Boutelou, Esteban. 1807. *Memoria sobre el cultivo de la vid en Sanlúcar de Barrameda y Xerez de la Frontera*. Madrid: Villalpando.

Bouvet, Eric, and Chelsea Roberts. 2004. "Early French Migration to South Australia:

Preliminary Findings on French Vignerons." In *The Regenerative Spirit*, edited by Sue Williams et al. Adelaide: Lythrum.

Bowley, Arthur L. 1900. *Wages in the United Kingdom in the Nineteenth Century*. Cambridge: Cambridge University Press.

Bradford, Sarah 1969. *The Englishman's Wine: The Story of Port*. London: Macmillan.

Bragoni, Beatriz. 2008. "El estímulo del mercado en la transformación empresaria. Consideraciones a raíz del negocio vitivinícola en Mendoza, 1880–1940." In *El vino y sus revoluciones*, edited by A. M. Mateu and S. Stein. Mendoza.

Braudel, Fernand. 1986. *L'identité de la France. Les hommes et les chose*. Paris: Flammarion.

Brennan, Thomas. 1988. *Public Drinking and Popular Culture in Eighteenth-Century Paris*. Princeton: Princeton University Press.

———. 1997. *Burgundy to Champagne: The Wine Trade in Early Modern France*. Baltimore: Johns Hopkins University Press.

Briggs, Asa. 1985. *Wine for Sale: Victoria Wine and the Liquor Trade, 1860–1984*. London: Batsford.

Bringas Gutiérrez, Miguel Ángel. 2000. *La productividad de los factores en la agricultura española (1752–1935)*. Madrid: Banco de España.

Brutails, Jean-Auguste. 1909. *Rapport. Commission d'enquête pour la délimitation de la région "Bordeaux."* Bordeaux.

Bunge, Alejandro E. 1929. *Informe sobre el problema vitivinícola. Sociedad Vitivinícola de Mendoza*. Buenos Aires: Impresora Argentia.

Burnett, John. 1999. *Liquid Pleasure: A Social History of Drinks in Modern Britain*. London: Routledge.

Busby, James. 1825/1979. *Treatise on the Culture of the Vine*. Hunters Hill: Reprinted David Ell Press.

———. 1833. *Journal of a Tour through Some of the Vineyards of Spain and France*. Sydney: Stephens and Stokes.

Bustamente, Nicolás de. 1890. *Arte de hacer vinos*. 4th ed. Barcelona.

Cabral Chamorro, Antonio. 1987a. "Observaciones sobre la regulación y ordenación del mercado del vino de Jerez de la Frontera 1850–1935: Los antecedentes del Consejo Regulador de la Denominación de Origen 'Jerez-Xérés-Sherry.'" *Agricultura y Sociedad* (44):171–99.

———. 1987b. "Un estudio sobre la composición social y arraigo del anarquismo en Jerez de la Frontera, 1869–1923." *Estudios de Historia Social*, 42–43.

California Board of State Viticultural Commissioners. 1888. *Annual Report of the Board of State Viticultural Commissioners for 1880*.

———. 1888. *Annual Report of the Board of State Viticultural Commissioners for 1887*.

———. 1891. *Directory of the Grape Growers, Wine Makers and Distillers of California, and of the Principal Grape Growers and Wine Makers of the Eastern States*. Sacramento: Johnson.

———. 1892. *Annual Report of the Board of State Viticultural Commissioners for 1891-2*.

———. 1914. *Bulletin. Viticulture in California*. Sacramento: State Office.

California State Agricultural Society. 1915. *Report of the California State Agricultural Society for 1914*. Sacramento.

California State Board of Agriculture. 1912. *Report of the California State Agricultural Society for the Year 1911*. Sacramento: The Society.

Campbell, Christy. 2004. *Phylloxera. How Wine Was Saved for the World*. London: HarperCollins.

Carmona, Juan, and James Simpson. 1999. "The 'Rabassa Morta' in Catalan Viticulture: The Rise and Decline of a Long-Term Sharecropping Contract, 1670s–1920s." *Journal of Economic History* 59:290–315.

———. 2003. *El laberinto de la agricultura española. Instituciones, contractos y organización entre 1850 y 1936*. Zaragoza: Prensas Universitarias de Zaragoza.

———. Forthcoming. "Explaining Contract Choice: Vertical Co-ordination, Sharecropping, and Wine, Europe 1850–1950." *Economic History Review*.

Carosso, Vincent P. 1951. *The California Wine Industry*. Berkeley: University of California Press.

Carrión, Pascual. 1932/1975. *Los latifundios de España*. Madrid: Ariel.

Carter, Susan B., et al. 2006. *Historical Statistics of the United States: Earliest Times to the Present*. New York: Cambridge University Press.

Casson, Mark. 1994. "Brands: Economic Ideology and Consumer Society." In *Adding Value: Brands and Marketing*, edited by G. Jones and N. Morgan. London: Routledge.

———. 1997. *Information and Organization: A New Perspective on the Theory of the Firm*. Oxford: Oxford University Press.

Castella, Francois de. 1922. "Preliminary Maceration." *Journal of Viticulture* (January): 36–44.

Castella, Hubert de. 1886. *John Bull's Vineyard*. Melbourne: Sands & McDougall.

Caupert, Maurice. 1921. *Essai sur la C.G.V.* Montpellier: Economiste meridional.

Cavoleau, Jean Alexandre. 1827. *Oenologie française*. Paris: Madame Huzard.

Cazeux-Cazalet. 1909. *Rapport de la Commission de délimitation de la région Bordeaux à Monsieur le Ministre de l'Agriculture fait au nom de la sous-commission Cazeaux-Cazalet*. Cadillac: Labarthe.

Caziot, Pierre. 1952. *La Valeur de la terre en France*. 3rd ed. Paris: Baillière.

Centro Vitivinícola Nacional. 1910. *La viti-vinicultura en 1910*. Buenos Aires: Coll.

Chaffal, Georges. 1908. *Les Crises viticoles modernes et la dépréciation actuelle du vignoble*. Lyon.

Chaptal, Jean-Antoine. 1819. *De l'Industrie françoise*. 2 vols. Paris: Renouard.

Cipolla, Carlo M. 1975. "European Connoisseurs and California Wines, 1875–1895." *Agricultural History* 49 (1):294–310.

Clarence-Smith, W.G. 1995. "Cocoa plantations in the Third World, 1870s–1914: The Political Economy of Inefficiency." In *The New Institutional Economics and Third World Development*, edited by J. Harriss, J. Hunter, and C. Lewis. London: Routledge.

Clique, Hubert 1931. *Les caves coopératives de vinification en Bourgogne*. Paris: Recueil Sirey.

Cockburn, Ernest. 1945. *Port Wine and Oporto*. London: Wine and Spirit Publications.

Cocks, Charles. 1846. *Bordeaux: Its Wines and the Claret Country*. London: Longman.

———. 1850. *Bordeaux, ses environs et ses vins classés par ordre de mérite: Guide de l'étranger à Bordeaux*. Bordeaux: Féret et Fils.

———. 1969. *Bordeaux et ses vins*. 12th ed. Bordeaux: Féret.

Cocks, Charles, and Édouard Féret. 1868. *Bordeaux et ses vins: Classés par ordre de mérite*. 2nd ed. Paris: Masson.

———. 1883. *Bordeaux and Its Wines*. 2nd ed. (translation of French 4th edition of 1881) Paris: Masson.

———. 1898. *Bordeaux et ses vins classés par ordre de mérite.* 7th ed. Bordeaux: Féret & Fils.

———. 1908. *Bordeaux et ses vins classés par ordre de mérite.* 8th ed. Bordeaux: Féret & Fils.

Coria López, Luis. 2006. "El siglo anterior al boom vitivinícola mendocino (1780/1883)." *Revista Universum* 2 (21):100–24.

———. 2008. "Producto bruto geográfico de Mendoza para 1914. Determinación y resultados." In *XLIII Asociación Argentina de Economía Política.* Córdoba.

Coste-Floret, Paul. 1894. *Procédés modernes de vinification.* Montpellier and Paris: Coulet and Masson.

Crampton, Charles Albert. 1888. "Composition of American Beers, Wines, and Ciders, and the Substances Used in Their Adulteration." In *U.S. Department of Agriculture, Yearbook 1887.* Washington, D.C.

Creuss, William V. 1937. "Knowing the Condition of Your Wine." *Wines and Vines* (May).

La Crisis Agrícola y Pecuaria, La. 1887–89. *Actas y dictámenes de la comisión creada por el Real Decreto de 7 de julio de 1887 para estudiar la crisis que atraviesa la agricultura y la ganadería.* 6 vols. Madrid.

Croft-Cooke, Rupert. 1957. *Port.* London: Putnam.

Cullen, Louis M. 1998. *The Brandy Trade under the Ancien Régime: Regional Specialisation in the Charente.* Cambridge: Cambridge University Press.

Davis, Ralph. 1972. "The English Wine Trade in the Eighteenth and Nineteenth Centuries." *Annales Cisalpines d'histoire sociale* 3:87–106.

Degrully, Paul. 1910. *Essai historique et économique sur la production et le marché des vins en France* Paris: Giard & Brière.

Demonet, Michel. 1990. *Tableau de l'agriculture Française. L'enquête de 1852.* Paris: Éditions de l'École de Hautes Études en Sciences Sociales.

Denis, Pierre. 1922. *The Argentina Republic: Its Development and Progress.* London: Fisher Unwin.

Denman, James L. 1876. *Wine and Its Counterfeits.* London: Spottiswoode.

De Turk, Issac. 1890. *Sonoma District. Report of the Board of State Viticultural Convention,* San Francisco: State Office.

Devoto, Fernando. 2004. *Historia de la inmigración en la Argentina.* 2nd. ed. Buenos Aires: Sudamericana.

De Vries, Jan. 1984. *European Urbanization, 1500–1800.* Harvard Studies in Urban History. Cambridge: Harvard University Press.

Dion, Roger. 1959: 1977. *Histoire de la vigne et du vin au France.* Paris: Flammarion.

Dovaz, Michael. 1983. *L'encyclopédie des vins de Champagne.* Paris: Julliard.

Drummond, Jack C., and Anne Wilbraham. 1958. *The Englishman's Food: A History of Five Centuries of English Diet.* London: Cape.

Dubois, Raymond, and W. Percy Wilkinson. 1901a. *New Methods of Grafting and Budding.* Melbourne: Rutherglen Viticultural Station.

———. 1901b. *Trenching and Subsoiling for American Vines.* Melbourne: Rutherglen Viticultural Station.

Du Breuil, Alphonse. 1867. *Vineyard Culture Improved and Cheapened, with Notes and Adaptations to American Culture.* Cincinnati: Clarke.

Duguid, Paul. 2003. "Developing the Brand: The Case of Alcohol, 1800–1880." *Enterprise & Society* 4 (3):405–41.

———. 2005a. "Introduction [to special issue on wine and networks]." *Business History Review* 79 (3):453–66.

———. 2005b. "Networks and Knowledge: The Beginning and End of the Port Commodity Chain, 1703–1860." *Business History Review* 79 (3):499–526.

Duguid, Paul, and Teresa da Silva Lopes. 1999. "Ambiguous Company: Institutions and Organizations in the Port Wine Trade, 1814–1834." *Scandinavian Economic History Review* 47:84–102.

———. 2001. "Divide and Rule: Regulation in the Port Wine Trade, 1812–1840." In *Business History Year Book 3*, edited by T. Gourvish. Aldershot: European Society for Business History.

Dunstan, David. 1994. *Better than Pommard!: A History of Wine in Victoria*. Melbourne: Australian Scholarly.

Dye, Alan. 1998. *Cuban Sugar in the Age of Mass Production: Technology and the Economics of the Sugar Central, 1899–1929*. Stanford: Stanford University Press.

Elías de Molins, José. 1904. *Los trigos y los vinos en España*. Barcelona.

Espejo, Zoilo. 1879–80. *Conferencias agrícolas de la provincia de Madrid*. Madrid.

Études et Conjoncture. 1955. *La strucuture et les rendements du vignoble du Bas-languedoc et du Roussillon* (6):515–60.

Exposición Vinícola Nacional. 1878–79. *Estudio sobre la . . . de 1877*.

Faith, Nicholas. 1978. *The Winemasters*. London: Hamilton.

———. 1983. *Victorian vineyard. Château Loudenne and the Gilbeys*. London: Constable in association with Christie's Wine Publications.

———. 1988. *The Story of Champagne*. London: Hamish Hamilton.

———. 2003. *Australia's Liquid Gold*. Chatham: Mitchell Beazley.

———. 2004. *Cognac*. London: Mitchell Beazley.

Federico, Giovanni. 1994. "Commercialization and Economic Development in Italy, 1869–1940." In *The Economic Development of Italy since 1870*, edited by G. Federico. Aldershot: Edward Elgar.

Feinstein, Charles. 1990. "New Estimates of Average Earnings in the United Kingdom, 1880–1913." *Economic History Review* 43 (4):595–632.

Féret, Édouard. 1874–89. *Statistique générale de la Gironde*, vol. 2 vinicole. Bordeaux: Féret.

Fernández-Pérez, Paloma 1999. "Challenging the Loss of an Empire: González & Byass of Jerez." *Business History* 41:72–87.

Fernández de la Rosa, G. 1909. "El viñedo y los vinos jerezanos." *Boletín de Agricultura Técnica y Económica*.

Fernández, Eva. 2010. "Unsuccessful Response to Quality Uncertainty: Brands in Spain's Sherry Industry, 1920–1990. *Business History* 52 (1):100–119.

Fernández García, Eva. 2008. *Productores, comerciantes y el estado: Regulación y redistribución de rentas en el mercado de vino en España, 1890–1990*, Ph.D. thesis, Universidad Carlos III de Madrid.

Ferrouillat, Paul, and M. Charvet. 1896. *Les celliers: Construction et matériel vinicole avec la description des principaux celliers du Midi, du Bordelais, de la Bourgogne et de l'Algérie*. Montpellier and Paris: Coulet and Masson.

Foex, Gustave. 1902. *Manuel of Modern Viticulture: Reconstruction with American Vines*. Melbourne: Brain.

Forbes, Patrick 1967. *Champagne: The Wine, the Land, and the People*. London: Gollancz.

Ford, Richard. 1846/1970. *Gatherings from Spain*. London: Dent.

Frader, Laura Levine 1991. *Peasants and Protest: Agricultural Workers, Politics, and Unions in the Aude, 1850–1914*. Berkeley: University of California Press.

France. 1934. *Annuaire statistique année 1933*, edited by Direction Générale. Paris: Imprimerie Nationale.

France. 1939. *Annuaire statistique année 1938*, edited by Direction Générale. Paris: Imprimerie Nationale.

France, Chambre des députés. 1909. *Enquête sur la situation critique de la viticulture, 4e rapport*. Vol. Annexe 2512.

France, Direction Générale des douanes. *Tableau Général du Commerce de la France*.

France, Ministère de l'Agriculture. 1913. *Statistique agricole annuelle 1912*.

———. 1937. *Statistique agricole de la France. Annexe à l'enquête de 1929, monographie du département de la Gironde*. Paris.

Francis, Alan D. 1972. *The Wine Trade*. London: Adam & Charles Black.

Franck, Rafaël, Noel D. Johnson, and John V. C. Nye. 2010. "Trade, Taxes, and Terroir: Excise Taxes and Market Integration in Third Republic France." Paper given at the 4th AAWE meeting. Davis, California.

Franck, William. 1824. *Traité sur les vins du Médoc et les autres vins rouges et blancs du département de la Gironde*. Bordeaux: Laguillotière et Compe.

French, Michael, and Jim Phillips. 2000. *Cheated Not Poisoned? Food Regulation in the United Kingdom, 1875–1938*. Manchester: Manchester University Press.

Gabriel, Pierre. 1913. "La viticulture dans le département de l'Aube," Ph.D. thesis, Paris.

Galanti, Arminio N. (1900). *La industria vitivinícola argentina*. Buenos Aires: Ostwald.

———. 1915. *Estudio crítico sobre la cuestión vitivínicola; estudios y pronósticos de otros tiempos*. Buenos Aires: Perrotti.

Galassi, Francesco. 1992. "Tuscans and Their Farms: The Economics of Share Tenancy in Fifteenth Century Florence." *Rivista di storia economica* 9:77–94.

Galet, Pierre. 1988. *Cépages et vignobles de France*. Montpellier: Déhan.

———. 2004. *Cépages et vignobles de France. Tome III Les vignobles de France*. 2 vols. Paris: Tec & Doc.

Gallo, Ezequiel. 1993. "Society and Politics, 1880–1916." In *Argentina since Independence*, edited by L. Bethell. Cambridge: Cambridge University Press.

García de los Salmones, Nicolás. 1893. *La invasión filoxérica en España y las cepas americanas*. Vol. 1. Barcelona.

Garrabou, Ramon, and Enric Tello. 2002. "Salario como coste, salario como ingreso: El precio de los jornales agrícolas en la Cataluña contemporánea, 1727–1930." In *El nivel de vida en la España rural, siglos XVIII–XX*, edited by J. M. M. Carrión.

Garrier, Gilbert. 1998. *Histoire sociale et culturelle*. Paris: Larousse.

Gasparin, Count Adrien Étienne Pierre de. 1848. *Cours de Agriculture*. Vol. 4. Paris.

Gayon, Ulysse. 1901. *Studies on Wine-Sterilizing Machines*. Melbourne: Brain.

———. 1904. "Pasteurisation des vins." *Revue de Viticulteu*, March 17.

Génieys, Pierre 1905. *La Crise viticole méridionale*. Toulouse: Privat.

Gereffi, Gary. 1994. "The Organisation of Buyer-Driven Global Commodity Chains." In *Commodity Chains and Global Capitalism*, edited by G. Gereffi and M. Korzeniewicz. Westport: Greenwood.

Gervais, Charles. 1903. *Indicateur des vignobles méridionaux, comprenant l'Hérault, l'Aude, le Gard, les Pyrénées-Orientales, les Bouches-du-Rhône, le Vaucluse, la Drôme, le Var, l'Algérie (départements d'Alger, d'Oran, de Constantine), la Tunisie.* 2nd ed. Montpellier.

Gervais, Misaël. 1913. *La coopération en viticulture.* Paris: Baillière.

Gide, Charles. 1901. "La crise du vin en France et les associations de vinification." *Revue d'économie politique* 15:217–35.

———. 1907. "The Wine Crisis in South France." *Economic Journal* 17 (September): 370–75.

———. 1926. *Les associations coopératives agricoles.* Paris: Association pour l'enseignement de la coopération.

González Gordon, Manuel. 1972. *Jerez, Xerez, Sherry.* London: Cassell.

González Inchaurraga, Íñigo. 2006. *El Marqués que reflotó el Rioja.* Madrid: LID Editorial Empresarial.

Gonzalez y Álvarez, Francisco. 1878. *Apuntes sobre los vinos españoles.* Madrid.

Gough, J. B. 1998. "Winecraft and Chemistry in 18th-Century France: Chaptal and the Invention of Chaptalization." *Technology and Culture* 39 (1):74–104.

Gourvish, Terry, and Richard Wilson. 1994. *The British Brewing Industry, 1830–1980.* Cambridge: Cambridge University Press.

Guy, Kolleen M. 1996. "Wine, Work, and Wealth: Class Relations and Modernization in the Champagne Wine Industry, 1870–1914, Ph.D. diss., Indiana University.

———. 2003. *When Champagne Became French: Wine and the Making of a National Identity.* Baltimore: Johns Hopkins University Press.

Guyot, Jules 1861. *Culture de la vigne et vinification.* 2nd ed. Paris: Librairie Agricole de la Maison rustique.

———. 1868. *Étude des vignobles de France pour servir a l'enseignement mutuel de la viticulture et de la vinification françaises.* Paris: Masson.

Hancock, David. 2009. *Oceans of Wine: Madeira and the Emergence of American Trade and Taste.* New Haven: Yale University Press.

Hansard's Parliamentary Debates. 1860, 156. *House of Commons.*

Harrison, Brian. 1971. *Drink and the Victorians.* London: Faber.

Harrison, Godfrey. 1955. *Bristol Cream.* London: Batsford.

Hayami, Yujiro, and Keijiro Otsuka. 1993. *The Economics of Contract Choice.* Oxford: Clarendon.

Heintz, William F. 1977. "The Role of Chinese Labor in Viticulture and Winemaking in Nineteenth Century California," Master's thesis, Sonoma State College.

———. 1990. *Wine Country: A History of Napa Valley: The Early Years, 1838–1920.* Santa Barbara: Capra.

———. 1999. *California's Napa Valley: One Hundred and Sixty Years of Wine Making.* San Francisco: Scottwall Associates.

Henderson, Alexander. 1824. *The History of Ancient and Modern Wines.* London: Baldwin, Cradock and Joy.

Hidalgo Tablada, José de. 1880. Tratado de la Fabricación de Vinos. Madrid: Cuesta.

Higounet, Charles. 1993. *Château Latour: The History of a Great Vineyard, 1331–1992.* Kingston upon Thames, UK: Segrave Foulkes.

———, ed. 1974. *La Seigneurie et le vignoble de Château Latour: Histoire d'un grand cru du Médoc, XIVe–XXe siècle.* 2 vols. Bordeaux: Fédération historique du Sud-Ouest.

Hilgard, Eugene W. 1884. "The future of grape-growing in California." *The Overland Monthly* (January): 1–6.

———. 1886. "Wine Fermentation." *Report of the Viticultural Work during the Seasons 1883–4 and 1884–5, Sacramento.* University of California.

Hoffman, Elizabeth, and Gary D. Libecap. 1991. "Institutional Choice and the Development of U.S. Agricultural Policies in the 1920s." *Journal of Economic History* 51:397–411.

Hoffman, Phillip. 1984. "The Economic Theory of Sharecropping in Early Modern France." *Journal of Economic History* 42:155–62.

Hume, David. 1752. *Political Discourses.* 2nd. ed. Edinburgh: Kincaid & Donaldson.

Huret, Jules. 1913. *La Argentina, del Plata a la Cordillera de los Andes.* Paris: Fasquelle.

Husmann, George. 1883. *American Grape Growing and Wine Making.* New York: Orange Judd Co.

———. 1888. *Grape Culture and Wine-Making in California; A Practical Manual for the Grape-Grower and Wine-Maker.* San Francisco: Payot, Upham & Co.

———. 1896. *American Grape Growing and Wine Making.* 4th ed. New York: Orange Judd Co.

———. 1899. "The Present Condition of Grape Culture in California." In *U.S. Department of Agriculture, Yearbook 1888.*

———. 1903. "Grape, Raisin, and Wine Production in the United States." In *U.S. Department of Agriculture, Yearbook 1902.* Washington, D.C.

International Institute of Agriculture. 1911. *Monographs on Agriculture. Co-operation in Various Countries.* Rome: IIA.

———. 1915a. *Monographs on Agriculture. Co-operation in Various Countries.* Vol. 2. Rome: IIA.

———. 1915b. *International Yearbook of Agricultural Statistics, 1913 and 1914.* Rome: IIA.

———. 1927. *International Yearbook of Agricultural Statistics, 1926–7.* Rome: IIA.

Isnard, Hildebert. 1954. *La vigne en Algérie; étude géographique.* 2 vols. Gap: Ophrys: Gap.

Italy, Istituto Centrale di Statistica. 1958. *Sommario di statistiche storiche italiane, 1861–1855.* Roma.

Italy, Ministero dell'Agricoltura e delle Foreste. 1932. *Per la tutela del vino Chianti.* Bologna: Brunelli.

Italy, Ministero di Agricoltura, Industria e Commerico (MAIC). 1914. *Il vino in Italia: Produzione-Commercio con l'estero-Prezzi.* Roma.

Jeffreys, James B. 1954. *Retail Trading in Britain, 1850–1950.* Cambridge: Cambridge University Press.

Jeffs, Julian. 2004. *Sherry.* London: Faber.

Joslyn, Maynard A., and W. V. Cruess. 1934. *Elements of Wine Making.* Berkeley: University of California Press.

Jouannet, Francois. 1839. *Statistique du département de la Gironde.* 3 vols. Paris: Dupont.

Judt, Tony. 1979. *Socialism in Provence 1871–1914. A Study in the Origins of the Modern French Left.* Cambridge: Cambridge University Press.

Jullien, Andre. 1816. *Topographie de tous les vignobles connus . . . : Suivie d'une classification générale des vins.* Paris: L'auteur.

———. 1824. *The Topography of All the Known Vineyards; Containing a Description of the Kind and Quality of Their Products, and a Classification. Tr. from the French*. London: Whittaker.

———. 1826. *Manuel du sommelier ou instruction pratique sur la manière de soigner les vins*. 3rd. ed. Paris.

Kaerger, Karl. 1901. *La agricultura y la colonización en Hispanoamérica*. Buenos Aires.

Kaplan, Teresa. 1977. *Anarchists of Andalusia, 1868–1903*. Princeton: Princeton University Press.

Kehrig, Henri. 1884. *Le privilège des vins à Bordeaux*. Paris and Bordeaux.

Kelly, Alexander. 1861. *The Vine in Australia*. Melbourne: Sands.

Klein, Benjamin, and Keith B. Leffler. 1981. "The Role of Market Forces in Assuring Contractual Performance." *Journal of Political Economy* 89 (4):615–41.

Knox, Trevor M. 1998. "Organization Change and Vinification Cooperatives in France's Midi." Working papers from University of Connecticut, Department of Economics, 1998–2004.

Labrousse, Ernest. 1933. *Esquisse du mouvement des prix et des revenue en France au XVIIIe siècle*. Paris: Dalloz.

———. 1944. *La crise de l'economie francaise a la fin de l'ancien regime et au debut de la revolution*. Paris: Presses universitaries de France.

Lachiver, Marcel. 1982. *Vin, vigne et vignerons en la région parisienne du XVIIe au XIXe siècle*. Pontoise.

———. 1988. *Vins, vignes et vignerons. Histoire du vignoble français*. Paris: Fayard.

Lachman, Henry. (1903). "The Manufacture of Wines in California." *U.S. Department of Agriculture, Bureau of Chemistry*, 1 (72):25–40.

Laffer, Henry E. 1949. *The Wine Industry in Australia*. Adelaide: Australian Wine Board.

Lafforgue, G. 1954. Cent cinquante ans de production viticole en Gironde. *Bulletin Technique des Ingenieurs des Services Agricoles*, 293–301.

Lains, Pedro. 1992. "Foreign Trade and Economic Growth in the European Periphery: Portugal, 1851–1913." Instituto de Ciências Sociais, Universidade de Lisboa, Lisboa.

Lains, Pedro, and P. Sousa. 1998. "Estatística e produção agrícola em Portugal, 1848–1914." *Análise Social* 33:935–68.

Lamoreaux, Naomi R. 1985. *The Great Merger Movement in American Business, 1895–1904*. Cambridge: Cambridge University Press.

Lana, José Miguel. 1999. *El sector agrario Navarro (1785–1935)*. Pamplona: Gobierno de Navarra.

———. 2002. "La aventura exterior de la agricultura Navarra (1850–1990)." *Estudios Agrosociales y Pesqueros*.

Lapsley, James T. 1996. *Bottled Poetry: Napa Winemaking from Prohibition to the Modern Era*. Berkeley: University of California Press.

Latham, Robert, and William Matthews, eds. 1970. *The Diary of Samuel Pepys*. Vol. 4. Berkeley: University of California Press.

Law, Marc. 2003. "The Origins of State Pure Food Regulation." *Journal of Economic History*, 1103–30.

Le guide ou conducteur de l'étranger à Bordeaux. 1825. Bordeaux: Fillastre et neveu.

Le Roy Ladurie, Emmanuel. 1976. *The Peasants of Languedoc*. Urbana: University of Illinois Press.

Leggett, Herbert B. 1941. *Early History of Wine Production in California*. San Francisco: Wine Institute.

Levi, Leone. 1872. "On the Limits of Legislative Interference with the Sale of Fermented Liquors." *Journal of the Statistical Society* 35 (1):25–56.

Lévy-Leboyer, Maurice. 1971. "La décélération de l'économie française." *Revue d'Histoire Economique et Sociale* (49).

Lewis, W. Arthur. 1978. *Growth and Fluctuations, 1870–1913*. London: Allen & Unwin.

Lheureux, Lucien. 1905. *Les syndicats dans la viticulture champenoise*. Paris: Librarie generale de droit et de jurisprudence.

Liaudat, Héctor C. 1934. *Analisis del problema vitivinícola*. Buenos Aires: Ministerio de Agricultura de la Nación.

Libecap, Gary D. 1992. "The Rise of the Chicago Packers and the Origins of Meat Inspection and Antitrust." *Economic Inquiry* 30 (2):242–62.

Lopes, Teresa da Silva. 2007. *Global Brands: The Evolution of Multinationals in Alcoholic Beverages*: Cambridge: Cambridge University Press.

López Estudillo, Antonio. 1992. "La vid y los viticultores de Jerez, la crisis comercial y el impacto de la filoxera: Un campo abierto a la investigación." *Revista de Historia de Jerez* 1:43–71.

———. 2001. *Republicanismo y Anarquismo en Andalucía*. Cordoba: La Posada.

Loubère, Leo A. 1974. *Radicalism in Mediterranean France: Its Rise and Decline, 1848–1914*. Albany: State University of New York Press.

———. 1978. *The Red and the White. The History of Wine in France and Italy in the Nineteenth Century*. Albany: State University of New York Press.

Maddison, Angus. 1995. *Monitoring the World Economy, 1820–1992*. Paris: OECD.

Maizière, Armand. 1848. *Origine et développement du commerce du vin de Champagne*. Reims: Jacquet.

Maldonado Rosso, Javier 1999. *La formación del capitalismo en el marco del Jerez. De la vitivinicultura tradicional a la agroindustria vinatera moderna (siglos xviii–xix)*. Madrid: Huerga & Fierro.

Malefakis, Edward. 1970. *Agrarian Reform and the Peasant Revolution in Spain: Origins of the Civil War*. New Haven: Yale University Press.

Mandeville, Léonce. 1914. *Étude sur les sociétés coopératives de vinification du midi de la France*. Toulouse: Douladoure.

Manuel de l'agriculteur du Midi. 1831. Avignon.

Marescalchi, Arturo. 1924. *Come si abbassa el costo del vino*. Casale Monferrato: Marescalchi.

Marcilla Arrazola, Juan. 1922. *Vinficación en países cálidos*. Madrid: Espasa-Calpe.

———. 1949–50. *Tratado práctico de viticultura y enología españolas*. Madrid: Sociedad Anónima Española de Traductores y Autores.

Markham, Dewey. 1998. *1855. A History of the Bordeaux Classification*. New York: Wiley.

Martins Pereira, Gaspar, and João Nicolau de Almeida. 1999. *Porto Vintage*. Porto: Institito do Vinho do Porto.

Mateu, Ana María. 2007. "Los caminos de construcción del cooperativismo vitivinícola en Mendoza, Argentina (1900–1920)." In *Documentos de Trabajo*. Universidad de Belgrano.

————. 2009. "Estudio y análisis de la modalidad empresarial de los Arizu en Mendoza. 1880–1930," Ph.D. thesis, Universidad Nacional de Cuyo, Mendoza.

Mateu, Ana María, and Steve Stein. 2006. "Diálogos entre sordos. Los pragmáticos y los técnicos en la época inicial de la industria vitivinícola argentina." *Historia Agraria* (39):267–92.

Maxwell, Herbert. 1907. *Half-a-Century of Successful Trade*. London: Pantheon.

Mazade, Marcel. 1900. *First Steps in Ampelography*. Melbourne: Brain.

McCulloch, John Ramsey. 1845. *Commercial Dictionary*. London.

Mendelson, Richard. 2009. *From Demon to Darling: A Legal History of Wine in America*. Berkeley: University of California Press.

Mills, S. A. 1908. *The Wine Story of Australia*. Sydney: Attkins.

Mingay, Gordon E., ed. 1975. *Arthur Young and His Times*. London: Macmillan.

Mitchell, Brian R. 1992. *International Historical Statistics, Europe, 1750–1988*. 3rd ed. Basingstoke and New York: Macmillan and Stockton Press.

————. 1995. *International Historical Statistics: Africa, Asia & Oceania, 1750–1988*. 2nd ed. New York: Stockton Press.

————. 2007. *International Historical Statistics: The Americas, 1750–2005*. 6th ed. London: Stockton.

Mokyr, Joel. 2004. *The Gifts of Athena: Historical Origins of the Knowledge Economy*: Princeton: Princeton University Press.

Mondini, Salvatore. 1900. *La viticoltura e l'enologia in Italia*. Roma.

Montañés, Enrique. 1997. *Transformación agrícola y conflictividad campesina en Jerez de la Frontera, 1880–1923*. Cádiz: Universidad de Cádiz.

————. 2000. *La empresa exportadora del Jerez. Historia Económica de González Byass, 1835–1885*. Cádiz: Universidad de Cádiz.

Moreau-Bérillon, Jules Camille. 1922. *Au pays du Champagne. Le Vignoble. Le Vin*. Reims: Michaud.

Moreira, Vital. 1998. *O Governo de Baco: A organização institucional do Vinho do Porto*. Oporto: Afrontamento.

Morgan, Percy T. (1902). "The Outlook for the Wine Interest." *Pacific Rural Press*, August 9.

————. (1904). "The California Wine Industry." *Pacific Rural Press*, January 16, pp. 36–37.

Morilla Critz, José. 1995. "La irrupción de California en el mercado de productos vitícolas y sus efectos en los países mediterráneos (1865–1925)." In *California y el Mediterráneo*, edited by J. Morilla Critz. Madrid: MAPA.

Morilla Critz, José, Alan L. Olmstead, and Paul W. Rhode. 1999. "'Horn of Plenty': The Globalization of Mediterranean Horticulture and the Economic Development of Southern Europe, 1880–1930." *Journal of Economic History* 59 (2):316–52.

Navarro Soler, Diego. 1875. *Guía razonada del cultivador de viñas y cosechero de vinos*. Valencia: Domenech.

Nourrisson, Didier. 1990. *Le buveur du XIXè siècle*. Paris: Albin Michel.

Nye, John V. C. 1991. "The Myth of Free Trade Britain and Fortress France: Tariffs and Trade in the Nineteenth Century." *Journal of Economic History* 51 (1):23–46.

————. 1994. "The Unbearable Lightness of Drink: British Wine Tariffs and French Economic Growth, 1689–1860." In *Political Economy of Protectionism and Commerce,*

Eighteenth–Twentieth Centuries, edited by Peter Lindert, J. Nye and J.-M.Chevet. Milan: Università Bocconi.

———. 2007. *War, Wine, and Taxes: The Political Economy of Anglo-French Trade, 1689–1900*. Princeton: Princeton University Press.

O'Brien, Denis P. 2004. *The Classical Economists Revisited*. Princeton: Princeton University Press.

Olmstead, Alan L., and Paul W. Rhode. 2002. "The Red Queen and the Hard Reds: Productivity Growth in American Wheat, 1800–1940." *Journal of Economic History* 62 (4):929–66.

———. 2010. "Quantitative Indices on the Early Growth of the California Wine Industry." In *Homenaje a Gabriel Tortella*, edited by J. Morilla Critz et al. Alcalá: Universidad de Alcalá.

Olson, Mancur. 1965. *The Logic of Collective Action*. Cambridge: Harvard University Press.

Ordish, George 1972. *The Great Wine Blight*. London: Dent.

Osmond, Robert, and Kym Anderson. 1998. *Trends and Cycles in the Australian Wine Industry, 1850–2000*. Adelaide: Centre for International Economic Studies, University of Adelaide.

Pacottet, Pablo. 1924. *Vinificación*. Barcelona: Salvat (translation of French 3rd ed., 1915).

Pacottet, Paul. 1911. *Vinificación en la provincia de Mendoza*. Paris: Bailliere.

Paguierre, M. 1828. *Classification and Description of the Wines of Bordeaux*. Edinburgh: Blackwood.

———. 1829. *Classification et description des vins de Bordeaux*. Paris: Audot.

Palmer, Hans Christian. 1965. "Italian Immigration and the Development of California Agriculture." University of California, Berkeley.

Pan-Montojo, Juan L. 1994. *La Bodega del Mundo. Historia de la vitivincultura en España, 1800–1936* Madrid: Alianza.

———. 2003. "Las industrias vinícolas Españolas: Desarrollo y diversificación productiva entre el siglo xviii y 1960." In *Las industrias agroalimentarias en Italia y España durante los siglos XIX y XX*, edited by C. Barciela López and A. Di Vittoriio. Alicante.

Parada y Barreto, Diego. 1868. *Noticias sobre la historia y estado actual del cultivo de la vid y del comercio vinatero de Jerez de la Frontera*. Jerez de la Frontera.

Paul, Harry W. 1996. *Science, Vine, and Wine in Modern France*. Cambridge: Cambridge University Press.

Pech, Rémy. 1975. *Entreprise viticole et capitalisme en Languedoc Roussillon*. Toulouse: L'Universite de Toulouse-Le Mirail.

Pedrocco, Giorgio. 1994. "Un caso e un modello: Viticoltura e industria enologica, inStudi sull'agricoltura Italiana," edited by P. D'Attore and A. De Bernardi. Milan: Fondazione Giangiacomo Feltrinelli.

Peninou, Ernest P., and Gail G. Unzelman. 2000. *The California Wine Association and Its Member Wineries, 1894–1920*. Santa Rosa, CA: Nomis.

Peninou, Ernest P., Gail G. Unzelman, and Michael M. Anderson. 1998. *History of the Sonoma Viticultural District: The Grape Growers, the Wine Makers, and the Vineyards*. Santa Rosa, CA: Nomis.

Penning-Rowsell, Edmund. 1973. *The Wines of Bordeaux*. Harmondsworth: Penguin.

Pérez Romagnoli, Eduardo. 2006. "Las industrias inducidas y derivadas de la vitivinicultura en Mendoza y San Juan (1885–1914)." In *La región vitivinícola argentina*, edited by R. A. Richard Jorba. Buenos Aires: Quilmes.

Perpinyá i Grau, Román. 1932. *La crisi del Priorat*. Barcelona.

Petit-Lafitte, Auguste. 1868. *La vigne dans le Bordelais; histoire–histoire naturelle–commerce–culture*. Paris: Rothschild.

Piard, Paul 1937. *L'Organisation de la Champagne viticole. Des syndicats vers la corporation*. Paris: Chancelier.

Pijassou, R. 1980. *Le Médoc: Un grand vignoble de Qualité*. 2 vols. Paris: Tallandier.

Pinchemel, Philippe. 1987. *France: A Geographical, Social and Economic Survey*. Cambridge: Cambridge University Press.

Pinilla, Vicente, and Maria-Isabel Ayuda. 2002. "The Political Economy of the Wine Trade: Spanish Exports and the International Market, 1890–1935." *European Review of Economic History* 6:51–85.

Pinney, Thomas. 1989. *A History of Wine in America: From the Beginnings to Prohibition*. Berkeley: University of California Press.

———. 2005. *A History of Wine in America: From Prohibition to the Present*. Berkeley: University of California Press.

Ponsot, Pierre. 1986. *Atlas de historia económica de la baja Andalucía: siglos XVI y XIX*. Granada.

Pope, David. 1971. "Viticulture and Phylloxera in North-East Victoria." *Australian Economic History Review* 10 (1).

Porter, George R. 1847. *The Progress of the Nation*. 2nd. ed. London: Murray.

Postel-Vinay, Gilles. 1989. "Debt and Agricultural Performance in the Languedocian Vineyard, 1870–1914." *Research in Economic History* (supplement 5):161–86.

Postel-Vinay, Gilles, and Jean-Marc Robin. 1992. "Eating, Working, and Saving in an Unstable World: Consumers in Nineteenth- Century France." *Economic History Review* 45 (3).

Pouget, Roger 1990. *Histoire de la lutte contre le Phylloxéra de la vigne en France*. Paris: INRA.

Prados de la Escosura, Leandro. 1982. *Comercio exterior y crecimiento en España, 1826–1913*. Madrid: Banco de España

Prest, Alan Richmond. 1954. *Consumers' Expenditure in the United Kingdom, 1900–1919*. Cambridge: Cambridge University Press.

Preston, Paul. 1984. "The Agrarian War in the South." In *Revolution and War in Spain, 1931–1939*, edited by P. Preston. London: Methuen.

Prestwich, Patricia. 1988. *Drink and the Politics of Social Reform: Anti-alcoholism in France*. Palo Alto: Society for the Promotion of Science and Scholarship.

Price, Roger. 1983. *The Modernization of Rural France*. London: Hutchinson.

Quevedo y García Lomas, José. 1904. *Memoria . . . al concurso de 1903. El socialismo agrario en Andalucía y la reforma del servicio agronómico del Estado*. Madrid: Minuesa de los Ríos.

Ramon Muñoz, Ramon. 2009. "Migration and Trade: The Case of Southern-European Immigration and Olive Oil Imports in the Americas, 1875–1930." In *Fourth Iberian Cliometrics Workshop*. Lisbon.

Redding, Cyrus. 1833. *A History and Description of Modern Wines*. London: Whittaker, Treacher, & Arnot.

———. 1851. *A History and Description of Modern Wine*. London: H. G. Bohn.

Rhode, Paul W. 1995. "Learning, Capital Accumulation, and the Transformation of California Agriculture." *Journal of Economic History* 55 (4):773–800.

Richard, Eugène 1934. *Le marché du vin à Paris*. Paris: Editions Domat-Montchrestien.

Richard Jorba, Rodolfo A. 1994. "Estado y empresarios regionales en los cambios economicos y espaciales. La modernización en Mendoza (1870–1910)." *Siglo XIX Cuadernos de Historia* (9):69–99.

———. 1998. *Poder, economía y espacio en Mendoza, 1850–1900: Del comercio ganadero a la agroindustria vitivinícola*. Mendoza: Universidad Nacional de Cuyo.

———. 2006. "Transiciones económicos-sociales: Inmigración y mundo del trabajo." In *La región vitivinícola argentina*, ed. R. A. Richard Jorba. Buenos Aires: Quilmes

———. 2007. "Sumando esfuerzos y conocimientos: La inmigración europea en el desarrollo de la viticultura capitalista en la provincia de Mendoza, 1870–1910." *Centro de Estudios Históricos, Segreti* (6):163–89.

———. 2008. "Los empresarios y la construcción de la vitivinicultura capitalista en la provincia de Mendoza (Argentina), 1850–2006." *Scripta Nova* 12 (271).

Richard Jorba, Rodolfo A., et al. 2006. *La región vitivinícola argentina: Transformaciones del territorio, la economía y la sociedad, 1870–1914*. Buenos Aires: Quilmes.

Rivera Medina, Ana María. 2006. *Entre la Cordillera y la Pampa: La Vitivinicultura en Cuyo, Argentina (sxviii)*. San Juan: Universidad Nacional de San Juan.

Rivera y Casanova, Julián. 1897. *La tierra labrantía y el trabajo agrícola en la provincia de Zaragoza*. Madrid.

Roberts, Edwards. 1889. "California Wine-Making." *Harper's Weekly* 33 (March 9): 197–200.

Robinson, Jancis. 2006. *The Oxford Companion to Wine*. 3rd ed. Oxford: Oxford University Press.

———. 2009. "Bordeaux versus the Rest of the World." *Financial Times*, May 16 and 17, p. 4.

Rocchi, Fernando. 2006. *Chimneys in the Desert: Industrialization in Argentina during the Export Boom Years, 1870–1930*. Stanford: Stanford University Press.

Rock, David. 1993. "Argentina in 1914: The Pampas, the Interior, Buenos Aires." In *Argentina since Independence*, edited by L. Bethell. Cambridge: Cambridge University Press.

Roos, L. 1900. *Wine Making in Hot Climates*. Melbourne: Victoria Dept. of Agriculture.

Rosés, Joan, and Blanca Sánchez-Alonso. 2004. "Regional Wage Convergence in Spain, 1850–1930." *Explorations in Economic History*.

Roudié, Philippe. 1994. *Vignobles et vignerons du Bordelais (1850–1980)*. 2nd ed. Bordeaux: Presses Universitaires de Bordeaux.

Rowntree, Joseph, and Arthur James Sherwell. 1900. *The Temperance Problem and Social Reform*. 7th ed. London: Hodder & Stoughton.

Roxas Clemente y Rubio, Simón de. 1807. *Ensayo sobre las variedades de la vid común que vegetan en Andalucía*. Madrid: Villalpando.

Salavert, Jan. 1912. "Le Commerce des Vins de Bordeaux," Ph.D. thesis, l'Université et des Facultés, Bordeaux.

Salvatore, Ricardo. (1986). "Control del trabajo y discriminación: El sistema de contratista en Mendoza, Argentina, 1880–1920." *Desarrollo Económico* 102:229–53.

Samuel, H. 1919. "The Taxation of the Various Classes of the People." *Royal Statistical Society* 82.

Sánchez Alonso, Blanca. 1992. *La inmigración española en Argentina. Siglos XIX y XX.* Barcelona: Jucar.

———. 1995. *Las causas de la emigración española 1880–1930.* Madrid: Alianza.

Seff, James, and John Cooney. 1984. "The Legal and Political History of California Wine." In *The Book of California Wine*, edited by D. Muscatine, M. A. Amerine and B. Thompson. Berkeley and London: University of California Press and Sotheby Publications.

Segarra Blasco, A. 1994. *Aiguardent i mercat a la Catalunya del segle XVII.* Vic.

Sempé, Henri. 1898. *Régime économique du vin. Production, consommation, échange.* Bordeaux and Paris.

Shaw, L.M.E. 1998. *The Anglo-Portuguese Alliance and the English Merchants in Portugal, 1654–1810.* Farnham; Ashgate.

Shaw, Thomas George. 1864. *Wine, the Vine and the Cellar.* 2nd ed. London: Longman.

Shear, Sherwood, and Gerald Pearce. 1934. "Supply and Price Trends in the California Wine-Grape Industry." Giannini Foundation mimeo, 34.

Sheingate, Adam D. 2001. *The Rise of the Agricultural Welfare State: Institutions and Interest Group Power in the United States, France, and Japan.* Princeton: Princeton University Press.

Sicsic, Pierre. 1991. "Labour Markets and Establishment Size in Nineteenth Century France," Ph.D. diss., Harvard University.

Siebel, John E., and Anton Schwarz. 1933. *History of the Brewing Industry and Brewing Science in America.* Chicago: Peterson.

Simois, Domingo, and José Lavenir. 1903. "El cultivo de la viña y la elaboración del vino en la provincia de Mendoza." In *Investigación Vinícola*, edited by P. Arata. Buenos Aires.

Simon, André. 1905. *The History of the Champagne Trade in England.* London: Wyman.

———. 1919. *Wine and Spirits: The Connoisseur's Textbook.* London: Duckworth.

———. 1920. *The Blood of the Grape: The Wine Trade Text Book.* London: Duckworth.

———. 1934a. *Champagne.* London: Constable.

———. 1934b. *Port.* London: Constable.

Simpson, James. 1985a. "Agricultural Growth and Technological Change: The Olive and Wine in Spain, 1860–1936." University of London.

———. 1985b. "La Producción de vinos en Jerez de la Frontera." In *La Nueva Historia Económica en España*, edited by P. Martin Aceña and Leandro Prados de la Escosura. Madrid.

———. 1986. *La producción agraria en 1866–1890: Un enfoque de la agricultura española del siglos xix.* Madrid: Banco de España.

———. 1995. *Spanish Agriculture: The Long Siesta, 1765–1965.* Cambridge: Cambridge University Press.

———. 2000. "Cooperation and Cooperatives in Southern European Wine Production." In *Advances in Agricultural Economic History*, edited by K. Kauffman. Stamford, CT: JAI Press.

———. 2004. "Selling to Reluctant Drinkers: The British Wine Market, 1860–1914." *Economic History Review* 57 (1):80–108.

———. 2005. "Cooperation and Conflicts: Institutional Innovation in France's Wine Markets, 1870–1911." *Business History Review* 79 (Autumn): 527–58.

———. 2005. "Too Little Regulation? The British Market for Sherry, 1840–90." *Business History* 47 (3):367–82.

———. 2010. "The British Market and the International Wine Trade: The Limits to Export-Led Growth before 1914." In *Homenaje a Gabriel Tortella*, edited by J. Morilla et al. Alcalá: Universidad de Alcalá.

Singer-Kérel, Jeanne. 1961. *Le coût de la vie à Paris de 1840 à 1954*. Paris: Colin

Smith, Harvey. 1975. "Work Routine and Social Structure in a French Village: Cruzy, Hérault in the Nineteenth Century." *Journal of Interdisciplinary History* 5:357–82.

———. 1978. "Agricultural Workers and the French Wine-Growers' revolt of 1907." *Past and Present*, 79:101–125.

South Australia. 1901. *Proceedings of the Parliament. Report of the Wine and Produce Depôt*.

South Australia. Parliamentary Papers no. 43. 1915. "Report of the Minister of Agriculture, Year Ended June 30th 1914." Adelaide.

Spain, Dirección General de Aduanas. *Estadística del comercio exterior de España* Madrid.

Spain, Dirección General de Agricultura Industria y Comercio. 1891. *Avance estadístico sobre el cultivo y producción de la vid en España formado por la Junta Consultiva Agronómica*. Madrid.

Spain, Ministerio de Agricultura. 1933. "Anuario estadístico de las producciones agrícolas. Año 1932." Madrid.

Spain, Ministerio de Fomento. 1909. *La invasión filoxérica*. Madrid.

———. 1886. *Información vinícola*. Madrid.

Stanziani, Alessandro. 2003. "La falsification du vin en France, 1880–1905: Un cas de fraude agro-alimentaire. " *Revue d'histoire moderne e contemporaine* 50 (2):154–86.

Stein, Steve. 2007. "Grape Wars: Quality in the History of Argentine Wine." In *Wine, Society and Globalization: Multidisciplinary Perspectives on the Wine Industry*, edited by G. Campbell and N. Guibert. Basingstoke: Palgrave Macmillan.

Sturzenegger, Adolfo, and Ramiro Moya. 2003. "Economic Cycles." In *A New Economic History of Argentina*, edited by G. della Paolera and A. Taylor. Cambridge: Cambridge University Press.

Suárez, Leopoldo. 1914. "La crisis vitivinícola." In *La Industria*.

———. 1922. *La acción del Estado en la Industria Vitivinícola de Mendoza*. Mendoza: Italia.

Sullivan, Charles L. 1998. *A Companion to California Wine: An Encyclopedia of Wine and Winemaking from the Mission Period to the Present*. Berkeley: University of California Press.

———. 2003. *Zinfandel: A History of a Grape and Its Wine*. Berkeley: University of California Press.

Tait, Geoffrey Murat. 1936. *Port; from the Vine to the Glass*. London: Harper.

Tennent, Sir James Emerson. 1855. *Wine, Its Use and Taxation*. London: Madison.

Thudichum, John L.W., and Auguste Dupre. 1872. *A Treatise on the Origin, Nature, and Varieties of Wine*. London: Macmillan.

Topik, S. C. 1988. "The Integration of the World Coffee Market." In *Integration of Com-*

modity Markets in History. B4 Twelfth International Economic History Congress, edited by C. E. Nuñez. Sevilla: Universidad de Sevilla.

Toutain, Jean-Claude. 1961. "Le produit de l'agriculture française de 1700 à 1958: La croissance." *Cahiers de l'ISEA, AF 4.*

Tovey, Charles. 1870. *Champagne: Its History, Manufactures, Properties . . . with Some Remarks upon Wine and Wine Merchants.* 1st. ed. London: Hotten.

———. 1877. *Wine and Wine Countries: A Record and Manual for Wine Merchants and Wine Consumers.* London: Whittaker.

Tovey, Charles. 1883. *Wine Revelations.* London: Whittaker.

Trebilcock, Clive. 1981. *The Industrialization of the Continental Powers 1780–1914.* London: Longman.

Trianes, José. 1908. *Por la viticultura argentina y por el consumidor argentino.* Buenos Aires.

Trianes, Rafael. 1935. Tres estudios sobre la cuestión vinícola. *Eco.*

Turner, Michael E. 2000. "Agricultural Output, Income and Productivity." In *The Agrarian History of England and Wales*, edited by E. J. T. Collins. Cambridge: Cambridge University Press.

United Kingdom. Parliamentary Papers. 1835. *Second Report on the Commercial Relations between France and Great Britain. Silks and Wine.* By John Bowring.

———. 1852. *Report from the Select Committee on Import Duties on Wines.* Vol. 17.

———. 1854–55. *First Report from the Select Committee on Adulteration of Food, &c.; with the Minutes of Evidence, and Appendix. (432).* Vol. 8.

———. 1854–55. *Second Report from the Select Committee on Adulteration of Food (480–1).*

———. 1859. *Reports of H.M.'s Secretaries of Embassy and Legation on the Effects of Vine Disease on Commerce of the Countries in Which They Reside.* Vol. 30.

———. 1872. *Report from the Select Committee on Adulteration of Food Act.*

———. 1878/79. *Report from the Select Committee on Wine Duties.*

———. (1927). *Report of the Committee on National Debt and Taxation (Colwyn Report).* xi.

Unwin, Tim. 1991. *Wine and the Vine: An Historical Geography of Viticulture and the Wine Trade.* London: Routledge.

Valentin, Jean. 1977. *La revolution viticole dans l'Aube, 1789–1907.* 2 vols. Carcassonne.

Valls Junyent, Francesc 1996. *La dinámica del canvi agrari a la Catalunya interior: L'Anoia, 1720–1860.* Barcelona: Ajuntament d'Igualada.

Vamplew, Wray, ed. 1987. *Australians: Historical Statistics.* Broadway, NSW: Fairfax, Syme & Weldon Associates.

Victoria. 1900. *Royal Commission on Refrigerating Stores and Central Wine Depôt. Report on Central Wine Depôt.* Melbourne: Government Printer.

Victoria, Department of Agriculture. 1900. Report for the Year 1900. Melbourne.

Vilar, Pierre. 1962. *La Catalogne dans l'Espagne moderne.* 3 vols. Paris: SEVPEN.

———. 1980. "Algunos puntos de historia de la viticultura mediterránea." In *Crecimiento y desarrollo: Economía e historia. Reflexiones sobre el caso español*, ed. Vilar. Barcelona: Ariel.

Vitu, Henri. 1912. *La question des délimitations régionales.* Paris.

Vizetelly, Henry. 1876. *Facts about Sherry: Gleaned in the Vineyards and Bodegas of the*

Jerez, Seville, Moguer & Montilla Districts during the Autumn of 1875. London: Ward, Locke, and Tyler.

——. 1879. *Facts about Champagne and Other Sparkling Wines Collected during Numerous Visits to the Champagne and Other Viticultural Districts of France and the Principal Remaining Wine-Producing Countries of Europe.* London: Ward, Lock.

——. 1880. *Facts about Port and Madeira.* London: Ward, Lock.

Walsh, Jacobo. 1844. *Memoria sobre el solo y único arbitrio que puede exponerse para que prospere la riqueza viñera en la ciudad de Jerez de la Frontera.* Sevilla: Diario de Comercio.

Warburton, Clark. 1932. *The Economic Results of Prohibition.* New York: Columbia University Press.

Ward, Ebenezer. 1864/1980. *The Vineyards of Victoria.* Adelaide.

Warner, Charles K. 1960. *The Winegrowers of France and the Government since 1875.* New York: Columbia University Press.

Waugh, Alec. 1957. *Merchants of Wine: House of Gilbey.* London: Cassell.

Weir, Ron B. 1984. "Obsessed with Moderation: The Drink Trades and the Drink Question (1870–1930)." *British Journal of Addiction* 79:93–107.

Wetmore, Charles A. (1885). "Viticulture and Viniculture in California." 2nd annual report. Board of State Viticultural Commissioners.

——. 1894. *Treatise on Wine Production in the Report of the California. Board of State Viticultural Commissioners.* Sacramento: State Office, A. J. Johnston, Supt. of State Printing.

Whitington, Ernest. 1903/1997. *The South Australian Vintage 1903.* Adelaide: Friends of the State Library of South Australia.

Wilson, George B. 1940. *Alcohol and the Nation.* London: Nicholson and Watson.

Williamson, Jeffrey G. 1995. "The Evolution of Global Labor Markets since 1830: Background Evidence and Hypotheses." *Explorations in Economic History* 32 (2):141–96.

Wright, Gordon. 1964. *Rural Revolution in France: The Peasantry in the Twentieth Century.* Stanford: Stanford University Press.

Yates, P. Lamartine. 1959. *Forty Years of Foreign Trade: A Statistical Handbook with Special Reference to Primary Products and Under-developed Countries.* London: Allen and Unwin.

Young, Arthur. 1794. *Travels during the Years 1787, 1788 and 1789, Undertaken . . . the Kingdom of France.* 2 vols. Bury St. Edmunds: Richardson.

Zoido Naranjo, Florencio. 1978. "Observations sur la crise du phylloxéra et ses conséquences dans la vignoble de Xeres." In *Géographie Historique. Colloque de Bordeaux.* Paris.

Index